Karl J. Smith

Student Solutions Manual for Calculus with Applications

2ND EDITION

Brooks/Cole Publishing Company
Pacific Grove, California

Brooks/Cole Publishing Company
A Division of Wadsworth, Inc.

Printed in the United States of America

10 9 8 7 6 5 4 3 2 1

ISBN 0-534-16886-8

Sponsoring Editor: *Faith B. Stoddard*
Editorial Assistants: *Sarah Wilson and Carol Ann Benedict*
Production Coordinator: *Dorothy Bell*
Cover Design: *Vernon T. Boes*
Cover Photo: *San Francisco Convention and Visitors Bureau*
Printing and Binding: *Malloy Lithographing, Inc.*

PREFACE

I wrote this solutions manual to accompany my book *Calculus with Applications*, *Second Edition*, to help you through the course. I have provided the complete solutions for the odd-numbered problems in the text. I have included the steps that I would include if I were working the problem on the board in class. I hope you will find this manual beneficial.

I used **EXP** 2.0 and CONVERGE 1.2 in preparing this manual, and found both of these pieces of software very easy to use. **EXP** is a scientific word processing language that is available from Brooks/Cole Publishing Company, and CONVERGE is a mathematics graphing program that is available from JEM Software.

Karl J. Smith
Sebastopol

CONTENTS

CHAPTER 1

Functions

1.1 What is Calculus?

1. (Answers vary) A mathematical model is a body of mathematics based on certain assumptions and data about the real world which is used to predict future occurrences.

3. (Answers vary) A function of a variable is a rule that assigns to each value of x exactly one value for $f(x)$.

5. a. y is a function of x, since for each year, x, there will be one and only one closing price, y. **b.** y is not a function of x, since for each closing price, x, we may not be able to uniquely determine the year, y.

7. **a.** $g(1954) = \$\,.29$ **b.** $g(1974) = \$\,.53$

9. **a.** $f(8) = 2(8) - 5 = 11$ **b.** $f(-3) = 2(-3) - 5 = -11$

11. **a.** $f(2) = 5(2) - 3 = 7$ **b.** $f(10) = 5(10) - 3 = 47$

c. $f(-15) = 5(-15) - 3 = -78$ **d.** $f(100) = 5(100) - 3 = 497$

13. **a.** $h(0) = 6 - 4(0) = 6$ **b.** $h(8) = 6 - 4(8) = -26$

c. $h(-7) = 6 - 4(-7) = 34$ **d.** $h(100) = 6 - 4(100) = -394$

15. **a.** $m(0) = 0^2 - 3(0) + 1 = 1$ **b.** $m(1) = 1^2 - 3(1) + 1 = -1$

c. $m(2) = 2^2 - 3(2) + 1 = -1$ **d.** $m(3) = 3^2 - 3(3) + 1 = 1$

17. **a.** Since any number can be substituted for x, the domain is all real numbers.

b. Since any number can be substituted for x, the domain is all real numbers.

19. **a.** Since the denominator can not be zero, $x + 5 \neq 0$.

The domain is all real numbers except 5.

b. Since the denominator can not be zero, $2x + 1 \neq 0 \Rightarrow x \neq -\frac{1}{2}$.

Since the square root of a negative number is not a real number,

$2x - 1 \geq 0 \Rightarrow x \geq \frac{1}{2}$. The domain is $[\frac{1}{2}, \infty)$.

21. a. $f(t) = 5t - 2$ b. $f(w) = 5w - 2$

23. a. $f(t + h) = 5(t + h) - 2 = 5t + 5h - 2$

b. $f(s + t) = 5(s + t) - 2 = 5s + 5t - 2$

25. $f(t + h + 8) = 5(t + h + 8) - 2 = 5t + 5h + 38$

27.
$$\begin{aligned} g(3 + h) &= 2(3 + h)^2 - 4(3 + h) - 5 \\ &= 2(9 + 6h + h^2) - 4(3 + h) - 5 \\ &= 18 + 12h + 2h^2 - 12 - 4h - 5 \\ &= 2h^2 + 8h + 1 \end{aligned}$$

29. $f(2x^2) = 5(2x^2) - 2 = 10x^2 - 2$

31.
$$\begin{aligned} g[f(x)] &= g[5x - 2] \\ &= 2(5x - 2)^2 - 4(5x - 2) - 5 \\ &= 2(25x^2 - 20x + 4) - 4(5x - 2) - 5 \\ &= 50x^2 - 40x + 8 - 20x + 8 - 5 \\ &= 50x^2 - 60x + 11 \end{aligned}$$

33.
$$\begin{aligned} f([g(5x)] &= f[2(5x)^2 - 4(5x) - 5] \\ &= f(50x^2 - 20x - 5) \\ &= 5(50x^2 - 20x - 5) - 2 \\ &= 250x^2 - 100x - 27 \end{aligned}$$

35.
$$\begin{aligned} g[g(x)] &= g(2x^2 - 4x - 5) \\ &= 2(2x^2 - 4x - 5)^2 - 4(2x^2 - 4x - 5) - 5 \\ &= 2(4x^4 - 8x^3 - 10x^2 - 8x^3 + 16x^2 + 20x - 10x^2 + 20x + 25) \\ &\qquad - 4(2x^2 - 4x - 5) - 5 \\ &= 8x^4 - 32x^3 - 8x^2 + 80x + 50 - 8x^2 + 16x + 20 - 5 \\ &= 8x^4 - 32x^3 - 16x^2 + 96x + 65 \end{aligned}$$

37. $$\frac{f(x + h) - f(x)}{h} = \frac{2(x + h) - 2x}{h} = \frac{2x + 2h - 2x}{h} = \frac{2h}{h} = 2$$

39.
$$\begin{aligned} \frac{f(x + h) - f(x)}{h} &= \frac{2(x + h)^2 - 2x^2}{h} \\ &= \frac{2(x^2 + 2xh + h^2) - 2x^2}{h} \\ &= \frac{4xh + 2h^2}{h} \\ &= 4x + 2h \end{aligned}$$

41. $\dfrac{f(x+h)-f(x)}{h}=\dfrac{[2(x+h)^2-3]-[2x^2-3]}{h}$

$=\dfrac{2(x^2+2xh+h^2)-3-2x^2+3}{h}$

$=\dfrac{4xh+2h^2}{h}$

$=4x+2h$

43. $\dfrac{f(x+h)-f(x)}{h}=\dfrac{[(x+h)^2-2(x+h)+1]-[x^2-2x+1]}{h}$

$=\dfrac{[x^2+2xh+h^2-2x-2h+1]-[x^2-2x+1]}{h}$

$=\dfrac{2xh+h^2-2h}{h}$

$=2x+h-2$

45. **a.** Locating 1954 in the left column, the price for round steak was $r(1954)=\$\,.92$.

b. Locating 1954 in the left column, the price for $\frac{1}{2}$ gallon milk was $m(1954)=\$\,.45$.

47. $s(1984)-s(1944)=1.48-.34=\$1.14$

49. **a.** $1.15-.64=\$\,.51$ **b.** $e(1984)-e(1944)$

51. **a.** $\dfrac{f(1944+40)-g(1944)}{40}=\dfrac{g(1984)-g(1944)}{40}$

$=\dfrac{1.52-.21}{40}$

$=.03275\approx\$\,.03$

b. The change in price (over the last 40 years) is $g(1984)-g(1944)$.

The average change in price *per year* is: $\dfrac{g(1984)-g(1944)}{40}$

53. **a.** $\dfrac{s(1954)-s(1944)}{1954-1944}=\dfrac{\$\,.52-\$\,.34}{10}=.018\approx\$\,.02$

b. $\dfrac{s(1964)-s(1944)}{1964-1944}=\dfrac{\$\,.59-\$\,.34}{20}=.0125\approx\$\,.01$

c. $\dfrac{s(1974)-s(1944)}{1974-1944}=\dfrac{\$1.08-\$\,.34}{30}=.024\overline{6}\approx\$\,.02$

d. $\dfrac{s(1984)-s(1944)}{1984-1944}=\dfrac{\$1.48-\$\,.34}{40}=.0285\approx\$\,.03$

e. $\dfrac{s(1944+h)-s(1944)}{h}$

55. $P(20)=249.6(1.02)^{18}\approx 356.5$. The population will be about 356,500,000

57. $P(-9) = 249.6(1.02)^{-9} \approx 208.9$. The population was about 208,900,000

59. a. $\dfrac{M(1987) - M(1982)}{5} = \dfrac{2{,}421{,}000 - 2{,}495{,}000}{5} = -14{,}800$

b. The average change in number of marriages from 1982 to year 1982 + h.

1.2 Functions and Graphs

1. A picture of the ordered pairs for the function.

3. The x-intercepts are the points where the graph crosses (or touches) the x-axis. To find the x-intercept, set $y = 0$. The y-intercept is the point where the graph crosses the y-axis. To find the y-intercept, set $x = 0$. **5.** $(2, f(2))$

7. $(x_0, f(x_0))$ **9.** $(x_0, g(x_0))$ **11.** $(3, h(3))$ **13.** $(x_0 + t, h(x_0 + t))$

15. Since all points on the vertical line have an x-coordinate of $x = -3$, the domain is $x = -3$. Since the y-coordinates of all points on the line range through all values from $-\infty$ to ∞, the range is all real numbers. The x-intercept is $(-3, 0)$. There is no y-intercept. Since there is a vertical line that contains more than one point on the graph, this relation is not a function.

17. For all points on the graph, the x-coordinates vary from -2 to 5, so the domain is $-2 \leq x \leq 5$. For all points on the graph, the y-coordinates vary from -5 to 3, so the range is $-5 \leq y \leq 3$. The x-intercepts are $(-\frac{3}{2}, 0)$ and $(\frac{15}{4}, 0)$. The y-intercept is $(0, 3)$. Since there is no vertical line that contains more than one point on the graph, this relation is a function.

19. Since the x- and y-coordinates of all points on the line range through all values from $-\infty$ to ∞, the domain and the range are all real numbers. The x-intercepts are $(-3, 0)$, $(-1, 0)$, and $(2, 0)$. The y-intercept is $(0, \frac{5}{2})$. Since there is no vertical line that contains more than one point on the graph, the relation is a function.

21. A few points are:

$x = -3\ f(-3) = 2(-3) + 3 = 3$

$x = -1,\ f(-1) = 2(-1) + 3 = 1$

$x = 0,\ f(0) = 2(0) + 3 = 3$

$x = 2,\ f(2) = 2(2) + 3 = 7$

The graph is shown on the right.

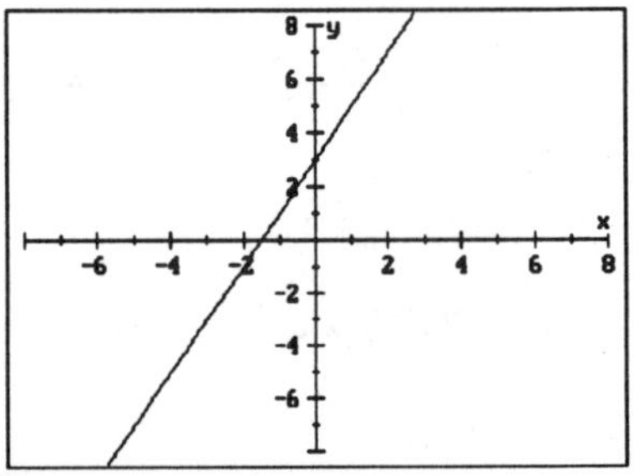

23. A few points are:

$x = -2,\ f(-2) = 6 - 2(-2) = 10$

$x = 0,\ f(0) = 6 - 2(0) = 6$

$x = 1,\ f(1) = 6 - 2(1) = 4$

The graph is shown on the right.

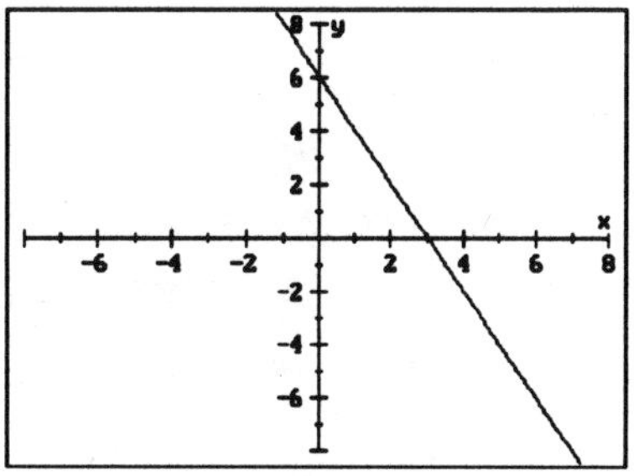

25. A few points are:

$x = -1,\ f(-1) = -3(-1) - 1 = 2$

$x = 0,\ f(0) = -3(0) - 1 = -1$

$x = 2,\ f(2) = -3(2) - 1 = -7$

The graph is shown on the right.

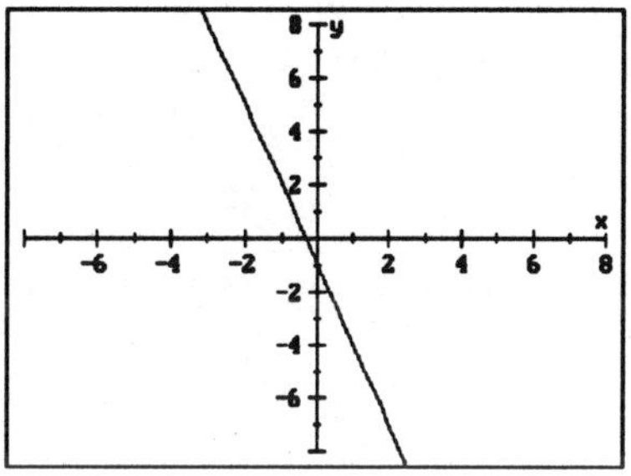

27. A few points are:

$x = -4,\ g(-4) = (\frac{1}{2})(-4)^2 = 8$

$x = -2,\ g(-2) = (\frac{1}{2})(-2)^2 = 2$

$x = -1,\ g(-1) = (\frac{1}{2})(-1)^2 = \frac{1}{2}$

$x = 0,\ g(0) = (\frac{1}{2})(0^2) = 0$

$x = 2,\ g(2) = (\frac{1}{2})(2^2) = 2$

$x = 4,\ g(4) = (\frac{1}{2})(4^2) = 8$

The graph is shown on the right.

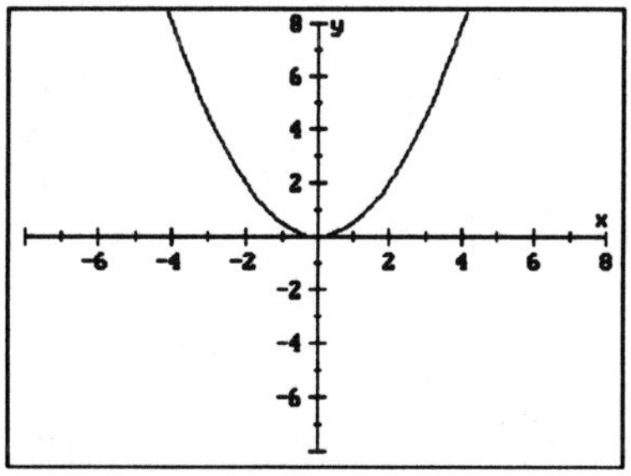

29. A few points are:

$x = -5,\ g(-5) = (-5)^2 + 4(-5) + 4 = 9$

$x = -3,\ g(-3) = (-3)^2 + 4(-3) + 4 = 1$

$x = -2,\ g(-2) = (-2)^2 + 4(-2) + 4 = 0$

$x = -1,\ g(-1) = (-1)^2 + 4(-1) + 4 = 1$

$x = 0,\ g(0) = (0)^2 + 4(0) + 4 = 4$

The graph is shown on the right.

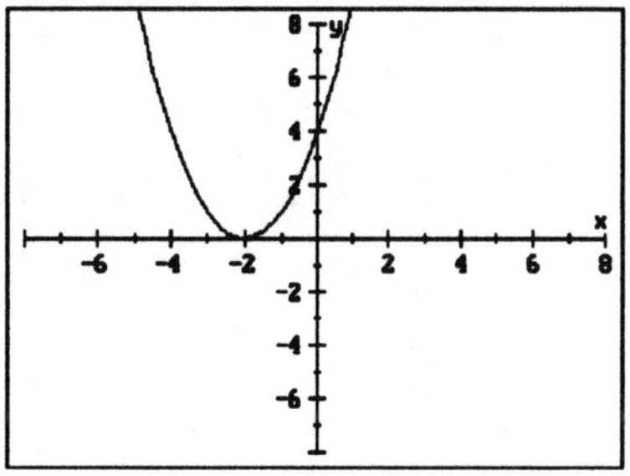

31. A few points are:

$x = 0,\ g(0) = 0^2 - 6(0) + 9 = 9$

$x = 2,\ g(2) = 2^2 - 6(2) + 9 = 1$

$x = 3,\ g(3) = 3^2 - 6(3) + 9 = 0$

$x = 4,\ g(4) = 4^2 - 6(4) + 9 = 1$

$x = 5,\ g(5) = 5^2 - 6(5) + 9 = 4$

The graph is shown on the right.

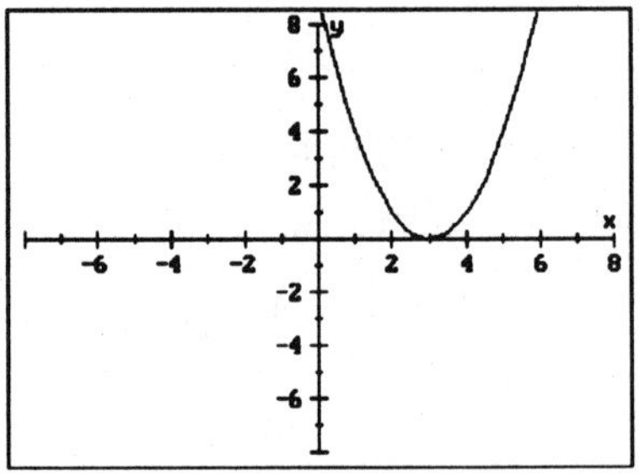

33. A few points are:

$x = -2,\ g(-2)$

$= 2(-2)^2 - 4(-2) + 5 = 21$

$x = 0,\ g(0) = 2(0)^2 - 4(0) + 5 = 5$

$x = 1,\ g(1) = 2(1)^2 - 4(1) + 5 = 3$

$x = 2,\ g(2) = 2(2)^2 - 4(2) + 5 = 5$

$x = 3,\ g(3) = 2(3)^2 - 4(3) + 5 = 11$

The graph is shown on the right.

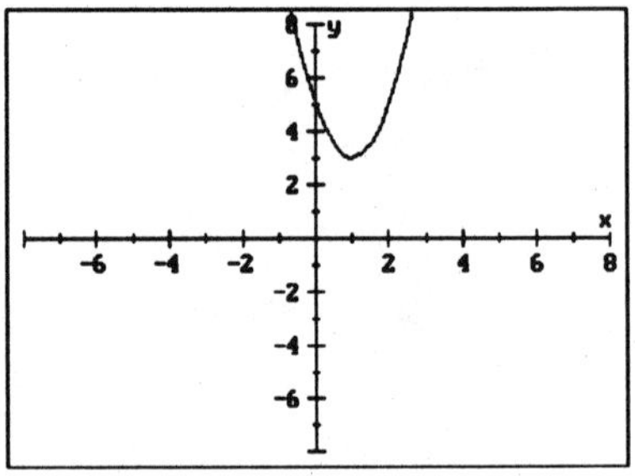

35. A few points are:

$x = -2,\ h(-2) = \frac{-1}{-2} = \frac{1}{2}$

$x = -1,\ h(-1) = \frac{-1}{-1} = 1$

$x = -\frac{1}{2},\ h(-\frac{1}{2}) = \frac{-1}{-1/2} = 2$

$x = 0$, $h(0)$ is undefined.

$x = \frac{1}{2}\ ,\ h(\frac{1}{2}) = -\frac{1}{1/2} = -2$

$x = 1,\ h(1) = -\frac{1}{1} = -1$

The graph is shown at the right.

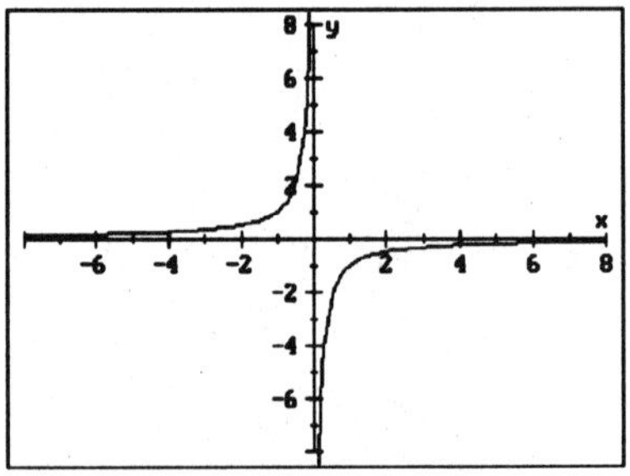

37. A few points are:

$x = -2,\ h(-2) = \frac{-2}{-2} = 1$

$x = -1,\ h(-1) = \frac{-2}{-1} = 2$

$x = -\frac{1}{2}\ ,\ h(-\frac{1}{2}\ ) = \frac{-2}{-1/2} = 4$

$x = 0$, $h(0)$ is undefined

$x = \frac{1}{2}\ ,\ h(\frac{1}{2}) = \frac{-2}{1/2} = -4$

$x = 1,\ h(1) = \frac{-2}{1} = -2$

The graph is shown on the right.

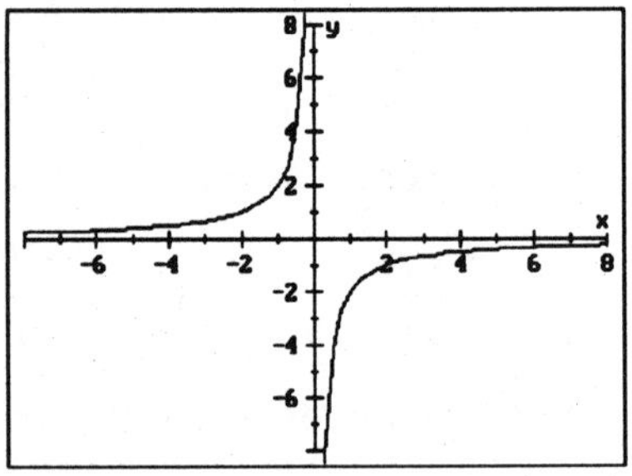

39. A few points are:

$x = -1,\ h(-1) = \frac{2}{-1-2} = -\frac{2}{3}$

$x = 0,\ h(0) = \frac{2}{0-2} = -1$

$x = 1,\ h(1) = \frac{2}{1-2} = -2$

$x = 2,\ h(2)$ is undefined.

$x = 3,\ h(3) = \frac{2}{3-2} = 2$

$x = 4,\ h(4) = \frac{2}{4-2} = 1$

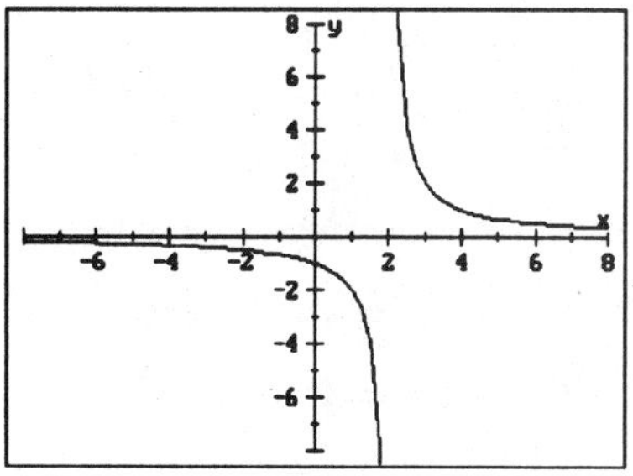

41. A few points are:

$x = 0,\ y = 2$

$x = 1,\ y = 1 - 3 + 2 = 0$

$x = 2,\ y = 8 - 6 + 2 = 4$

$x = 3,\ y = 27 - 9 + 2 = 20$

$x = 1,\ y = -1 + 3 + 2 = 4$

$x = -2,\ y = -8 + 6 + 2 = 0$

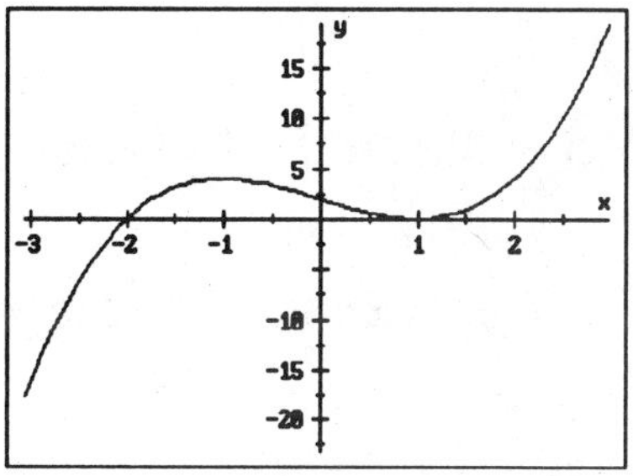

43. The endpoints are:

$x = 19{,}451,\quad T = .28(19{,}451) - 3{,}250$
$= 2196.28$
(open point)

$x = 47{,}050\quad T = .28(47{,}050) - 2411.5$
$= 10{,}762.50$
(closed point)

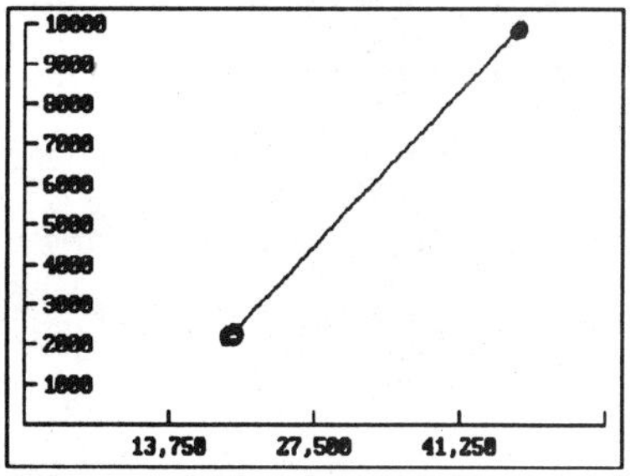

45. A few points are:

$d = 0,\ P = .4(0) + 7.6 = 7.6$

$d = 5,\ p = .4(5) + 7.6 = 9.6$

$d = 10,\ p = .4(10) + 7.6 = 11.6$

The graph is shown on the right.

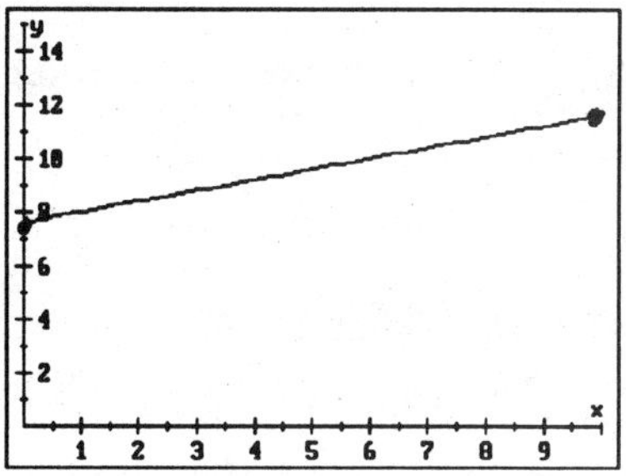

47. A few points are:

$t = 0,\ n = 150 - (0 - 10)^2 = 50$

$t = 4,\ n = 150 - (4 - 10)^2 = 114$

$t = 8,\ n = 150 - (8 - 10)^2 = 146$

$t = 10,\ n = 150 - (10 - 10)^2 = 150$

$t = 12,\ n = 150 - (12 - 10)^2 = 146$

$t = 22,\ n = 150 - (22 - 10)^2 = 6$

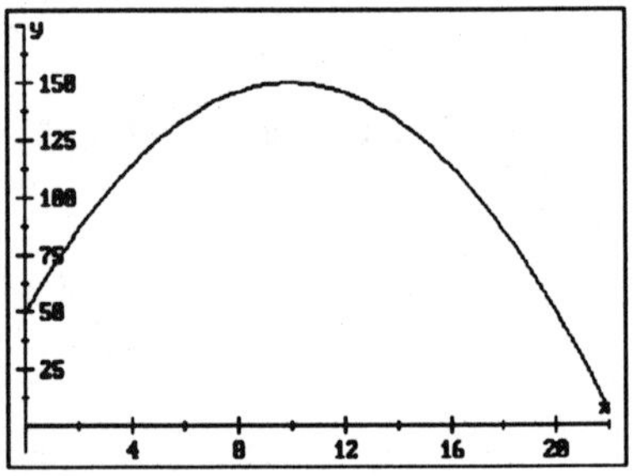

1.3 Linear Functions

1. To find the x-intercept, set $y = 0$: $0 = 2x + 4 \Rightarrow x = -2$. To find the y-intercept, set $x = 0$: $y = 2(0) + 4 \Rightarrow y = 4$. The x-intercept is $(-2, 0)$; the y-intercept is $(0, 4)$.

3. To find the x-intercept, set $y = 0$: $4x + 3(0) + 4 = 0 \Rightarrow x = -1$. To find the y-intercept, set $x = 0$: $4(0) + 3y + 4 = 0 \Rightarrow y = -\frac{4}{3}$. The x-intercept is $(-1, 0)$; the y-intercept is $(0, -\frac{4}{3})$.

5. First divide both sides of the equation by 50; $2x - 5y + 10 = 0$. To find the x-intercept, set $y = 0$: $2x - 5(0) + 10 = 0 \Rightarrow x = -5$. To find the y-intercept, set $x = 0$: $2(0) - 5y + 10 = 0 \Rightarrow y = 2$. The x-intercept is $(-5, 0)$; the y-intercept is $(0, 2)$.

7. Since we cannot set $y = 0$, there is no x-intercept. To find the y-intercept, set $x = 0$: $y + 2 = 0 \Rightarrow y = -2$. (If this bothers you, write the equation as $0x + y + 2 = 0$. Now set $x = 0$.) The y-intercept is $(0, -2)$.

9. The slope is $\frac{4 - 3}{5 - 2} = \frac{1}{3}$.

11. The slope is $\frac{-3 - 2}{-2 - 5} = \frac{-5}{-7} = \frac{5}{7}$.

13. The slope is $\frac{-2 - (-3)}{-1 - (-2)} = \frac{1}{1} = 1$.

15. This is in the form $y = mx + b$, where m is the slope and b is the y-intercept. The slope is $m = 2$; the y-intercept is $b = 4$ or $(0, 4)$.

17. This is in the form $y = mx + b$, where m is the slope and b is the y-intercept. The slope is $m = 9$; the y-intercept is $b = 1$ or $(0, 1)$.

19. First isolate y to get the form $y = mx + b$:

$2x - 3y + 5 = 0$ *Subtract 2x and 5 from both sides.*

$-3y = -2x - 5$ *Divide both sides by − 3.*

$y = \frac{2}{3}x + \frac{5}{3}$

The slope is $m = \frac{2}{3}$; the y-intercept is $b = \frac{5}{3}$ or $(0, \frac{5}{3})$.

21. Write the equation in the form $y = mx + b$:

$y - 5 = 0$

$y = 5$

$y = 0x + 5$

The slope is $m = 0$; the y-intercept is $b = 5$ or $(0, 5)$.

23. This a vertical line; hence it has no slope nor a y-intercept.

25. Start with the y-intercept, $(0, -5)$. Since the slope is $m = 2 = \frac{2}{1}$, from the y-intercept move 2 units up and 1 unit right. The coordinates of this point are $(1, -3)$. Draw a line through these two points.

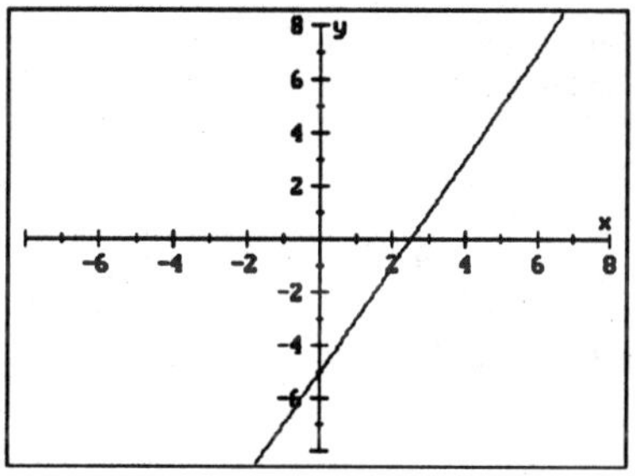

27. Start with the y-intercept, $(0, 2)$. Since the slope is $m = -\frac{1}{4} = \frac{-1}{4}$, from the y-intercept move 1 unit down and 4 units right. [The coordinates of this point are (4, 1).] Now draw a line through these two points.

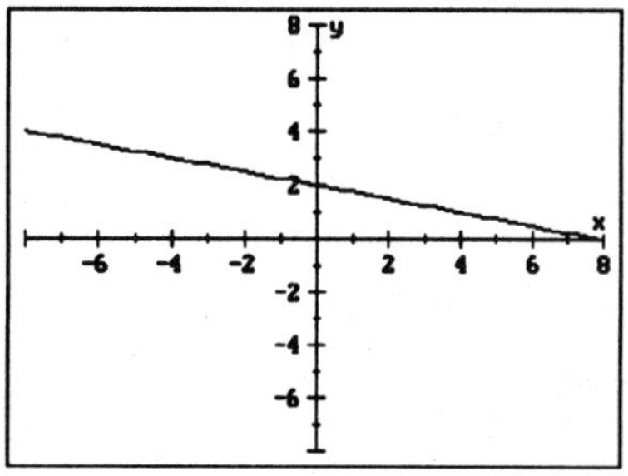

29. Start with the y-intercept, $(0, \frac{2}{5})$. Since the slope is $m = \frac{3}{5}$, from the y-intercept move 3 units up and 5 units right. Now draw a line through these two points.

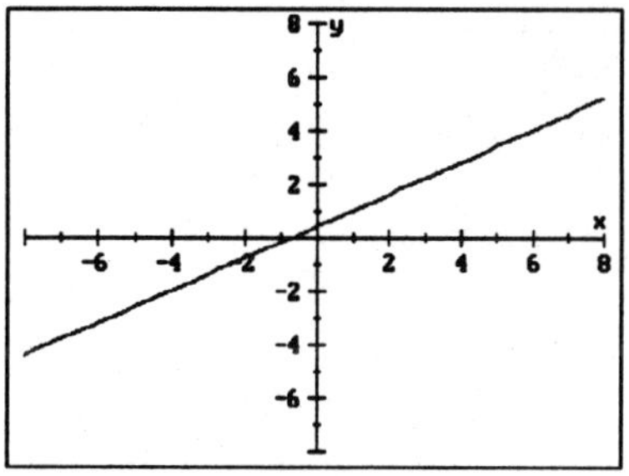

31. First write $x + 3y - 9 = 0$ in the form $y = mx + b \Rightarrow y = -\frac{1}{3}x + 3$. Start with the y-intercept, (0, 3). Since the slope is $m = -\frac{1}{3} = \frac{-1}{3}$, from the y-intercept move 1 unit down and 3 units right. Now draw a line through these two points.

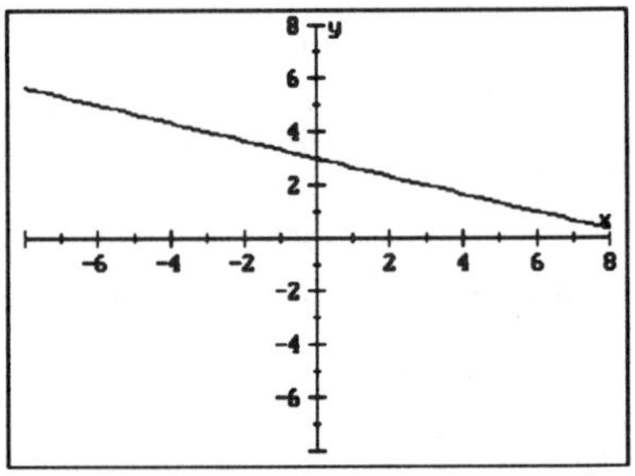

33. First write $x = \frac{2}{3}y$ in the form $y = mx + b \Rightarrow y = \frac{3}{2}x + 0$. Starting with the y-intercept, (0, 0) move 3 units up and 2 units right. Draw a line through these two points.

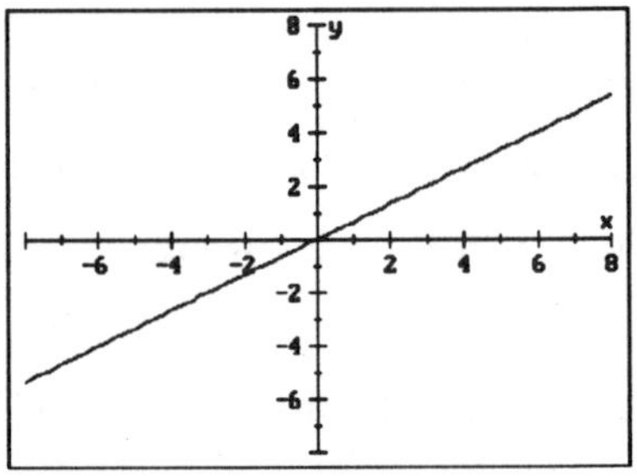

35. This equation has no slope or y-intercept. It is the vertical line $x = -\frac{5}{2}$.

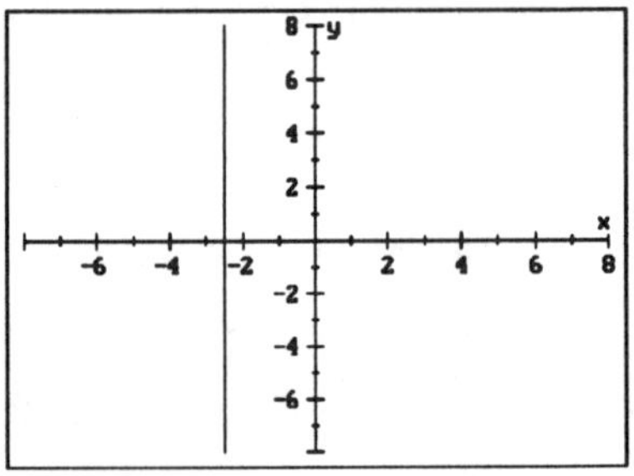

37.

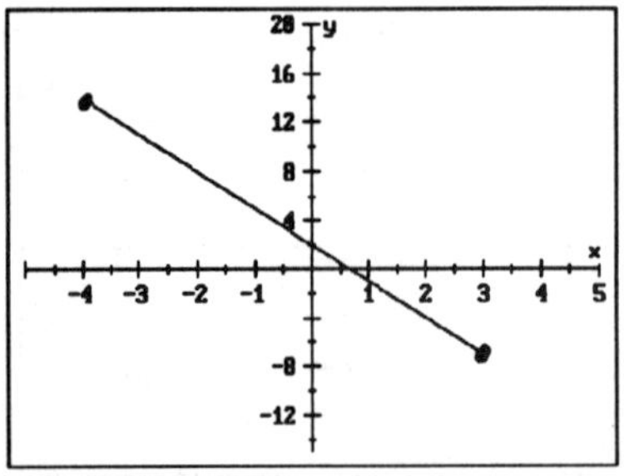

39.

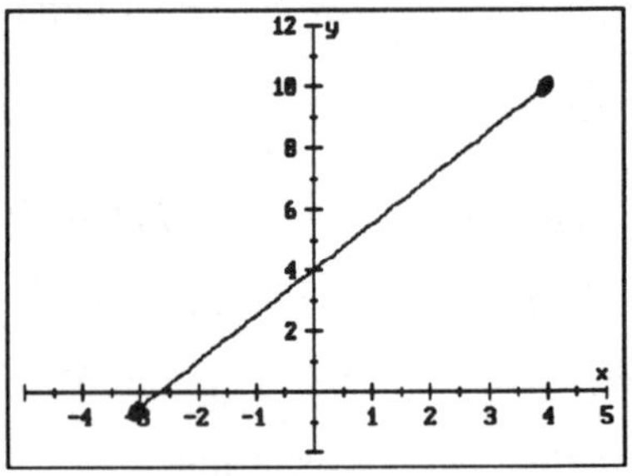

41.

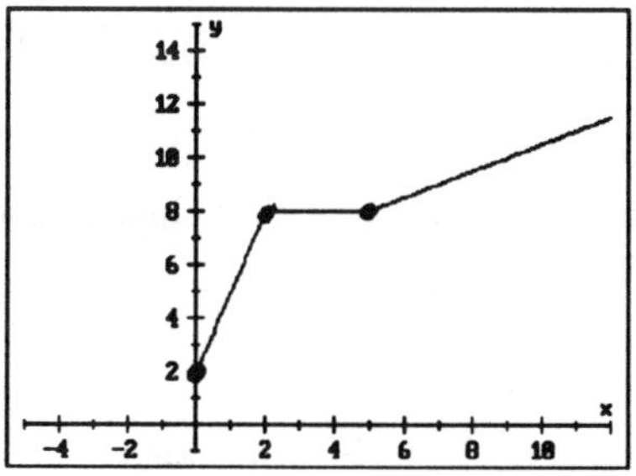

43.

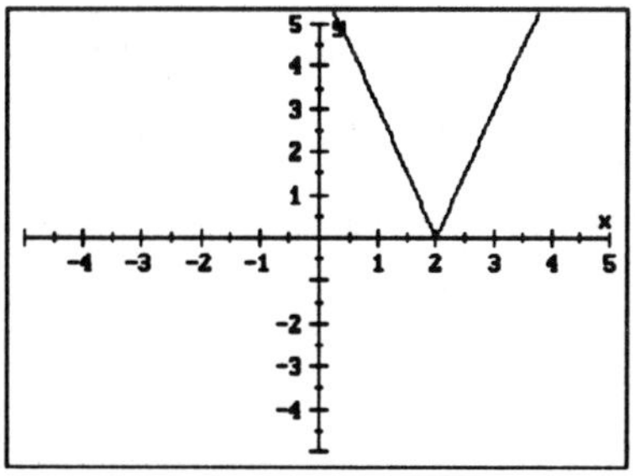

45.

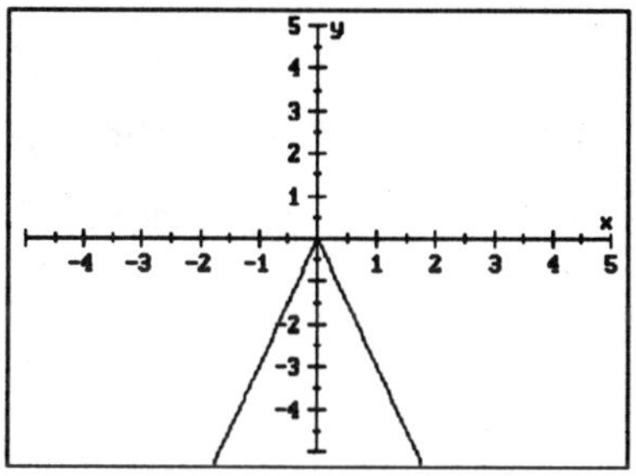

47.

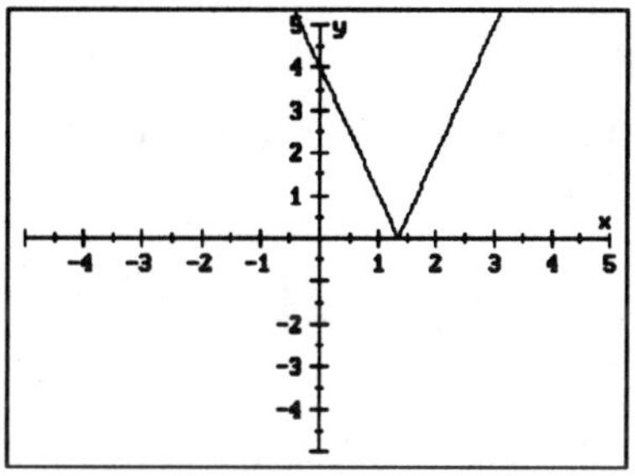

49. Using the form $y = mx + b$ where $m = -2$ is the slope and $b = -3$ is the y-intercept, $y = -2x - 3$. In standard form this is $2x + y + 3 = 0$.

51. Using the form $y = mx + b$, $m = 0$ and $b = 4$. In standard form this is $y - 4 = 0$.

53. Using the point-slope form, $y - y_1 = m(x - x_1)$ we obtain

$$y - 5 = -1[x - (-3)]$$

$$y - 5 = -x - 3.$$

In standard form this is $x + y - 2 = 0$.

55. Using the point-slope form $y - y_1 = m(x - x_1)$ we get

$$y - (-3) = \tfrac{3}{5}(x - 4)$$

$$y + 3 = \tfrac{3}{5}x - \tfrac{12}{5}.$$

Multiplying both sides by 5 gives $5y + 15 = 3x - 12$. In standard form this is $3x - 5y - 27 = 0$.

57. First we find the slope: $m = \dfrac{-2 - 6}{1 - 5} = 2$. Using either ordered pair, say (5, 6) in the point-slope form we get $y - 6 = 2(x - 5)$. In standard form this is $2x - y - 4 = 0$.

59. First we find the slope: $m = \frac{6 - 6}{7 - 5} = 0$. This must, therefore, be the horizontal line, $y = 6$. In standard form the equation is $y - 6 = 0$.

61. We want the equation of the line through (5, 10) and (10, 20). The slope is $\frac{20 - 10}{10 - 5} = 2$. Using the point-slope form $y - y_1 = m(x - x_1)$, $y - 10 = 2(x - 5)$. In standard form this is $2x - y = 0$. If $x = 2$, then $y = 4$; four boxes could be supplied.

63. We want the equation of the line through (20, 14.2) and (30, 17.2). The slope is $\frac{17.2 - 14.2}{30 - 20} = \frac{3}{10}$. The equation is $y - 14.2 = \frac{3}{10}(x - 20)$. In standard form this is $3x - 10y + 82 = 0$. To predict the population in 2000 let $x = 40$: $y = .3(40 - 20) + 14.2$; $y = 20.2$ (million).

65. We want the equation of the line through (50, 60) and (260, 60). The slope is 0. The equation is $y - 60 = 0(x - 50) \Rightarrow y - 60 = 0$ where y represents the cost.

67. $$m = \frac{400 - 100}{40 - 10} = \frac{300}{30} = 10$$

$$y - 100 = 10(x - 10)$$

$$y = 10(x - 10) + 100$$

If $x = 53.5$, then $y = 10(53.5 - 10) + 100 = 535$

The expected cost is \$535.

69. $$m = \frac{67.2 - 70.6}{4} = -.85$$

$$y - 70.6 = -.85(x - 3)$$

If $x = 12$, then $y = 62.95$ The expected percentage is 63.0%.

71. Let x = amount on Form 1040, line 37; $T = .15x$

73.

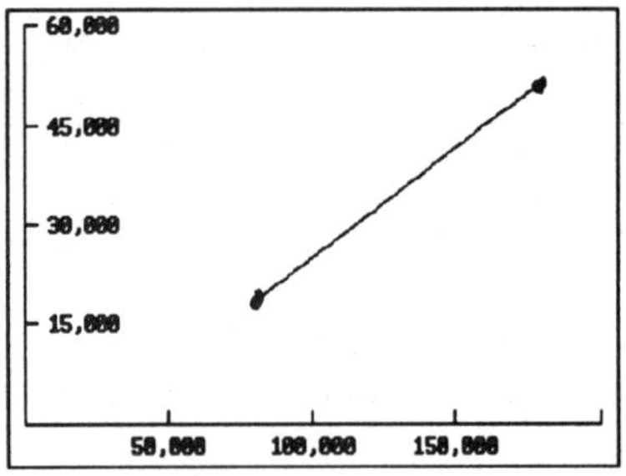

75. Let x = amount on Form 1040, line 37

$$T = \begin{cases} .15x \text{ for } 0 < x \le 32{,}450 \\ .28x + 4{,}867.50 \text{ for } 32{,}450 < x \le 78{,}401 \\ .33x + 17{,}733.50 \text{ for } 78{,}401 < x \le 185{,}730 \end{cases}$$

1.4 Quadratic and Polynomial Functions

1. See the discussion following Example 2.

This has the form $y = ax^2$ where $a = 1$.

The parabola opens up from the vertex

(0, 0) since $a > 0$. Some points are:

Let $x = 1$, then $y = (1)^2 = 1$

Let $x = 2$, then $y = (2)^2 = 4$

Let $x = 3$, then $y = (3)^2 = 9$

Using the symmetry with respect to the vertical axis, the graph is shown on the right.

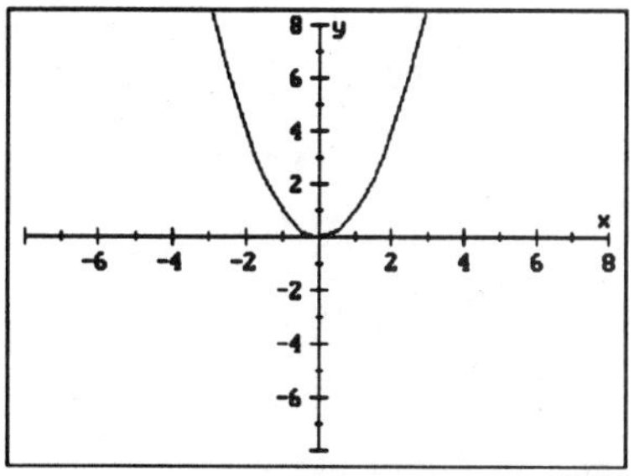

3. See the discussion following Example 2.

This has the form $y = ax^2$ where $a = -2$.

The parabola opens up from the vertex

(0, 0) since $a < 0$. Some points are:

Let $x = 1$, then $y = -2(1)^2 = -2$

Let $x = 2$, then $y = -2(2)^2 = -8$

Let $x = 3$, then $y = -2(3)^2 = -18$

Using the symmetry with respect to the vertical axis, the graph is shown on the right.

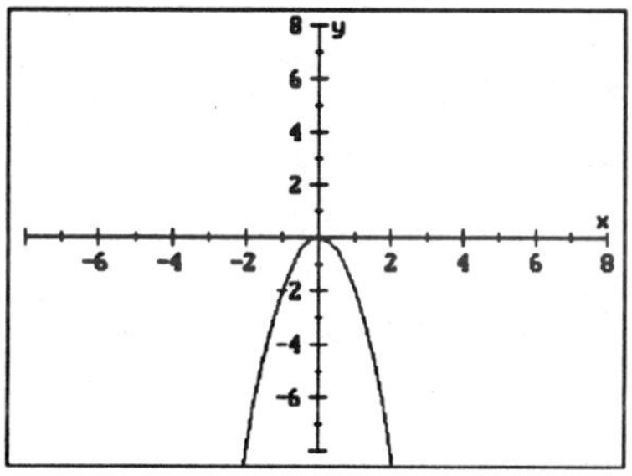

5. See the discussion following Example 2.

This has the form $y = ax^2$ where $a = -5$.

The parabola opens up from the vertex

(0, 0) since $a < 0$. Some points are:

Let $x = 1$, then $y = -5(1)^2 = -5$

Let $x = 2$, then $y = -5(2)^2 = -20$

Let $x = \frac{1}{2}$, then $y = -5(\frac{1}{2})^2 = -\frac{5}{4}$

Using the symmetry with respect to the vertical axis, the graph is shown on the right.

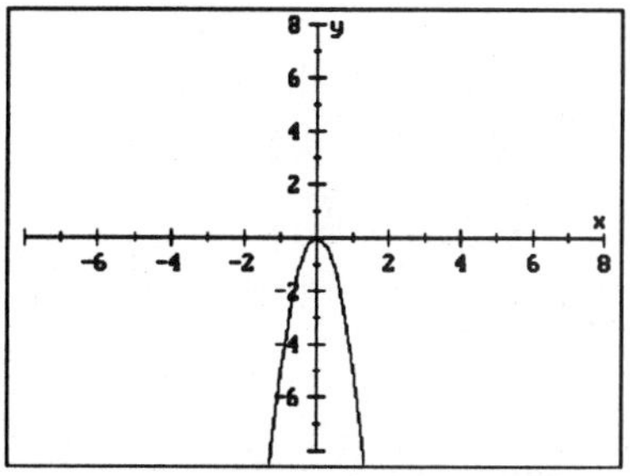

7. See the discussion following Example 2.

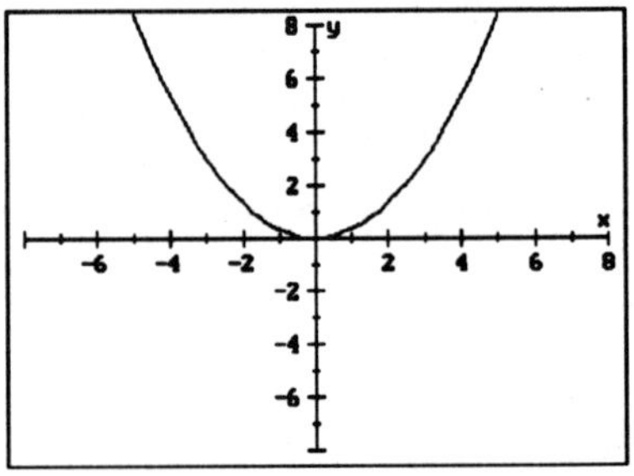

This has the form $y = ax^2$ where $a = \frac{1}{3}$.

The parabola opens up from the vertex

(0, 0) since $a > 0$. Some points are:

Let $x = 1$, then $y = (\frac{1}{3})(1)^2 = \frac{1}{3}$

Let $x = 3$, then $y = (\frac{1}{3})(3)^2 = 3$

Let $x = 6$, then $y = (\frac{1}{3})(6)^2 = 12$

Using the symmetry with respect to the vertical axis, the graph is shown on the right.

9. See the discussion following Example 2.

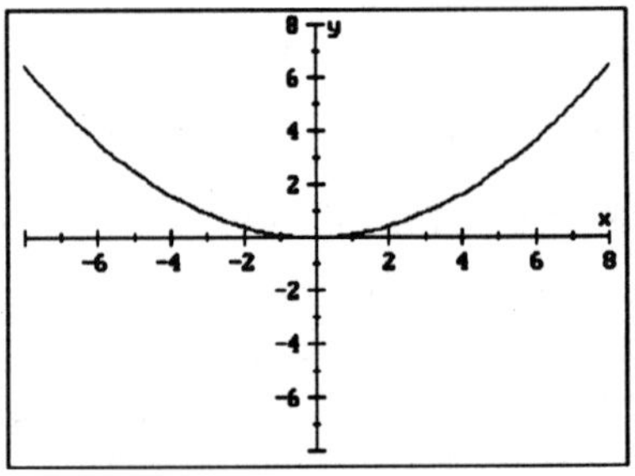

This has the form $y = ax^2$ where $a = \frac{1}{10}$.

The parabola opens up from the vertex

(0, 0) since $a > 0$. Some points are:

Let $x = 1$, then $y = \frac{1}{10}(1)^2 = \frac{1}{10}$

Let $x = 3$, then $y = \frac{1}{10}(3)^2 = \frac{9}{10}$

Let $x = 5$, then $y = \frac{1}{10}(5)^2 = \frac{5}{2}$

Using the symmetry with respect to the vertical axis, the graph is shown on the right.

11. See the discussion following Example 2.

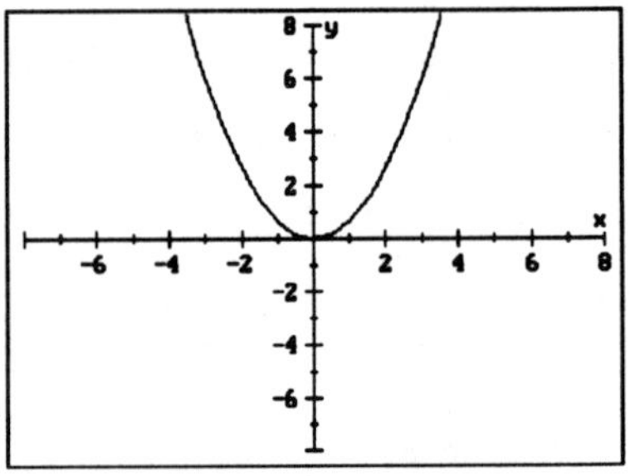

This has the form $y = ax^2$ where $a = \frac{2}{3}$.

The parabola opens up from the vertex

(0, 0) since $a > 0$. Some points are:

Let $x = 1$, then $y = \frac{2}{3}(1)^2 = \frac{2}{3}$

Let $x = 2$, then $y = \frac{2}{3}(2)^2 = \frac{8}{3}$

Let $x = 3$, then $y = \frac{2}{3}(3)^2 = 6$

Use the symmetry with respect to the vertical axis.

13. Vertex is (1, 0)

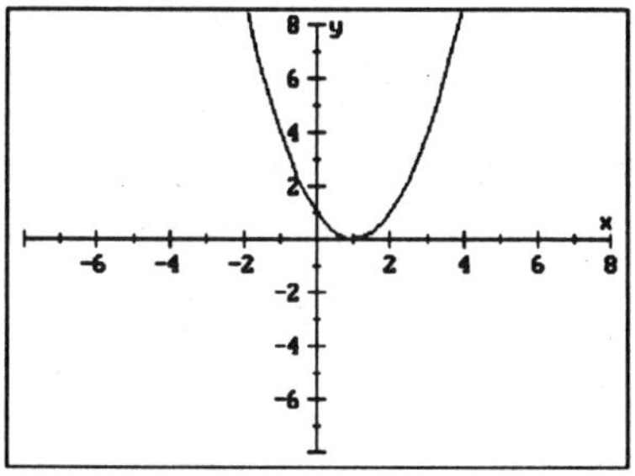

15. Vertex is (– 3, 0)

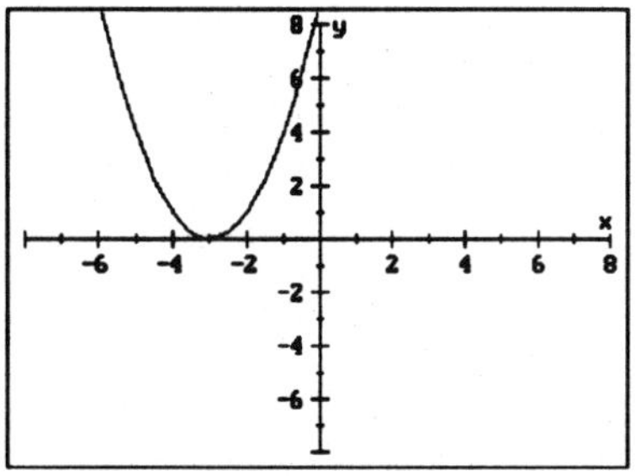

17. Vertex is (1, 0)

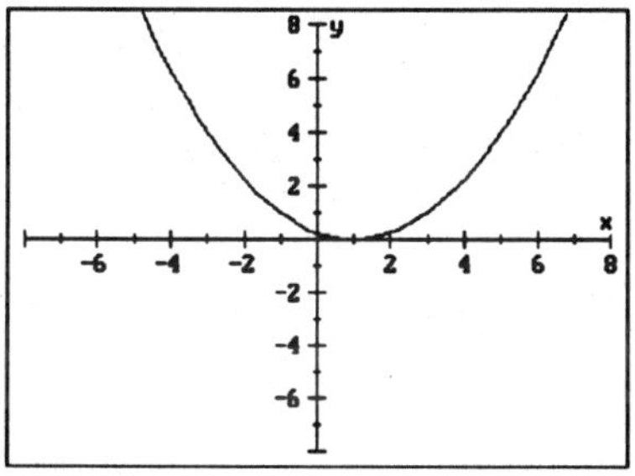

19. Vertex is (1, 2)

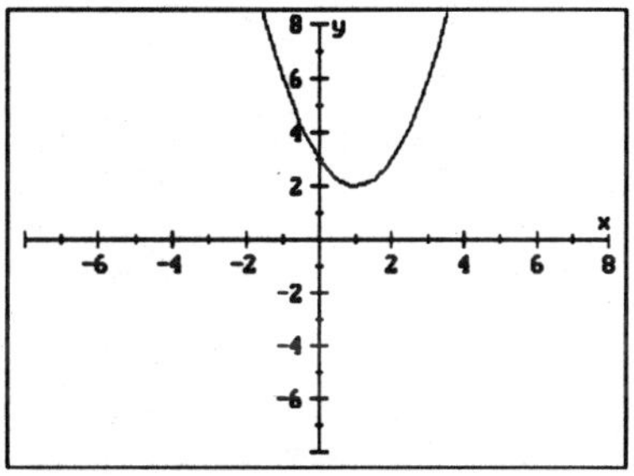

21. Vertex is (1, 2)

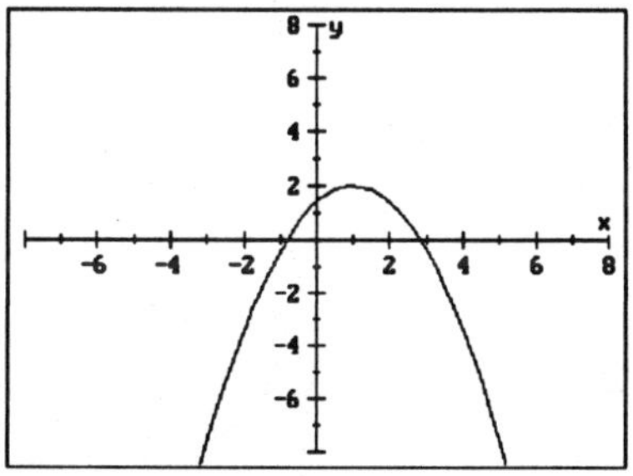

23. Vertex is $\left(-\frac{1}{3}, -\frac{2}{3}\right)$

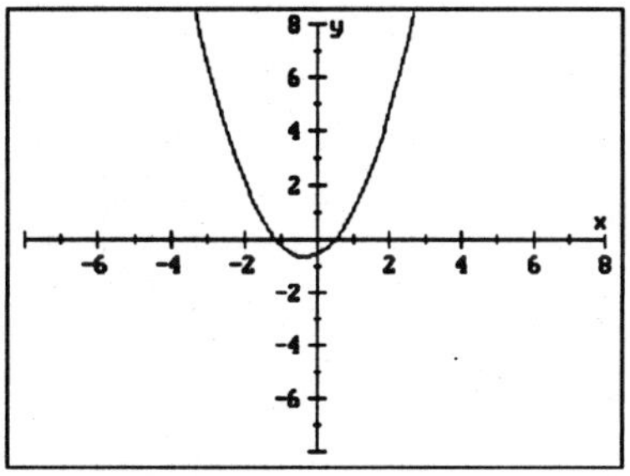

25. Vertex is $\left(\frac{3}{5}, -\frac{2}{5}\right)$

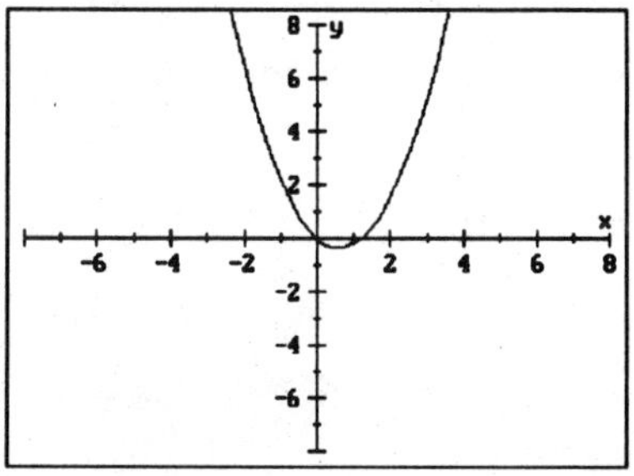

27. $y = (x - 2)^2$

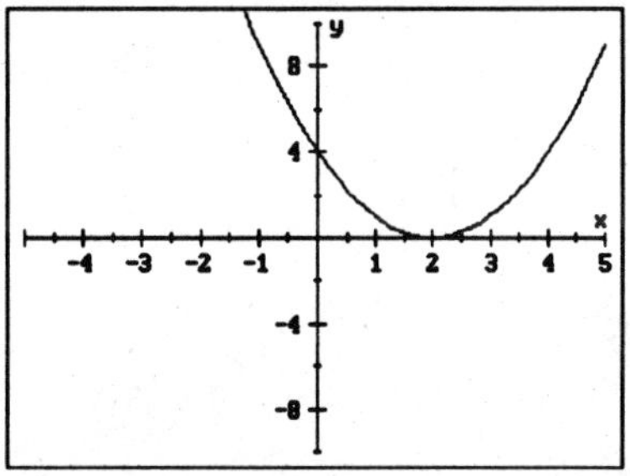

29. $y = -2(x + 1)$

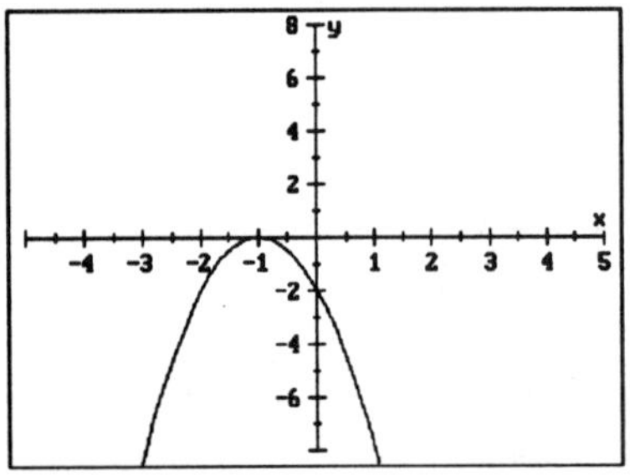

31. $y - 1 = (x + 2)^2$

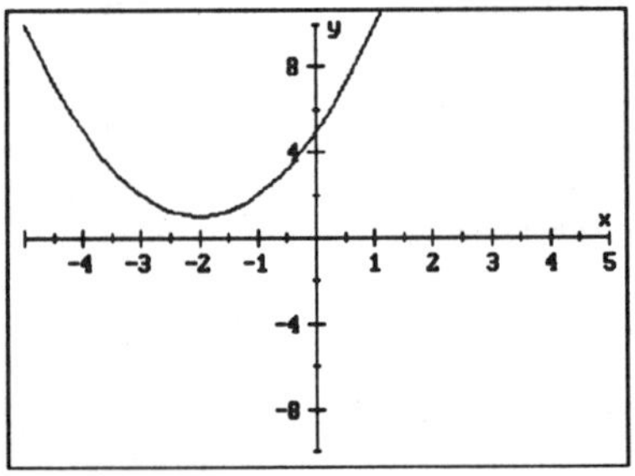

33. $y + 2 = 2(x + 2)^2$

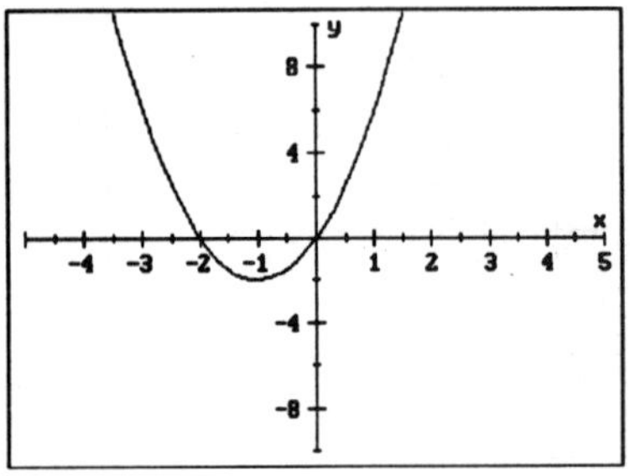

35. $y - 3 = -2(x + 1)^2$

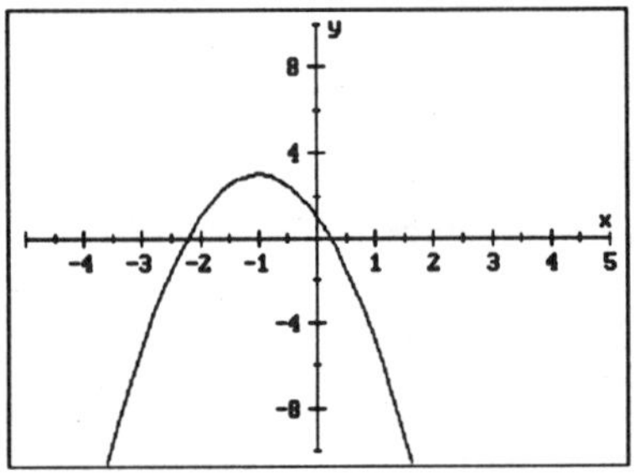

37. $y + 4 = -3(x - 2)^2$

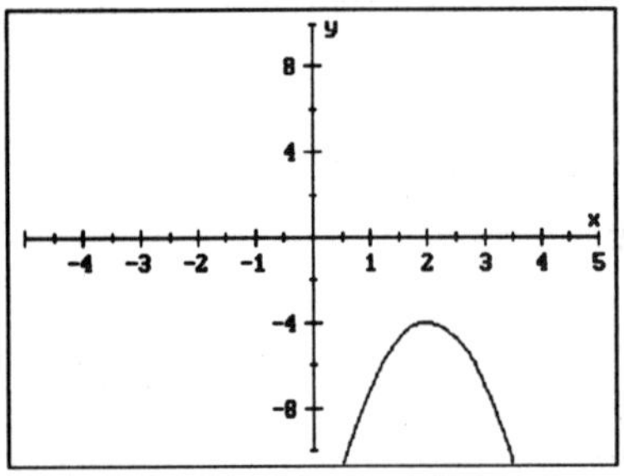

39. Subtract 3 from both sides to get $y - 3 = -4(x + 1)^2$. This has the form $y - k = a(x - h)^2$ where $a = -4$ and $(h, k) = (-1, 3)$. The graph is a parabola opening down (since $a < 0$) with vertex $(-1, 3)$. Hence, the maximum value of y is 3, occurring when $x = -1$.

41. Subtract 1,250 from both sides to get $y - 1{,}250 = -10(x - 450)^2$. This has the form $y - k = a(x - h)^2$ where $a = -10$ and $(h, k) = (450, 1250)$. The graph is a parabola opening down (since $a < 0$) with vertex $(450, 1250)$. Hence, the maximum value of y is 1,250, occurring when $x = 450$.

43. Add 1,400 to both sides to get $y + 1400 = 25(x - 560)^2$. This has the form $y - k = a(x - h)^2$ where $a = 25$ and $(h, k) = (560, -1400)$. The graph is a parabola opening up (since $a > 0$) with vertex $(560, -1400)$. Hence, the minimum value of y is $-1{,}400$, occurring when $x = 560$.

45. Complete the square to find $y + 14 = -3(x + 3)^2$. We see $a = -3$. The graph is a parabola opening down (since $a < 0$) with vertex $(-3, -14)$. Hence, the maximum value of y is -14, occurring when $x = -3$.

47. Complete the square to find $y - 13 = 2(x + 3)^2$. This has the form $y - k = a(x - h)^2$ where $a = 2$ and $(h, k) = (-3, 13)$. The graph is a parabola opening up (since $a > 0$) with vertex $(-3, 13)$. Hence, the minimum value for y is 13, occurring at $x = -3$.

49. Complete the square to find $y + 4 = -2(x - 1)^2$. The graph is a parabola opening down, since $a = -2$, with vertex $(1, -4)$. Hence, the maximum is $y = -4$, which occurs when $x = 1$.

51. a. $R(x) = (\text{Number of items})(\text{demand}) = x(1{,}040 - 10x) = -10x^2 + 1{,}040x$

b. The break-even points occur when $R(x) = C(x)$:

$$-10x^2 + 1{,}040x = 6{,}650 + 500x$$
$$-10x^2 + 540x - 6{,}650 = 0$$
$$x^2 - 54x + 665 = 0$$
$$(x - 19)(x - 35) = 0$$
$$x = 19, 35$$

$R(19) = 16{,}150$ and $R(35) = 24{,}150$ so the break-even points are (19, 16150) and (35, 24150).

53. a. First write the profit function in the form $P(x) - k = a(x - h)^2$. The graph of $P(x) - 1{,}156{,}250 = -10(x - 375)^2$ is a parabola which opens downward from the vertex (375, 1156250). The maximum profit is 1,156,250 which occurs when 375 boats are produced; that is 375 should be produced. **b.** The loss is \$250,000, since when $x = 0$, $P(0) = -10(0 - 375)^2 + 1{,}156{,}250 = -\$250{,}000$.

c. The maximum profit is \$1,156,150. (See part a.)

55.

$$
\begin{aligned}
P(x) &= \text{Revenue} - \text{Cost} \\
&= x(50 - x) - (200 + 20x) \\
&= 50x - x^2 - 200 - 20x \\
&= -x^2 + 30x - 200 \\
P(x) + 200 - 225 &= -(x^2 - 30x + 15^2) \\
P(x) - 25 &= -(x - 15)^2
\end{aligned}
$$

The maximum is 25 which occurs when $x = 15$.

57. The graphs of $R(x) = 50x - x^2$ and $C(x) = 200 + 20x$ are shown on the right. They intersect at (20, 600) and (10, 400).

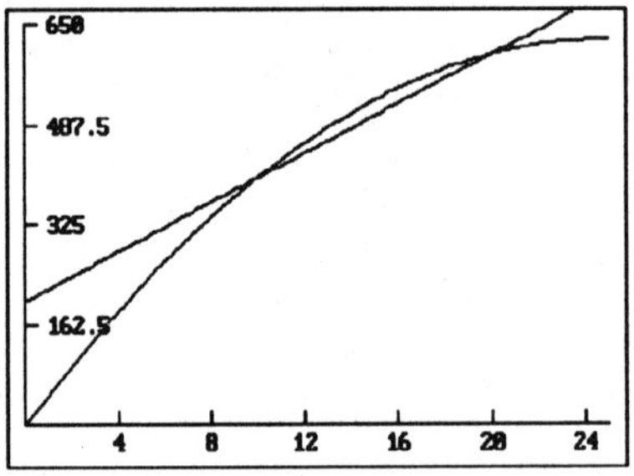

59. The graph of $d(x) = -.0005(x - 2{,}390)^2 + 3{,}456$ is a parabola with vertex at (2390, 3456). The maximum height was 3,456 units.

1.5 Rational Functions

1. A rational function is a function of the form $\dfrac{P(x)}{Q(x)}$ where $P(x)$ and $Q(x)$ are polynomials and $Q(x) \neq 0$.

3. A horizontal asymptote will be present if: (a) the degrees of the numerator and denominator are equal, in which case the ratio of leading coefficients gives the value of the asymptote or (b) the degree of the numerator is less than that of the denominator, in which case the x-axis will be the asymptote.

5. Set the denominator equal to zero and solve: $2x - 3 = 0 \Rightarrow x = \frac{3}{2}$. There is a vertical asymptote at $x = \frac{3}{2}$. Since the denominator has degree 1 which is greater than the degree of the numerator (which is 0), the horizontal asymptote is $y = 0$ (the x-axis).

7. $\dfrac{x+3}{x^2 + 2x - 3} = \dfrac{x+3}{(x+3)(x-1)} = \dfrac{1}{x-1}$; $x \neq -3$ Deleted point at $x = -3$. Vertical asymptote $x = 1$. Since the degree of the denominator (2) is greater than that of the numerator (0), the horizontal asymptote is $y = 0$ (the x-axis).

9. Set the denominator equal to zero: $6x^2 + 5x - 6 = 0 \Rightarrow (2x + 3)(3x - 2) = 0$ $\Rightarrow x -\frac{3}{2}$ or $x = \frac{2}{3}$. The vertical asymptotes are $x = -\frac{3}{2}$ and $x = \frac{2}{3}$. Since the degree of the denominator (2) is greater than the degree of the numerator (0), the horizontal asymptote is $y = 0$ (the x-axis).

11. Set the denominator equal to zero: $2x^2 + 4x + 1 = 0 \Rightarrow x = \dfrac{-2 \pm \sqrt{2}}{2}$. There is a vertical asymptote at $x = \dfrac{-2 \pm \sqrt{2}}{2}$. Since the degree of the denominator equals that of the numerator $(2 = 2)$, the horizontal asymptote is at the ratio of leading coefficients: $y = \frac{5}{2}$.

13. Set the denominator equal to zero: $x^2 + x + 2 = 0$. Since this quadratic equation has no solution, there is no vertical asymptote. Since the degree of the denominator (2) is greater than the degree of the numerator (1), the horizontal asymptote is $y = 0$ (the x-axis).

15. Set the denominator equal to zero: $x^2 + x - 2 = 0 \Rightarrow (x - 2)(x - 1) = 0 \Rightarrow$ $x -2$ or $x = 1$. The vertical asymptotes are $x = -2$ and $x = 1$. Since the degree of the denominator (2) is greater than that of the numerator (1) the horizontal asymptote is $y = 0$ (the x-axis).

17. Set the denominator equal to zero: $x = 0$ is the vertical asymptote. Since the degree of the denominator (1) is greater than that of the numerator (0), the horizontal asymptote is $y = 0$. A few points are

Let $x = -2$, then $y = -(\frac{1}{2}) = \frac{1}{2}$

Let $x = -1$, then $y = -(\frac{1}{-1}) = 1$

Let $x = -\frac{1}{2}$, then $y = \left(\frac{1}{-1/2}\right) = 2$

Let $x = \frac{1}{2}$, then $y = -\left(\frac{1}{1/2}\right) = -2$

Let $x = 1$, then $y = -(\frac{1}{1}) = -1$

Let $x = 2$, then $y = -(\frac{1}{2}) = -\frac{1}{2}$

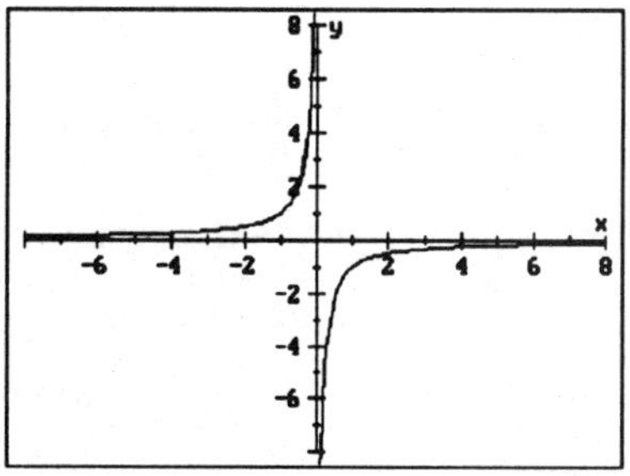

19. Set the denominator equal to zero:

$x - 4 = 0 \Rightarrow x = 4$ is the vertical asymptote.

Since the degree of the denominator (1) is greater than that of the numerator (0), the horizontal asymptote is $y = 0$. A few points are:

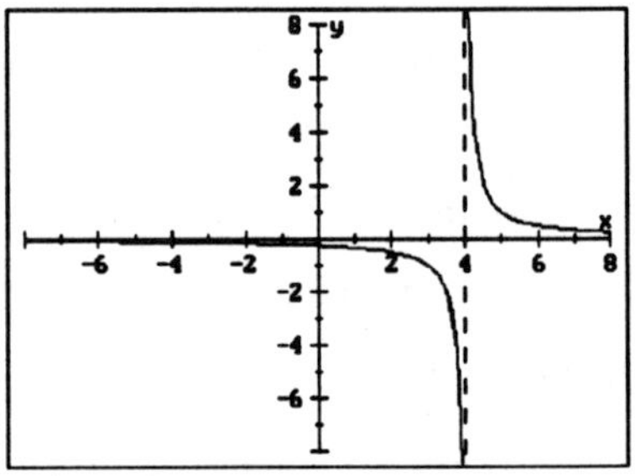

Let $x = 0$, then $y = \frac{1}{0-4} = -\frac{1}{4}$

Let $x = 2$, then $y = \frac{1}{2-4} = -\frac{1}{2}$

Let $x = 3$, then $y = \frac{1}{3-4} = -1$

Let $x = 3.5$ then $y = \frac{1}{3.5-4} = -2$

Let $x = 4.5$ then $y = \frac{1}{4.5-4} = 2$

Let $x = 6$, then $y = \frac{1}{6-4} = -\frac{1}{2}$

21. Add $\frac{1}{x} + 3 = \frac{1}{x} + \frac{3x}{x} = \frac{3x+1}{x}$. Setting the denominator equal to zero, the vertical asymptote is $x = 0$. Since the degree of the numerator equals that of the denominator, the horizontal asymptote is the ratio of leading coefficients: $y = \frac{3}{1}$.

A few points are:

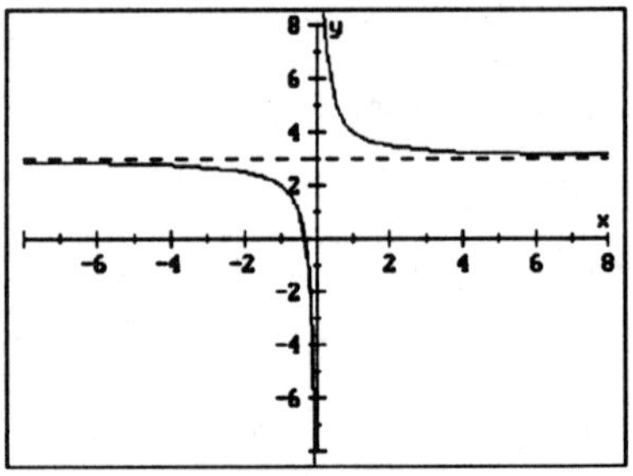

Let $x = -1$, then $y = \dfrac{3(-1)+1}{-1} = 2$

Let $x = -\frac{1}{3}$, then $y = \dfrac{3(-\frac{1}{3})+1}{-\frac{1}{3}} = 0$

Let $x = \frac{1}{3}$, then $y = \dfrac{3(\frac{1}{3})+1}{\frac{1}{3}} = 6$

Let $x = -1$, then $y = \dfrac{3(1)+1}{1} = 4$ You should generate more ordered pairs.

23. First factor and reduce: $y = \dfrac{2x^2 - 3x + 1}{x^2 + 2x - 3} = \dfrac{(2x-1)(x-1)}{(x+3)(x-1)} = \dfrac{2x-1}{x+3}$.

Note that $x \neq 1$. Now set the denominator equal to zero: $x + 3 = 0 \Rightarrow x = -3$ is the vertical asymptote. Since the degrees of numerator and denominator are equal, the horizontal asymptote is the ratio of leading coefficients: $y = \frac{2}{1} = 2$.

A few points are:

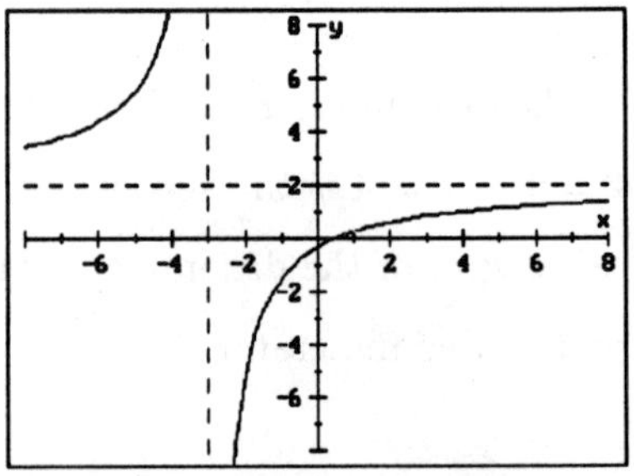

Let $x = -5$, then $y = \dfrac{2(-5)-1}{-5+3} = \frac{11}{2}$

Let $x = -4$, then $y = \dfrac{2(-4)-1}{-4+3} = 9$

Let $x = -2$, then $y = \dfrac{2(-2)-1}{-2+3} = -5$

25. $y = \dfrac{1}{x^2 - 9}$ vertical asymptote: $x^2 - 9 = 0$

$$(x - 3)(x + 3) = 0$$

$$x = 3,\ x = -3$$

horizontal asymptote $y = 0$

A few points are:

Let $x = \pm 4$, then y $= \dfrac{1}{16 - 9} = \dfrac{1}{7}$

Let $x = \pm 2$, then $y \dfrac{1}{4 - 9} = \dfrac{-1}{5}$

Let $x = 0$, then $y = -\dfrac{1}{9}$

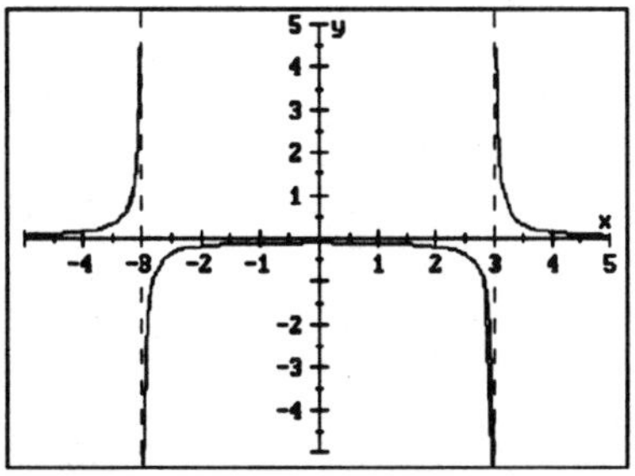

27. $y = \dfrac{1}{x^2 + 2x - 3} = \dfrac{1}{(x + 3)(x - 1)}$ vertical asymptotes $x = -3$, $x = 1$;

horizontal asymptote $y = 0$

A few points are:

$x = -4$, then $y = \dfrac{1}{5}$

$x = -2$, then $y = \dfrac{-1}{3}$

$x = 0$, then $y = \dfrac{-1}{3}$

$x = 2$, then $y = \dfrac{1}{5}$

$x = 3$, then $y = 12$

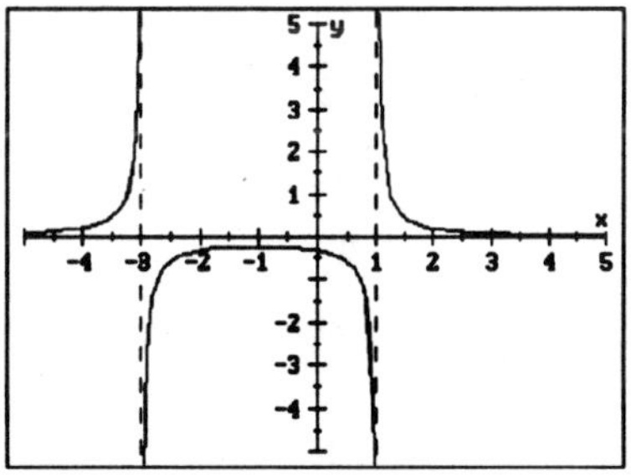

29. Let $p = 80$: $C = \dfrac{40{,}000(80)}{110 - 80} = \$106{,}667$.

31. Setting the denominator equal to zero, we get a vertical asymptote at $p = 110$. However, this is not in the restricted domain $0 \leq p \leq 100$. A few points are:

Let $p = 0$, $C = \dfrac{40{,}000(0)}{110 - 0} = 0$

Let $p = 20$, $C = \dfrac{40{,}000(20)}{110 - 20} \approx 8{,}889$

Let $p = 40$, $C = \dfrac{40{,}000(40)}{110 - 40} \approx 22{,}857$

Let $p = 60$, $C = \dfrac{40{,}000(60)}{110 - 60} = 48{,}000$

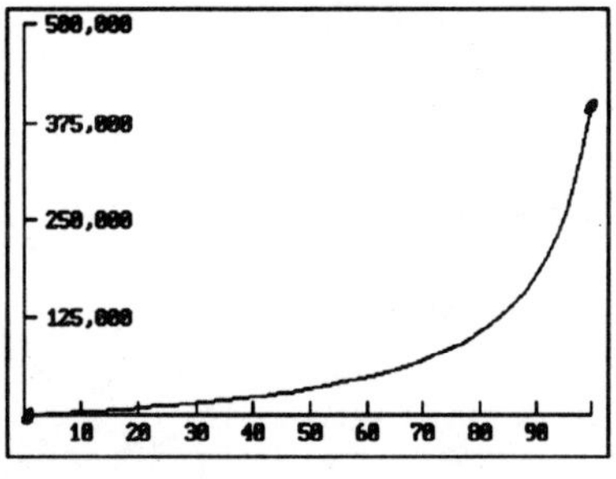

33. Let $p = 80$: $C = \dfrac{18{,}000(80)}{100 - 80} = \$72{,}000$

35. This is a rational function with vertical asymptote at $p = 100$.

A few points are:

Let $p = 0$, $C = \dfrac{18{,}000(0)}{100 - 0} = 0$

Let $p = 20$, $C = \dfrac{18{,}000(20)}{100 - 20} = 4500$

Let $p = 40$, $C = \dfrac{18{,}000(40)}{100 - 40} = 12{,}000$

Let $p = 60$, $C = \dfrac{18{,}000(60)}{100 - 60} = 27{,}000$

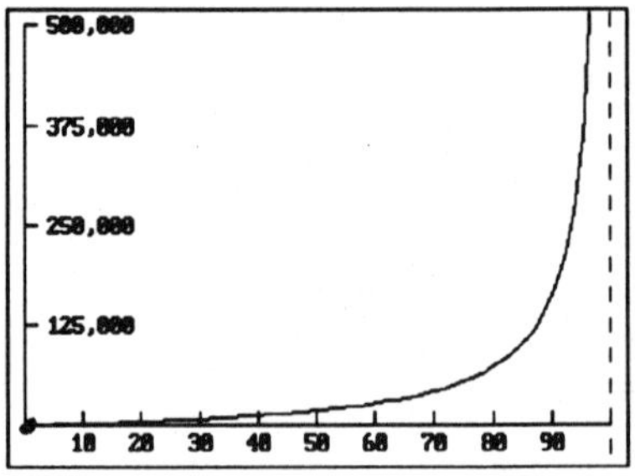

37. This is a rational function with vertical asymptote at $x = 0$. A few points are:

Let $x = 1$, $MA = \dfrac{256}{1^2} = 256$

Let $x = 2$, $MA = \dfrac{256}{2^2} = 64$

Let $x = 4$, $MA = \dfrac{256}{4^2} = 16$

Let $x = 6$, $MA = \dfrac{256}{6^2} = 7$

Let $x = 10$, $MA = \dfrac{256}{10^2} = 2.6$

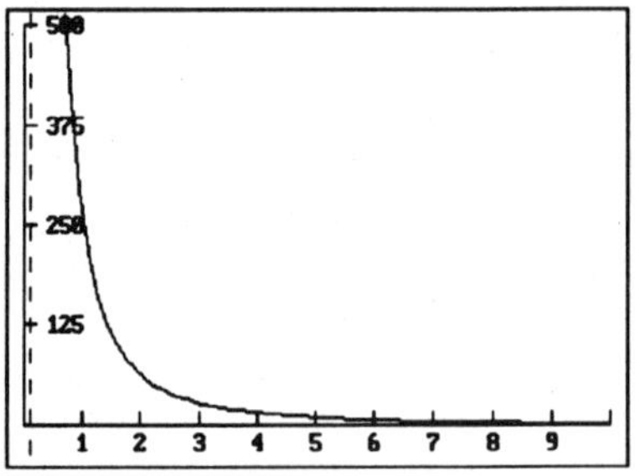

39. Using $\dfrac{\text{original pressure}}{\text{new pressure}} = \dfrac{\text{new volume}}{\text{original volume}}$,

replace original pressure with 2800 and original

volume with 400: $\dfrac{2800}{\text{new pressure}} = \dfrac{\text{new volume}}{400} \Rightarrow$

new pressure $= \dfrac{1{,}120{,}000}{\text{new volume}}$.

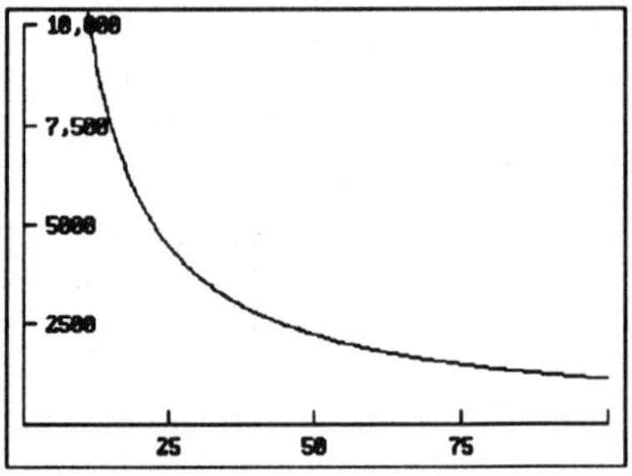

1.6 Review

Objective Questions

1. $f(10) = 5(10) - 3 = 47$

3. $g(-2) = 2(-2)^2 - 3(-2) + 1 = 15$

5. $f(s+3) = 3(s+3)^2 = 3s^2 + 18s + 27$

7. $f(x+h) - f(x) = 3(x+h)^2 - 3x^2 = 3x^2 + 6xh + 3h^2 - 3x^2 = 6xh + 3h^2$

9. The coordinates of A are (2, 3). Since for any value on the x-axis we choose there is a corresponding point on the graph, the domain is all real numbers. Since for any value on the y-axis less than or equal to 4 there is a corresponding point (in fact, 2 points) on the graph, the range is $y \le 4$. The x-intercepts are $(-1, 0)$ and (3, 0) and the y-intercept is (0, 3).

11. Coordinates of C are $(x_0 + h, f(x_0 + h))$. The domain is $s \le x \le t$; the range is $r \le y \le q$. The x-intercept is at d; the y-intercept is at q.

13.

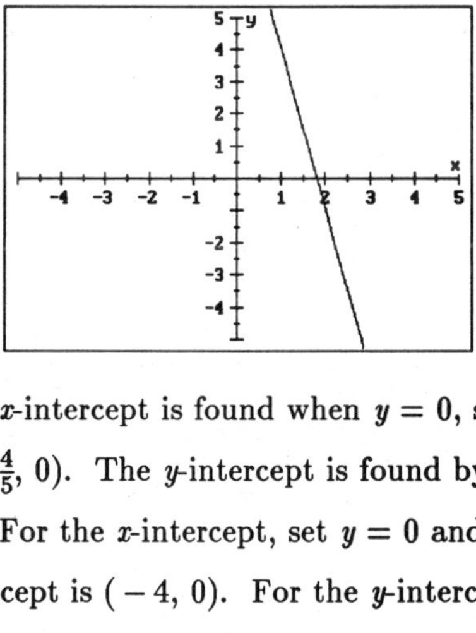

15.

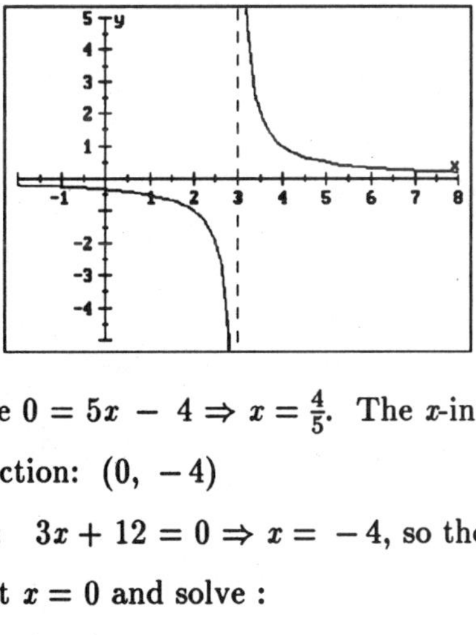

17. x-intercept is found when $y = 0$, so solve $0 = 5x - 4 \Rightarrow x = \frac{4}{5}$. The x-intercept is: $(\frac{4}{5}, 0)$. The y-intercept is found by inspection: $(0, -4)$

19. For the x-intercept, set $y = 0$ and solve: $3x + 12 = 0 \Rightarrow x = -4$, so the x-intercept is $(-4, 0)$. For the y-intercept, set $x = 0$ and solve : $-4y + 12 = 0 \Rightarrow y = 3$, so the y-intercept is (0, 3).

21. $m = \dfrac{y_2 - y_1}{x_2 - x_1} = \dfrac{-2-1}{-3-4} = \dfrac{-3}{-7} = \dfrac{3}{7}$

23. $m = \dfrac{y_2 - y_1}{x_2 - x_1} = \dfrac{-5-5}{-1-(-3)} = \dfrac{-10}{2} = -5$

25. $m = \frac{-2}{5}$, $b = 2$

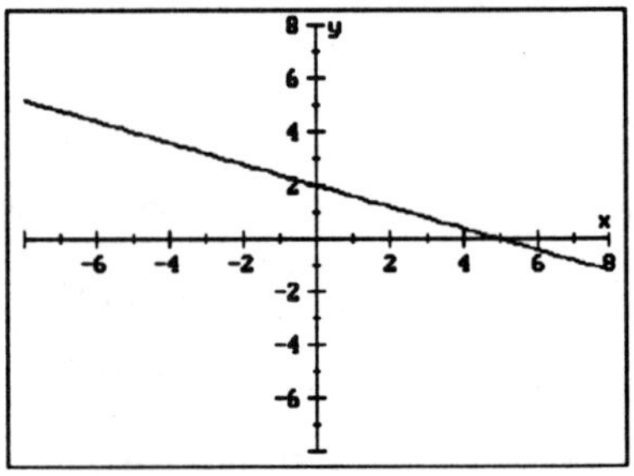

27. First isolate y:

$x + 2y = 8$

$2y = -x + 8$

$y = (-\frac{1}{2})x + 4$

So we start at 4 on the y-axis and count down 1, over 2.

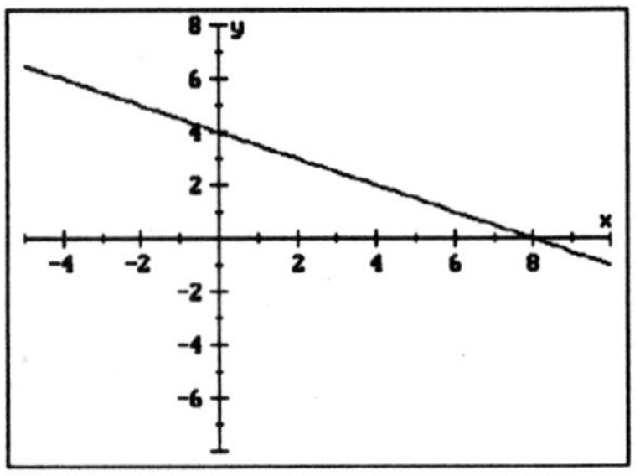

29.

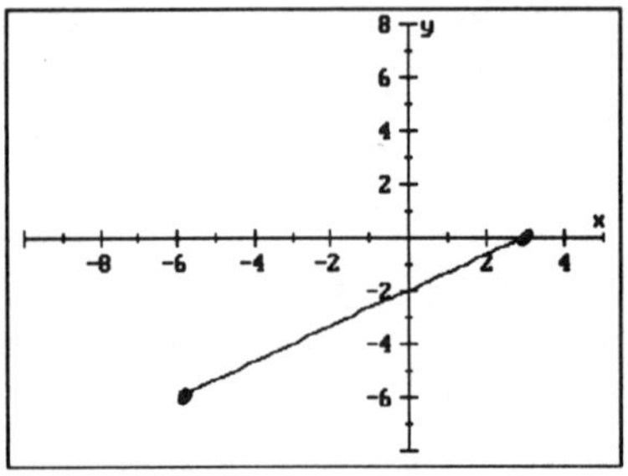

31.

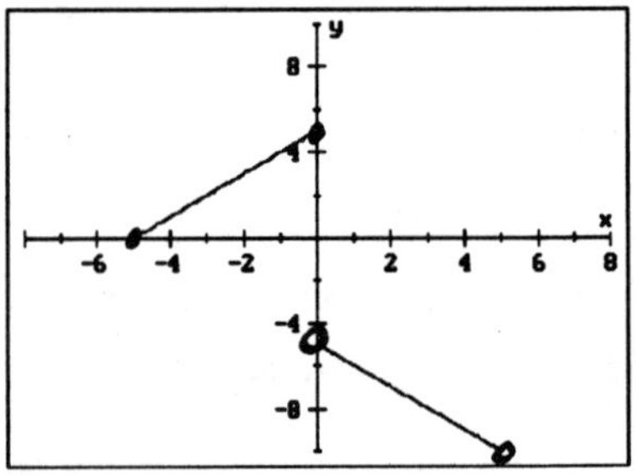

33. $m = \frac{25 - 10}{5 - 2} = \frac{15}{3} = 5$; $y - 10 = 5(x - 2)$ so $5x - y = 0$

35. Use the slope-intercept form

$m = \frac{1}{5}$, $b = -4$ so

$y = \frac{1}{5}x - 4$

$5y = x - 20$

$x - 5y - 20 = 0$

37.

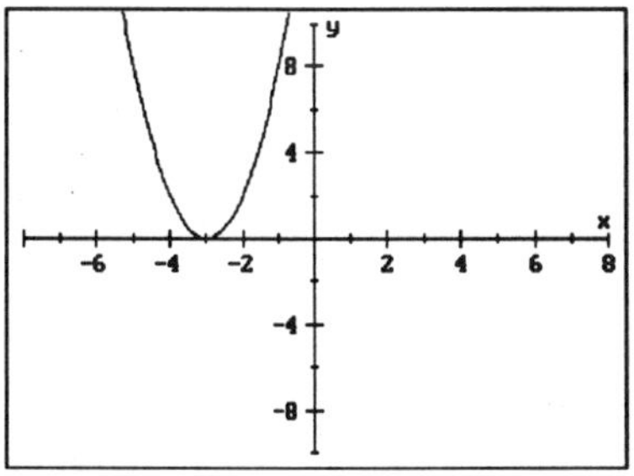

39. This has the form $y - k = a(x - h)^2$ where (h, k) is the vertex; thus the vertex is at $(-2, 4)$ and the related equation is $y = -\frac{2}{3}x^2$. Using $y = -\frac{2}{3}x^2$ some points *relative to the vertex* are:

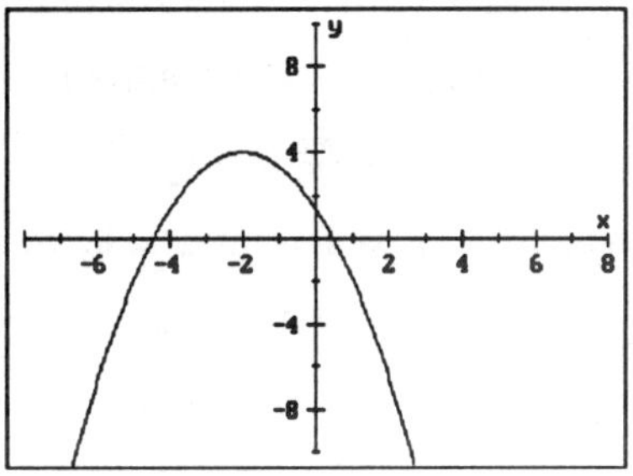

Let $x = 0$, then $y = -\frac{2}{3}(0)^2 = 0$; this is the vertex.
Let $x = 3$, then $y = -\frac{2}{3}(3)^2 = -6$; 3 right and 6 down from the vertex.
Let $x = -1$, then $y = -\frac{2}{3}(-1)^2 = -\frac{2}{3}$; 1 left and $\frac{2}{3}$ down from the vertex.
You should continue the above to generate more points. Also, use the symmetry of the parabola.

41. Maximum: $y = -250$; occurs when $x = 1{,}300$; these values are found by looking directly at the equation to find vertex.

43. Complete the square:

$$y - 268 = 2(x^2 - 6x)$$
$$y - 268 + 18 = 2(x^2 - 6x + 9)$$
$$y - 250 = 2(x - 3)^2$$

Minimum: $y = 250$; occurs when $x = 3$

45. The denominator is zero when $x = 5$, so there is a vertical asymptote at $x = 5$. Since the degree of the numerator is less than the degree of the denominator, there is a horizontal asymptote at $y = 0$.

47. Vertical asymptotes where the denominator is zero:

$$x^2 - x - 6 = 0$$
$$(x - 3)(x + 2) = 0$$
$$x = 3 - 2$$

The vertical asymptotes are at $x = 3$ and $x = -2$; horizontal asymptote at $y = 0$.

49.

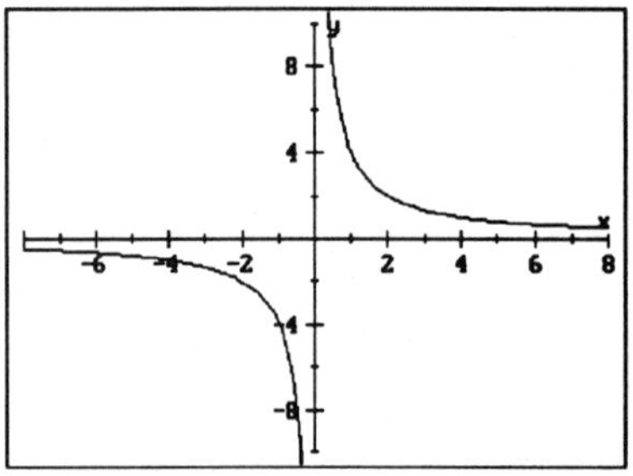

51. There are at least two ways to graph this. One way is to realize it is the same graph as in Problem 49 but shifted 3 units up. Another way is to add:

$g(x) = \frac{4}{x} + 3 = \frac{4}{x} + \frac{3x}{x} = \frac{3x + 4}{x}$

The denominator is zero at $x = 0$, which is equation of the vertical asymptote. Since the degree of $3x + 4$ is the same as the degree of x, there is a horizontal asymptote at $y = \frac{3}{1} = 3$. Some points are:

Let $x = 1$, then $y = \frac{4}{1} + 3 = 7$

Let $x = -1$, then $y = \frac{4}{(-1)} + 3 = -1$

Let $x = 2$, then $y = \frac{4}{2} + 3 = 5$

You should generate more points as above.

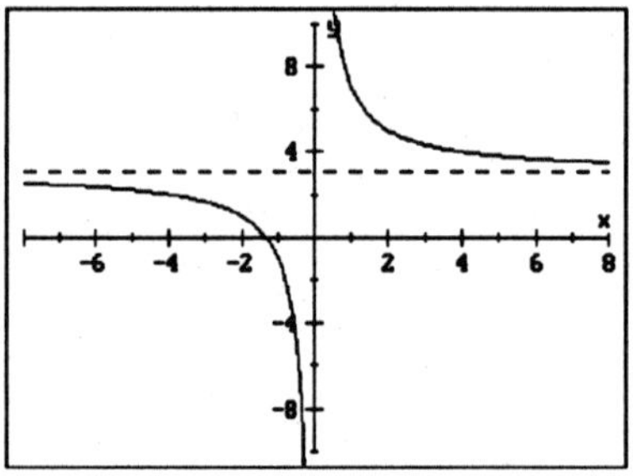

53. $\dfrac{Z(1990) - Z(1985)}{5}$

55. We have two points: (200, 2000) and (400, 8000). To find a linear equation relating supply (y) with cost (x), first compute the slope:

$m = \frac{y_2 - y_1}{x_2 - x_1} = \frac{8{,}000 - 2{,}000}{400 - 200} = 30$

Now use either ordered pair in the point-slope form; we will use (200, 2000):

$y - y_1 = m(x - x_1) \Rightarrow y - 2{,}000 = 30(x - 200)$ Now solve for y:

$y = 30x - 4{,}000$. In function notation, call $S(x)$ the supply function:

$S(x) = 30x - 4{,}000$

57. Break-even occurs when cost equals revenue:

$$25{,}000 + 1{,}000x = C(x) = R(x) = 4{,}000x - 50x^2$$

$$50x^2 - 3{,}000x + 25{,}000 = 0$$

$$x^2 - 60x + 500 = 0 \qquad \text{(divided by 50)}$$

$$(x - 10)(x - 50) = 0 \qquad \text{(factor)}$$

$$x = 10 \text{ or } x = 50$$

Break-even points occur at (10, 35000) and(50, 75000).

59. $C(95) = \dfrac{20{,}000(95)}{100 - 95} = \$380{,}000$

Sample Test

1. $f(3) = 2(3)^2 - 3(3) + 5 = 14$

3. $g(t + 1) = 1 - 4(t + 1) = -4t - 3$

5. $\dfrac{g(x+h) - g(x)}{h} = \dfrac{[1 - 4(x+h)] - [1 - 4x]}{h} = \dfrac{1 - 4x - 4h - 1 + 4x}{h}$

$$= \frac{-4h}{h} = -4$$

7. This is a linear equation.

A few points are:

Let $x = -1$, then $g(-1) = 1 - 4(-1) = 5$

Let $x = 0$, then $g(0) = 1 - 4(0) = 1$

Let $x = 1$, then $g(1) = 1 - 4(1) = -3$

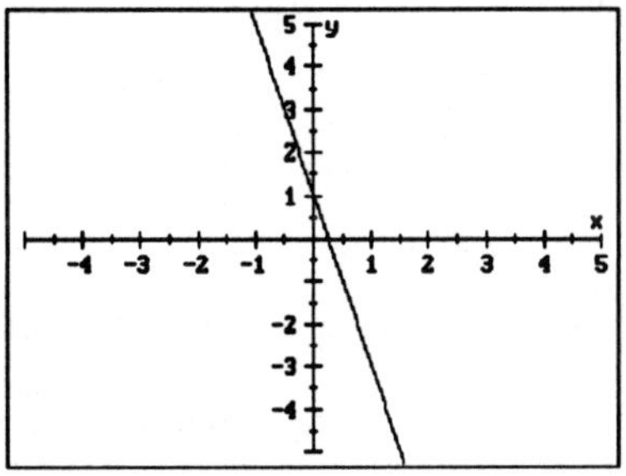

9. slope $= \dfrac{y_2 - y_1}{x_2 - x_1} = \dfrac{(-7) - (9)}{(-3) - (-5)} = \dfrac{-16}{2} = -8$

11. A few points are:

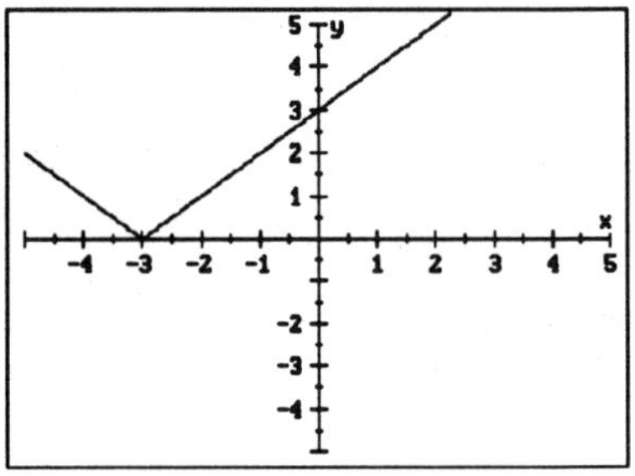

Let $x = -4$, then $y = |-4+3| = 1$

Let $x = -3$, then $y = |-3+3| = 0$

Let $x = -2$, then $y = |-2+3| = 1$

Also, remember that $|a| = \begin{cases} a \text{ if } a \geq 0 \\ -a \text{ if } a < 0 \end{cases}$

Therefore, $y = \begin{cases} x+3, \text{ if } x \geq 0 \\ -(x+3), \text{ if } x+3 < 0 \end{cases}$

This simplifies to

$$y = \begin{cases} x+3, \text{ if } x + \geq -3 \\ -x - 3, \text{ if } x < -3 \end{cases}$$

13. Using the point-slope form; $y - y_1 = m(x - x_1)$:

$$y - 20 = -5(x - 30)$$

$$y - 20 = -5x + 150$$

$$5x + y - 170 = 0.$$

15. Vertical lines have no slope. The vertical line passing through $(-3, -2)$ is $x = -3$ or $x + 3 = 0$.

17. This has the form $y - k = a(x - h)^2$, where $a = -\frac{2}{3}$ and $(h, k) = (40, 550)$. The graph is a parabola opening down (because $a < 0$) from the vertex (40, 550). Hence, the maximum value of y is 550, which occurs when $x = 40$.

19. Substitute $p = 90$: $C(p) = \dfrac{500(90)}{100 - 90} = \$4{,}500$

Modeling Application 1: Sample Essay

Gaining a Competitive Edge in Business; A Case Study

What might a consulting firm do for Solartex? In this chapter we have discussed mathematical models, formulas, graphs, lines, and systems. We will use all those ideas to analyze Solartex's situation, just as a consulting firm would. Let us begin by writing a cost function for Solartex. There are fixed costs of $8,500 per month and variable costs of \$95 + \$55 = \$150 per panel:

$$C = 150x + 8{,}500$$

where C is the monthly cost and x is the number of panels produced for the month. Does this cost function make sense?

Since Solartex is a small company, expenditures must be limited to \$20,000 per month. How many panels can be manufactured and not exceed this limit? Let $C = 20{,}000$ and solve for x:

$$20{,}000 = 150x + 8{,}500$$
$$11{,}500 = 150x$$
$$76\tfrac{2}{3} = x$$

This means that production must be limited to 76 panels per month. That is, the maximum number of panels that can be made per month is 76. The graph showing the cost function is given in Figure 1. Notice that the slope of the line is the cost per panel.

Table 1 Solartex's Monthly Costs

x	C
0	\$ 8,500
10	\$10,000
50	\$16,000
75	\$19,750
100	\$23,500

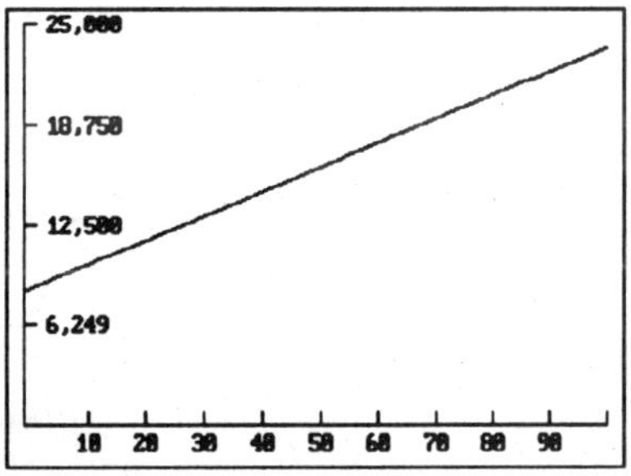

Figure 1 Cost function for Solartex solar panels for $0 \le x \le 76$

What is the minimum number of panels Solartex should make per month? To answer this, consider the number of panels manufactured as a function of the cost; that is, solve the equation for x:

$$C = 150x + 8{,}500$$
$$C - 8{,}500 = 150x$$
$$x = \frac{1}{150}C - \frac{8{,}500}{150}$$
$$x \approx .007C - 56.67$$

Consider the graph of this equation. Since $C \ge 8{,}500$ and $0 \le x \le 76$, we are interested only in that part of the graph shown in Figure 2.

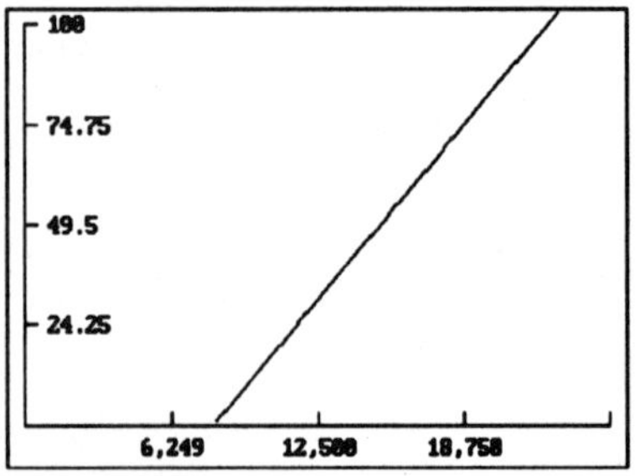

Figure 2 Number of panels as a function of cost

Notice how the number of panels produced depends on the amount of money spent. Also, since the slope of this line is about .007, we see that every dollar that Solartex spends buys about $\frac{7}{1,000}$ of a panel.

In order to make a profit, the revenue must exceed the cost. If R represents the revenue, then at a selling price of \$320 per panel,

$$R = 320x$$

The break-even point is shown in Figure 3.

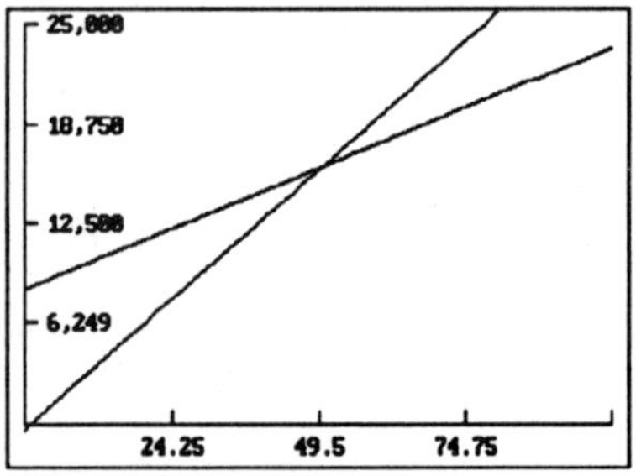

Figure 3 Break-even point is (50, 16000).

Check the point shown in Figure 3 algebraically:

$$\begin{cases} C = 150x + 8{,}500 \\ R = 320x \end{cases}$$

$$\begin{aligned} 150x + 8{,}500 &= 320x && \text{Solution by substitution;} \\ 8{,}500 &= 170x && R = C \text{ at break-even point} \\ 50 &= x \end{aligned}$$

In order to make a profit, Solartex must produce at least 50 panels per month. Thus, the production should be $50 \leq x \leq 76$.

Now, compare the monthly costs of Solartex with that of its competitor,

Sunenergy.
Let

$C =$ Monthly cost of Sunenergy
$x =$ Number of panels manufactured
$p =$ Cost per panel
$f =$ Fixed costs

then

$$C = px + f$$

We have been given the information in Table 2.

Table 2
Sunenergy's
Monthly Costs

x	C
85	\$20,475
60	\$17,100

Thus,
$$\begin{cases} 20{,}475 = p(85) + f \\ 17{,}100 = p(60) + f \end{cases}$$

Solve this system for (p, f):

$$\begin{array}{r} \\ -1 \end{array}\begin{cases} 85p + f = 20{,}475 \\ 60p + f = 17{,}100 \end{cases}$$

$$+\begin{cases} 85p + f = 20{,}475 \\ -60p - f = -17{,}100 \end{cases}$$
$$25p = 3{,}375$$
$$p = 135$$

If $p = 135$, then

$$17{,}100 = 135(60) + f$$
$$f = 17{,}100 - 8{,}100$$
$$f = 9{,}000$$

This means that the cost function for Sunenergy is $C = 135x + 9{,}000$. The overhead for the two companies is very similar, but Sunenergy's costs per panel are \$15 less than those for Solartex. This tells Solartex that the variable costs should be lowered, if possible.

Marginal cost is the amount that it cost a company to produce each additional item once they have set up shop and paid for the fixed costs. That is, Sunenergy has a marginal cost of $135 per panel compared to that of Solartex's marginal cost of $150. Solartex should try to lower the marginal cost.

Now, a consulting firm would also prepare a study of market demand. Suppose it is found that at $75 per panel, Solartex could sell 200 panels per month, but at $450, they could sell only 20 panels per month. On the other hand, if Solartex had to sell the panels at $225 each, they could find a supply of materials that would limit them to 30 panels per month, but if that price rose to $450 each, they could have sufficient materials to supply 100 panels per month. If the supply greatly exceeds demand, then money will be lost because of unsold items. On the other hand, if demand greatly exceeds supply, then money will be lost because of insufficient inventory. The most desirable situation is when the supply and demand are equal. If we assume that supply and demand are linear, then the solution to the system is called the equilibrium point and represents the point for which supply and demand are equal. We can use the graphs in Figure 4 to find the equilibrium point.

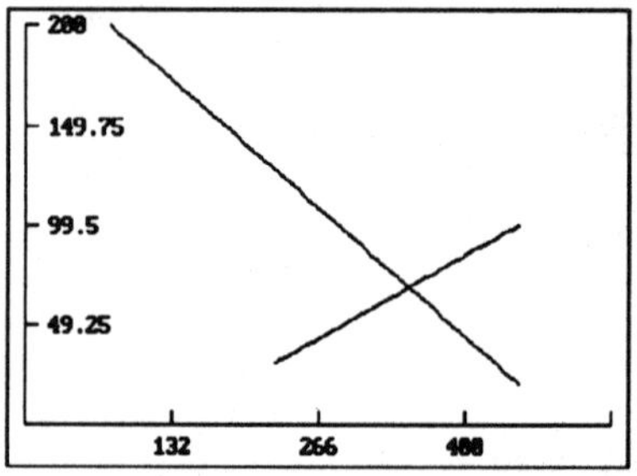

Figure 4 Equilibrium point for sales of solar panels for Solartex

From Figure 4 you can see that the equilibrium point is (350, 70). This suggests that Solartex might consider raising their price to $350 per panel and maintaining their production at 70 panels per month.

CHAPTER 2

The Derivative

2.1 Limits

1. On the graph, as x gets closer and closer to $x = 4$, the y-coordinates get closer and closer to $y = 0$. The limit is 0.

3. As x gets closer and closer to 0, the graph approaches 8, so the limit is 8.

5. On the graph, as x gets closer to -1 (approaching from the left), the y values are always $y = 7$, so the limit is 7.

7. On the graph, as x gets closer to 3 (approaching from the right), the y-values approach (but never reach) 6. The limit is 6.

9. As x approaches 2 from the right, the graph remains a constant value of 2, so the limit is 2.

11. On the graph, as x gets closer to $x = 4$, the y-coordinates approach $y = 2$, so the limit is 2.

13.

x	2	3	4	4.5	4.9	4.99
$f(x)$	3	7	11	13	14.6	14.96

limit is 15

15.

x	1.5	1.9	1.99	1.999	1.999	2.5	2.01	2.001
$f(x)$	-1	-2	-10	-100	$-1{,}000$	2	100	1,000

As x approaches 2, the $f(x)$ values appear to be increasing or decreasing without bound. The limit does not exist.

17. $\lim_{x\to 0} x^8 = 0^8 = 0$

19. $\lim_{x\to 3} (x^2 - 4) = 9 - 4 = 5$

21. $\lim_{x\to -3} \dfrac{1}{x - 3} = \dfrac{1}{-3 - 3} = -\frac{1}{6}$

23. $\lim_{x\to 0} \dfrac{1}{x^2 + 1} = \dfrac{1}{0^2 + 1} = 1$

25. $\lim_{x\to\infty} \frac{1}{x^2+1} = \lim_{x\to\infty} \frac{\frac{1}{x^2}}{1+\frac{1}{x^2}} = 0$

27. $\lim_{x\to -1} \frac{1}{x^2+1} = \frac{1}{(-1)^2+1} = \frac{1}{2}$

29. If we first factor and then reduce: $\frac{x^2-4}{x-2} = \frac{(x+2)(x-2)}{x-2} = x+2,\ x \neq 2.$ The restriction $x \neq 2$ is important! Fortunately, when evaluating the limit as x approaches 2 we do not include $x \neq 2$, but rather only values close to 2. So for our discussion, $\lim_{x\to 2} \frac{x^2-4}{x-2} = \lim_{x\to 2}(x+2) = 4$

31. $\lim_{x\to 3} \frac{x^2+3x-10}{x-2} = \lim_{x\to 3} \frac{(x+5)(x-2)}{x-2} = \lim_{x\to 3}(x+5) = 8$

33. $\lim_{x\to 4} \frac{\sqrt{x}-4}{x-16} = \lim_{x\to 4} \frac{\sqrt{x}-4}{x-16}\cdot\frac{(\sqrt{x}+4)}{(\sqrt{x}+4)} = \lim_{x\to 4} \frac{x-16}{(x-16)(\sqrt{x}+4)}$
$= \lim_{x\to 4} \frac{1}{\sqrt{x}+4} = \frac{1}{6}$

35. We can substitute $x = 9$ to find the limit: $\lim_{x\to 9} \frac{\sqrt{x}-3}{x-3} = \frac{\sqrt{9}-3}{9-3} = 0$

37. A table might help:

x	1.9	1.99	1.999	2.01	2.001
$f(x)$	−36.1	−296.01	−2996.001	304.01	3004.001

$f(x)$ seems to e increasing or decreasing without bound. As x approaches 2 the numerator approaches 3, but the denominator approaches 0, hence the limit does not exist.

39. $\lim_{x\to 2} \frac{x^3-8}{x^2+2x+4} = \lim_{x\to 2} \frac{(x-2)(x^2+2x+4)}{x^2+2x+4} = \lim_{x\to 2}(x-2) = 0.$ (The factoring was not really necessary; you could have substituted $x = 2$ to get $\frac{0}{12} = 0$.)

41. $\lim_{x\to 2} \frac{x+2}{x^3+8} = \lim_{x\to 2} \frac{x+2}{(x+2)(x^2-2x+4)} = \lim_{x\to 2} \frac{1}{x^2-2x+4} = \frac{1}{4}.$ (The factoring was not really necessary; you could have substituted $x = 2$ to get to get $\frac{4}{16}$ $= \frac{1}{4}$).

43. $\lim_{x\to\infty} \frac{2x^2-5x-3}{x^2-9}\cdot\frac{\frac{1}{x^2}}{\frac{1}{x^2}} = \lim_{x\to\infty} \frac{2-\frac{5}{x}-\frac{3}{x^2}}{1-\frac{9}{x^2}} = \frac{2-0-0}{1-0} = 2$

45. $\lim_{|x|\mapsto\infty} \frac{3x-1}{2x+3}\cdot\frac{\frac{1}{x}}{\frac{1}{x}} = \lim_{|x|\mapsto\infty} \frac{3-\frac{1}{x}}{2+\frac{3}{x}} = \frac{3-0}{2+0} = \frac{3}{2}$

47. $\lim_{x\to\infty} \frac{5x+10{,}000}{x-1}\cdot\frac{\frac{1}{x}}{\frac{1}{x}} = \lim_{|x|\mapsto\infty} \frac{5+\frac{10{,}000}{x}}{1-\frac{1}{x}} = \frac{5+0}{1-0} = 5$

49. As $x \to \infty$, $x + 2$ will grow without bound bound and $\frac{3}{x - 1}$ will approach 0:

$\lim_{x\to\infty} (x + 2 + \frac{3}{x - 1}) = \lim_{x\to\infty} (x + 2)$ which grows without bound; the limit does not exist.

51. $\lim_{x\to 0} \frac{1 - \frac{1}{x + 1}}{x} = \lim_{x\to 0} (\frac{x + 1 - 1}{x + 1} \cdot \frac{1}{x}) = \lim_{x\to 0} \frac{1}{x + 1} = \frac{1}{0 + 1} = 1$

53. $\lim_{x\to\infty} \frac{3x^2 - 5x + 15}{x + 3} \cdot \frac{\frac{1}{x^2}}{\frac{1}{x^2}} = \lim_{x\to\infty} \frac{3 - \frac{5}{x} + \frac{15}{x^2}}{\frac{1}{x} + \frac{3}{x^2}}$. Try a table of values to see that as x decreases without bound, the function is decreasing without bound.

55. $\lim_{x\to -\infty} \frac{4x^4 - 3x^3 + 2x + 1}{3x^4 - 9} \cdot \frac{\frac{1}{x^4}}{\frac{1}{x^4}} = \lim_{x\to -\infty} \frac{4 - \frac{3}{x} + \frac{2}{x^2} + \frac{1}{x^4}}{3 - \frac{9}{x^4}} = \frac{4 - 0 + 0 + 0}{3 - 0}$
$= \frac{4}{3}$

57. $\lim_{x\to 0^-} (x^2 + 3) = 3$ and $\lim_{x\to 0^+} (2x + 3) = 3$; $x \to 0$ thus, $\lim_{x\to 0} f(x) = 3$

59. Use $f(x) = 2x + 3$ since $1 > 0$; $\lim_{x\to 1} (2x + 3) = 5$

61. $\lim_{x\to 3^+} (x + 7) = 10$ and $\lim_{x\to 3^-} (x^2 + 1) = 10$; thus, $\lim_{x\to 3} f(x) = 10$.

63. $y = \lim_{|x|\to\infty} \frac{3x^2}{x^2 + 2} \cdot \frac{\frac{1}{x^2}}{\frac{1}{x^2}} = \lim_{|x|\to\infty} \frac{3}{1 + \frac{2}{x^2}} = 3$; horizontal asymptote $y = 3$

65. $y = \lim_{|x|\to\infty} \frac{2x^2 + 5}{3x^2 + 2} \cdot \frac{\frac{1}{x^2}}{\frac{1}{x^2}} = \lim_{x\to\infty} \frac{2 + \frac{5}{x^2}}{3 + \frac{2}{x^2}} = \frac{2}{3}$; horizontal asymptote $y = \frac{2}{3}$

67. $y = \lim_{x\to\infty} \frac{2x^2}{x^2 - 4} \cdot \frac{\frac{1}{x^2}}{\frac{1}{x^2}} = \lim_{x\to\infty} \frac{2}{1 - \frac{4}{x^2}} = 2$; horizontal asymptote $y = 2$

69. a. As n approaches 10 from the left, the graph drops closer to \$7.00 on the cost axis. So the limit is \$7.00. **b.** As n approaches 10 from the right, the graph rises closer and closer to \$12.00 on the cost axis. So the limit is \$12.00. **c.** As n approaches 13 from either side, the graph approaches \$7.00 on the cost axis, so the limit is \$7.00. **d.** Since in parts (a) and (b) we get different limiting values as we approach from both sides, the limit does not exist.

71. a. $\lim_{t \to t_1^-} P(t) = 40\%$ and $\lim_{t \to t_1^+} P(t) = 60\%$; since these are not equal, the limit does not exist. **b.** $\lim_{t \to t_2^-} P(t) = 80\%$, since the closer t gets to t_2 (from the left) the closer $P(t)$ gets to 80% (even though it never reaches 80%). **c.** $\lim_{t \to t_3} P(t) =$ 70% **d.** $\lim_{t \to \infty} \mathrm{P}(t) = 50\%$

73. $\lim_{h \to 0} \dfrac{f(1+h) - f(1)}{h} = \lim_{h \to 0} \dfrac{\sqrt{1+h} - \sqrt{1}}{h} = \lim_{h \to 0} \dfrac{\sqrt{1+h} - 1}{h} \cdot \dfrac{\sqrt{1+h}+1}{\sqrt{1+h}+1}$

$= \lim_{h \to 0} \dfrac{h}{h(\sqrt{1+h}+1)} = \lim_{h \to 0} \dfrac{1}{\sqrt{1+h}+1} = \dfrac{1}{\sqrt{1+0}+1} = \frac{1}{2}$

75. $\lim_{h \to 0} \dfrac{f(3+h) - f(3)}{h} = \lim_{h \to 0} \dfrac{\frac{1}{3+h} - \frac{1}{3}}{h} = \lim_{h \to 0} \left[\dfrac{3 - (3+h)}{3(3+h)} \cdot \dfrac{1}{h}\right] = \lim_{h \to 0}$

$\dfrac{-1}{3(3+h)} = -\frac{1}{9}$

2.2 Continuity

1. Suspicious point $x = 4$. Since the graph has a hole as $x = 4$, this is a point of discontinuity. (The function is not defined at $x = 4$.)

3. Suspicious point $x = 2$. Since the function is not defined at $x = 2$ (a hole), this is a point of discontinuity.

5. Suspicious points at $x = 1$ and $x = 4$; however, $f(1) = 6$ which matches the limiting value as x approaches 1 so the function is continuous at $x = 1$. The function does not exist at $x = 4$ (a hole), so $x = 4$ is a point of discontinuity.

7. Suspicious points at $x = 1$ and $x = 6$. Since the function is not defined at $x = 1$ or $x = 6$ (holes), so both of these are points of discontinuity.

9. Suspicious points at $x = -2$ and $x = 2$. Since the function is not defined at $x = -2$ (a hole), it is a point of discontinuity. Even though the function is defined at $x = 2$, $f(2) = 4$, the left- and right-hand limits do not match (jump), hence the $\lim_{x \to c} f(x)$ does not exist; $x = 2$ is also a point of discontinuity.

11. There are no suspicious points; hence the function is continuous for $x > 0$.

13. $\lim_{x \to 1} f(x) = f(1)$, yes.

15. Since $\lim_{x \to .001} f(x) = f(.001)$, yes.

17. Since there are no abrupt jumps or undefined values, it is continuous. The domain is $0 \le x < 24$, where x is the hour of the day.

19. If the selling price of IBM stock changes, even by only 1 cent, this would be a jump in the graph; hence, this function is not continuous (unless the price remains constant). The domain is $0 \le x < 24$ where x is the hour of the day.

21. As with the solution to Problem 19, even a 1 cent change is a sudden jump in the graph; hence, this function is not continuous. The domain is $t > 0$, where t is the number of minutes.

23. The numerator, 1, is a continuous function and the denominator, $x^2 + 5$, is continuous. All we need to do is check when the denominator is zero, but $x^2 + 5 \ne 0$ so $f(x) = \dfrac{1}{x^2 + 5}$ is continuous on $[-5, 5]$.

25. The numerator, 1, is a continuous function and the denominator, $x^2 - 9$, is continuous. All we need to do is check when the denominator is zero: $x^2 - 9 = 0$ so $x = \pm 3$ (suspicious points) But since neither of these values fall in the domain, the $f(x)$ is continuous on $[5, 10]$.

27. The function is not defined at $x = 1$ or -1, so the function has discontinuities at $x = 1$ and $x = -1$.

29. $f(x) =$ on $[-5, 5]$, $\dfrac{x+2}{x^2 - 6x - 16} = \dfrac{x+2}{(x-8)(x+2)}$. Although both $x = 8$ and $x = -2$ will cause division by zero, only $x = -2$ is in the domain; hence, $x = -2$ is the only discontinuity.

31. Suspicious point $x = 3$ is not in domain; continuous on $[0, 2]$.

33. Suspicious point $x = 8$ is not in domain; continuous on $[0,5]$.

35. $x = 3$ is a suspicious point. Now, $\lim_{x \to 3} f(x) = \lim_{x \to 3} \dfrac{1}{x - 3}$ which does not exist, hence $x = 3$ is a discontinuity.

37. $x = -2$ is the only suspicious point. We have two things to check: 1. $f(-2)$ exists, and 2. $\lim_{x \to -2} f(x) = f(-2)$. The first condition is easy: $f(-2) = -4$. The second condition is harder: $\lim_{x \to -2} f(x) = \lim_{x \to -2} \dfrac{x^2 - x - 6}{x + 2}$ $= \lim_{x \to -2} \dfrac{(x-3)(x+2)}{x+2} = \lim_{x \to -2} (x - 3) = -5$. Since this limit does not match the value for $f(-2)$, the function is discontinuous at $x = -2$.

39. $\dfrac{x^2 - 3x - 10}{x+2} = \dfrac{(x-5)(x+2)}{x+2}$ which is undefined at $x = -2$, but equal to

$x - 5$ for all other x. Since the domain is $0 \leq x \leq 5$, $f(x) = x - 5$ which is a polynomial and, therefore, continuous.

41. $x = -2$ is the only suspicious point. We have two things to check: 1. $f(-2)$ exists, and 2. $\lim_{x \to -2} f(x) = f(-2)$. The first is easy, $f(-2) = -3$. The second can be done as follows,

$$\lim_{x \to -2} f(x) = \lim_{x \to -2} \frac{(x - 5)(x + 2)}{x + 2} = \lim_{x \to -2} (x - 5) = 7 \neq f(-2).$$

So $x = -2$ is a discontinuity.

43. $f(x) = |x| = \begin{cases} -x \text{ if } -5 \leq x < 0 \\ x \text{ if } 0 \leq x \leq 5 \end{cases}$

Suspicious point at $x = 0$. $f(0) = 0$. $\lim_{x \to 0^+} = 0$ and $\lim_{x \to 0^-} (-x) = 0$ so $\lim_{x \to 0} f(x) = 0$. Continuous at $x = 0$. Thus, continuous on $[-5, 5]$.

45. $f(x) = |x + 3| = \begin{cases} -x - 3 \text{ if } -5 \leq x < -3 \\ x + 3 \text{ if } -3 \leq x \leq 5 \end{cases}$

Suspicious point at $x - 3$. $f(-3) = 0 - 3 + 3 = 0$. $\lim_{x \to -3} (x + 3) = 0$ and $\lim_{x \to -3^-} f(x) = 0$. Thus, continuous on $[-5, 5]$.

47. a. This is not a continuous function because, for example, at $w = 1$ there is a jump from $C = .25$ to $C = .45$, formally: $\lim_{x \to 1^+} C(w) = .45$ and $\lim C(w) = .25$; hence, the $\lim_{x \to 1} C(w)$ does not exist. The domain is $w > 0$. **b.** $C(1.9) = .29 + .23(1) = \$.52$; $C(2.01) = .29 + .23(2) = \$.75$; $C(2.89) = .29 + .23(2) = \$.75$

c.

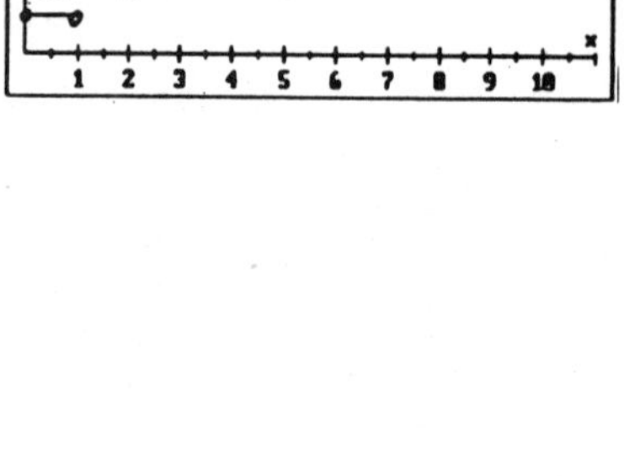

49. Answers vary.

A few examples are: $f(x) = \dfrac{1}{x - 2}$ or $g(x) = \dfrac{x^2 - 2x}{x - 2}$ or $h(x) = \dfrac{|x - 2|}{x - 2}$

2.3 Rates of Change

1. $\dfrac{\text{change in height}}{\text{change in time}} = \dfrac{80 - 40}{7 - 1} = \dfrac{20}{3}$ ft/sec

3. $\dfrac{\text{change in height}}{\text{change in time}} = \dfrac{80 - 40}{2 - 1} = 40$ ft/sec

5. $\dfrac{\text{change in output}}{\text{change in number of workers}} = \dfrac{2000 - 1000}{800 - 100} = \dfrac{10}{7}\,\dfrac{\text{units}}{\text{worker}}$

7. $\dfrac{\text{change in output}}{\text{change in number of workers}} = \dfrac{2000 - 1750}{800 - 500} = \dfrac{5}{6}\,\dfrac{\text{units}}{\text{worker}}$

9. $\dfrac{\text{change in scores}}{\text{change in time}} = \dfrac{530 - 580}{1991 - 1986} = \dfrac{-50}{5} = -10\,\dfrac{\text{points}}{\text{year}}$

11. $\dfrac{\text{change in scores}}{\text{change in time}} = \dfrac{530 - 548}{1991 - 1989} = -9\,\dfrac{\text{points}}{\text{year}}$

13. $\dfrac{\text{change in GNP}}{\text{change in time}} = \dfrac{4.5267 - .5153}{1987 - 1960} \approx .1486\,\dfrac{\text{trillion dollars}}{\text{year}}$

15. $\dfrac{\text{change in GNP}}{\text{change in time}} = \dfrac{4.5267 - 2.7320}{1987 - 1980} \approx .2564\,\dfrac{\text{trillion dollars}}{\text{year}}$

17. Average speed $= \dfrac{26.7}{6{:}36 - 6{:}09} = .99\,\dfrac{\text{miles}}{\text{minute}} \cdot \dfrac{60 \text{ minutes}}{\text{hour}} = 59.3\,\dfrac{\text{miles}}{\text{hour}}$

Note: $\dfrac{\text{miles}}{\text{hour}} = \text{mph}$

19. Average speed $= \dfrac{13.2}{7{:}28 - 7{:}06} = .6\,\dfrac{\text{miles}}{\text{minutes}} \cdot \dfrac{60 \text{ minutes}}{\text{hour}} = 36\,\dfrac{\text{miles}}{\text{hour}}$

21. slope $\to 0$ **23.** slope $= 2$ for *each* value of Q **25.** slope $\to 0$

27. $\dfrac{f(2) - f(-3)}{2 - (-3)} = \dfrac{[4 - 3(2)] - [4 - 3(-3)]}{5} = \dfrac{-15}{5} = -3$

29. $\dfrac{f(3) - f(-3)}{3 - (-3)} = \dfrac{5 - 5}{6} = 0$

31. $\dfrac{f(3) - f(1)}{3 - 1} = \dfrac{3(3)^2 - 3(1)^2}{2} = 12$

33. $\dfrac{[-2(4)^2 + (4) + 4] - [-2(1)^2 + (1) + 4]}{4 - 1} = \dfrac{-24 - 3}{3} = -9$

35. $\dfrac{\left[\dfrac{-2}{(5) + 1}\right] - \left[\dfrac{-2}{(1) + 1}\right]}{5 - 1} = \dfrac{-\frac{1}{3} + 1}{4} = \dfrac{1}{6}$

37. $\lim_{h \to 0} \frac{f(-3+h) - f(-3)}{h} = \lim_{h \to 0} \frac{[4 - 3(-3+h)] - [4 - 3(-3)]}{h}$

$= \lim_{h \to 0} \frac{49 - 3h - 13}{h}$

$= \lim_{h \to 0} (-3) = -3$

39. $\lim_{h \to 0} \frac{f(-3+h) - f(-3)}{h} = \lim_{h \to 0} \frac{5 - 5}{h} = 0$

41. $\lim_{h \to 0} \frac{f(1+h) - f(1)}{h} = \lim_{h \to 0} \frac{3(1+h)^2 - 3(1)^2}{h}$

$= \lim_{h \to 0} \frac{3 + 6h + 3h^2 - 3}{h}$

$= \lim_{h \to 0} (3h + 6) = 6$

43. $\lim_{h \to 0} \frac{f(1+h) - f(1)}{h}$

$= \lim_{h \to 0} \frac{[-2(1+h)^2 + (1+\text{h}) + 4] - [-2(1)^2 + (1) + 4]}{h}$

$= \lim_{h \to 0} \frac{-2 - 4h - 2h^2 + 1 + h + 4 - 3}{h}$

$= \lim_{h \to 0} (-2h - 3) = -3$

45. $\lim_{h \to 0} \frac{f(h) - f(0)}{h} = \lim_{h \to 0} \frac{\frac{-2}{h+1} - \frac{-2}{0+1}}{h}$

$= \lim_{h \to 0} \frac{-2 + 2(h+1)}{h+1} \cdot \frac{1}{h}$

$= \lim_{h \to 0} \frac{-2 + 2h + 2}{h(h+1)}$

$= \lim_{h \to 0} \frac{2}{h+1} = 2$

47. $\frac{P(30) - P(20)}{30 - 20} = \frac{[50(30) - 30^2] - 50(20) - 20^2]}{10} = 0$

49. $\frac{P(21) - P(20)}{21 - 20} = \frac{[50(21) - 21^2] - [50(20) - 20^2]}{1} = 9$

51. $\frac{C(200) - C(100)}{200 - 100} = \frac{[30(200)^2 - 100(200)] - [30(100)^2 - 100(100)]}{100} = 8{,}900$

53. $\frac{C(101) - C(100)}{101 - 100} = \frac{[30(101)^2 - 100(101)] - [30(100)^2 - 100(100)]}{1} = 5{,}930$

55. $\frac{C(x+h) - C(x)}{h} = \frac{[30(x+h)^2 - 100(x+h)] - [30(x)^2 - 100(x)]}{h}$

$= \frac{[30(x^2 + 2xh + h^2) - 100(x+h)] - [30x^2 - 100x]}{h}$

$$= \frac{30x^2 + 60xh + 30h^2 - 100x - 100h - 30x^2 + 100x}{h}$$

$$= \frac{60xh + 30h^2 - 100h}{h}$$

$$= 60x + 30h - 100$$

57. $\lim_{h \to 0} \frac{C(x+h) - C(x)}{h} = \lim_{h \to 0} (60x + 30h - 100)$

$$= 60x + 0 - 100 = 60x - 100$$

59. The instantaneous rate of change of cost at a given x.

2.4 Definition of Derivative

1. The "corner" point at (1, 0) has no derivative because $\lim_{h \to 0} \frac{f(1+h) - f(1)}{h}$ $= -2$ and $\lim_{h \to 0} \frac{f(1+h) - f(1)}{h} = 2$, so the $\lim_{h \to 0} \frac{f(1+h) - f(1)}{h}$ does not exist at $x = 1$.

3. Since the function is discontinuous at $x = 3$, the derivative does not exist at $x = 3$. The function is differentiable everywhere else.

5. Since the function is discontinuous at $x = 4$, the derivative does not exist at $x = 4$.

7. $f'(x) = \lim_{h \to 0} \frac{f(x+h) - f(x)}{h}$

$$= \lim_{h \to 0} \frac{2(x+h)^2 - 2x^2}{h}$$

$$= \lim_{h \to 0} \frac{2(x^2 + 2xh + h^2) - 2x^2}{h}$$

$$= \lim_{h \to 0} \frac{4xh + 2h^2}{h}$$

$$= \lim_{h \to 0} (4x + 2h)$$

$$= 4x$$

9. $\lim_{h \to 0} \frac{f(x+h) - f(x)}{h}$

$$= \lim_{h \to 0} \frac{-3(x+h)^2 - (-3x^2)}{h}$$

$$= \lim_{h \to 0} \frac{-3x^2 - 6xh - 3h^2 + 3x^2}{h}$$

$$= \lim_{h \to 0} (-6x - 3h)$$

$$= -6x$$

11. $f'(x) = \lim_{h\to 0} \frac{f(x+h) - f(x)}{h}$

$= \lim_{h\to 0} \frac{[4 - 5(x+h)] - [4 - 5x]}{h}$

$= \lim_{h\to 0} \frac{4 - 5x - 5h - 4 + 5x}{h}$

$= \lim_{h\to 0} (-5)$

$= -5$

13. $f'(x) = \lim_{h\to 0} \frac{f(x+h) - f(x)}{h}$

$= \lim_{h\to 0} \frac{[3(x+h)^2 + 4(x+h)] - [3x^2 + 4x]}{h}$

$= \lim_{h\to 0} \frac{3(x^2 + 2xh + h^2) + 4x + 4h - 3x^2 - 4x}{h}$

$= \lim_{h\to 0} \frac{6xh + 3h^2 + 4h}{h}$

$= \lim_{h\to 0} 6x + 3h + 4 = 6x + 4$

15. $f'(x) = \lim_{h\to 0} \frac{f(x+h) - f(x)}{h}$

$= \lim_{h\to 0} \frac{[-3(x+h)^2 - 50(x+h) + 125] - [-3x^2 - 50x + 125]}{h}$

$= \lim_{h\to 0} \frac{-6xh - 3h^2 - 50h}{h}$

$= \lim_{h\to 0} (-6x - 3h - 50)$

$= -6x - 50$

17. $f'(x) = \lim_{h\to 0} \frac{f(x+h) - f(x)}{h}$

$= \lim_{h\to 0} \frac{2(x+h)^3 - 2x^3}{h}$

$= \lim_{h\to 0} \frac{2x^3 + 6x^2h + h^3 - 2x^3}{h}$

$= \lim_{h\to 0} (6x + 6xh + h^2) = 6x^2$

19. $f'(x) = \lim_{h\to 0} \frac{f(x+h) - f(x)}{h}$

$= \lim_{h\to 0} \frac{\frac{-3}{x+h} - (-\frac{3}{x})}{h}$

$= \lim_{h\to 0} \frac{-3x + 3(x+h)}{x(x+h)} \cdot \frac{1}{h}$

$= \lim_{h\to 0} \frac{3}{x(x+h)}$

$= \frac{3}{x^2}$

21. $f'(x) = \lim_{h \to 0} \dfrac{f(x+h) - f(x)}{h}$

$= \lim_{h \to 0} \dfrac{2\sqrt{x+h} - 2\sqrt{x}}{h} \cdot \dfrac{2\sqrt{x+h} + 2\sqrt{x}}{2\sqrt{x+h} + 2\sqrt{x}}$

$= \lim_{h \to 0} \dfrac{4(x+h) - 4x}{h[2\sqrt{x+h} + 2\sqrt{x}]}$

$= \dfrac{1}{\sqrt{x}}$ or $\dfrac{\sqrt{x}}{x}$

23. The point of tangency is at $x = -3$, $y = 5x^2 = 5(-3)^2 = 45$. The slope of the tangent line at any x is

$$\begin{aligned} f'(x) &= \lim_{h \to 0} \frac{f(x+h) - f(x)}{h} \\ &= \lim_{h \to 0} \frac{5(x+h)^2 - 5x^2}{h} \\ &= \lim_{h \to 0} \frac{5(x^2 + 2xh + h^2) - 5x^2}{h} \\ &= \lim_{h \to 0} (10x + 5h) \\ &= 10x \end{aligned}$$

In particular, the slope at $x = -3$ is $f'(-3) = 10(-3) = -30$. Using the point-slope form with $m = -30$ and $(x_1, y_1) = (-3, 45)$: $y - 45 = -30(x+3)$ or $30x + y + 45 = 0$

25. The point of tangency is at $x = -2$, $y = 4 - 5x = 4 - 5x = 4 - 5(-2) = 14$. The slope of the tangent line is $f'(x) = -5$. (see Problem 11.) Using the point slope form with $m = -5$ and $(x_1, y_1) = (-2, 14)$: $y - 14 = -5(x+2)$ or $5x + y - 4 = 0$

27. The point of tangency is at $x = -1$, $y = 3 + 2x - 3x^2 = -2$. The slope of the tangent line is $f'(x) = \lim_{h \to 0} \dfrac{f(x+h) - f(x)}{h}$

$$\begin{aligned} &= \lim_{h \to 0} \frac{[3 + 2(x+h) - 3(x+h)^2] - [3 + 2x - 3x^2]}{h} \\ &= \lim_{h \to 0} \frac{3 + 2x + 2h - 3(x^2 + 2xh + h^2) - 3 - 2x + 3x^2}{h} \\ &= \lim_{h \to 0} \frac{2h - 6xh - 3h^2}{h} \\ &= \lim_{h \to 0} (2 - 6x - 3h) \\ &= 2 - 6x. \end{aligned}$$

So, at $x = -1$, the slope is $f'(-1) = 2 - 6(-1) = 8$. Using the point-slope form with $m = 8$ and $(x_1, y_1) = (-1, -2)$: $y + 2 = 8(x + 1)$ or $8x - y + 6 = 0$

29. $$C'(20) = \lim_{h \to 0} \frac{C'(20 + h) - C'(20)}{h}$$
$$= \lim_{h \to 0} \frac{[(20 + h)^2 - 40(20 + h) + 2{,}000] - [20^2 - 40(20 + 2{,}000]}{h}$$
$$= \lim_{h \to 0} \frac{400 + 40h + h^2 - 800 - 40h + 2{,}000 - 400 + 800 - 2{,}000}{h}$$
$$= \lim_{h \to 0} h = 0$$

The marginal cost is \$0.

31. $$\frac{C(21) - C(20)}{21 - 20} = \frac{[21^2 - 40(21) + 2000] - [20^2 - 40(20) + 2000]}{1} = 1$$

The actual cost of producing an extra item is \$1.

33. $$C'(x) = \lim_{h \to 0} \frac{C(x + h) - C(x)}{h}$$
$$= \lim_{h \to 0} \frac{[(x + h)^2 - 100(x + h) + 3{,}000] - [x^2 - 100x + 3{,}000]}{h}$$
$$= \lim_{h \to 0} (2x + h - 100) = 2x - 100$$

35. $C'(60) = 2(60) - 100$ from Problem 33, the marginal cost is \$20.

37. $\dfrac{C(61) - C(60)}{61 - 60} = [2(60) + 1 - 100]$ from Problem 33, the actual cost is \$21.

39. $$N'(3) = \lim_{h \to 0} \frac{N(3 + h) - N(3)}{h}$$
$$= \lim_{h \to 0} \frac{[2(3 + h)^2 - 200(3 + h) + 1{,}000] - [2(3)^2 - 200(3) + 1{,}000]}{h}$$
$$= \lim_{h \to 0} \frac{18 + 12h + 2h^2 - 600 - 200h + 1000 - 18 + 600 - 1{,}000}{h}$$
$$= \lim_{h \to 0} (-188 + 12h)$$
$$= -188$$

41. The velocity is $d'(t)$ so at $t = 3$ we want

$$d'(3) = \lim_{h \to 0} \frac{d(3 + h) - d(3)}{h}$$
$$= \lim_{h \to 0} \frac{[\frac{1}{5}(3 + h)^2 + 5(3 + h)] - [\frac{1}{5}(3)^2 + 5(3)]}{h}$$
$$= \lim_{h \to 0} \frac{\frac{9}{5} + \frac{6}{5}h + \frac{1}{5}h^2 + 15 + 5h - \frac{9}{5} - 15}{h}$$
$$= \lim_{h \to 0} (\tfrac{31}{5} + \tfrac{1}{5}h) = \tfrac{31}{5}$$; velocity is 6.2 mph.

43. The velocity at $t = 5$ is

$$\begin{aligned} m'(5) &= \lim_{h\to 0} \frac{m(5+h) - m(5)}{h} \\ &= \lim_{h\to 0} \frac{[(5+h)^2 + 10(5+h)] - [5^2 + 10(5)]}{h} \\ &= \lim_{h\to 0} \frac{25 + 10h + h^2 + 50 + 10h - 25 - 50}{h} \\ &= \lim_{h\to 0} h \\ &= 20;\text{ velocity is } 20\ \frac{\text{miles}}{\text{min}} \end{aligned}$$

2.5 Differentiation Techniques, Part I

1. $y' = 7x^6$ **3.** $y' = 12x^{11}$ **5.** $y = -5x^{-6}$ **7.** $y' = 0$

9. $y' = -4(-8x^{-9}) = 32x^{-9}$

11. First, $y = -\frac{1}{2}x^{1/2}$ so $y' = -\frac{1}{2}(\frac{1}{2}x^{-1/2}) = -\frac{1}{4}x^{-1/2} = -\frac{1}{4\sqrt{x}} = \frac{-\sqrt{x}}{4x}$

13. $y' = 5(-8x^{-9}) = -40x^{-9}$

15. $y' = 12(\frac{5}{4}x^{1/4}) = 15x^{1/4}$

17. Use the sum rule: $y' = (3x^2)' + (x)' = 6x + 1$

19. Use the sum and difference rules: $y' = (2x^2)' - (5x)' - (6)'$

$= 4x - 5 - 0 = 4x - 5$

21. Use the sum and difference rules:

$y' = (5x^3)' - (5x^2)' + (4x)' - (5)' = 15x^2 - 10x + 4 - 0 = 15x^2 - 10x + 4$

23. Use the sum rule:

$y' = (x^{-3})' + (x^2)' + (x^{-1})' = -3x^{-4} + 2x + (-1)x^{-2}$ or $-\frac{3}{x^4} + 2x - \frac{1}{x^2}$

$= \frac{-3 + 2x^5 - x^2}{x^4} = \frac{2x^5 - x^2 - 3}{x^4}$

25. First, $y = (x^4 + 2)^2 = x^8 + 4x^4 + 4$; so $y' = 8x^7 + 16x^3 + 0 = 8x^7 + 16x^3$

27. First, $y = \frac{2}{x} + \frac{5}{x^2} = 2x^{-1} + 5x^{-2}$; so $y' = -2x^{-2} + (-10)x^{-3}$

or $\frac{-2}{x^2} - \frac{10}{x^3} = \frac{-2x - 10}{x^3}$

29. First, $y = -5x^7 + 2x^{1/2} - 3x^{-1}$; so $y' = -35x^6 + 1x^{-1/2} - (-3)x^{-2}$

$= -35x^6 + x^{-1/2} + 3x^{-2}$ or $\dfrac{-35x^8 + x\sqrt{x} + 3}{x^2}$

31. $f(x) = \dfrac{x^3 + 4x^2 + 1}{x^2} = x + 4 + x^{-2}$ so $f'(x) = 1 - 2x^{-3}$

33. $f(x) = \dfrac{x^5 + 3x^3}{\sqrt{x}} = x^{5-1/2} + 3x^{3-1/2} = x^{9/2} + 3x^{5/2}$ so

$f'(x) = \frac{9}{2}x^{7/2} + \frac{15}{2}x^{3/2}$

35. $\frac{d}{dx}(\sqrt[3]{x} + \frac{5}{8}) = \frac{1}{3}x^{-2/3}$

37. $\frac{d}{dD}(\pi D) = \pi$

39. $\frac{dv}{dr} = \frac{d}{dr}(\frac{4}{3}\pi r^3) = 4\pi r^2$

41. $\frac{dK}{dC} = \frac{d}{dC}(\frac{C^2}{4\pi}) = \frac{2C}{4\pi} = \frac{C}{2\pi}$

43. $f'(x) = 5x^4$; $f'(1) = 5$

45. $f'(x) = -\frac{1}{2}x^{-3/2}$; $f'(1) = -\frac{1}{2}$

47. $f'(x) = 6x - 4$; $f'(0) = -4$

49. $f'(x) = 2x + 2$; $f'(0) = 2$; $y - 1 = 2(x - 0)$; $y - 1 = 2x$; $2x - y + 1 = 0$

51. $f(x) = 2 - 2x^{-2}$; $f'(x) = 4x^{-3}$; $f'(1) = 4$;

$$y - 0 = 4(x - 1)$$
$$y = 4x - 4$$
$$4x - y - 4 = 0$$

53. $f'(x) = \frac{1}{3}x^{-2/3}$; $f'(1) = \frac{1}{3}$;

$$y - 1 = \tfrac{1}{3}(x - 1)$$
$$3y - 3 = x - 1$$
$$x - 3y + 2 = 0$$

55. Marginal cost is $C'(x) = (20x^2)' + (500x)' + (250{,}000)'$

$$= 40x + 500 + 0$$
$$= 40x + 500$$

57. $m = \dfrac{400}{t} = 400t^{-1}$, so $\dfrac{dm}{dt} = 400(-1)t^{-2} = \dfrac{-400}{t^2}$

59. The rate of change of earnings at any t is $A'(t) = (.05t^2)' + (25t)' + (5)'$ $= .10t + 25$. Since 1986 is the initial time ($t = 0$), 1992 corresponds to $t = 6$, $A'(6) = .10(6) + 25 = 25.6$; that is, the rate of change of earnings is 25.6

(thousand dollars per year).

61. The instantaneous rate of change of number of tasks (N) per hour (unit change in x) is $N' = \frac{dN}{dt} = 25(\frac{1}{2})x^{-1/2} = \frac{25}{2\sqrt{x}}$. At the end of the fifth hour ($x = 5$), the subjects are learning at the rate of $\frac{25}{2\sqrt{5}} = 6$ tasks per hour.

63. The rate of change of P with respect to x is

$P'(x) = .0005 + .00001(2x)$

$= .0005 + .00002x$, so the population is changing at the rate of $P'(1{,}000{,}000)$

$= .0005 + (.00002)(1{,}000{,}000)$

$= 20$ foxes per rabbit.

65. **a.** $\downarrow t = 0$ (1990) and $h = 4$

$$\frac{P(4) - P(0)}{4} = \frac{[-.05(4)^2 + 5(4) + 250] - [250]}{4} = \frac{19.2}{4} = 4.8$$

The average rate of change in the CPI from 1990 to 1994 was 4.8

b. $\lim_{h \to 0} \frac{P(1+h) - P(1)}{h} = P'(1) = -.1(1) + 5 = 4.9$; $P'(x) = -.1t + 5$

67. Let $C(x) = kf(x)$. Prove $C'(x) = kf'(x)$

$$C'(x) = \lim_{h \to 0} \frac{C(x+h) - C(x)}{h} = \lim_{h \to 0} \frac{kf(x+h) - kf(x)}{h}$$

$$= \lim_{h \to 0} k\left[\frac{f(x+h) - f(x)}{h}\right] = k \lim_{h \to 0} \frac{f(x+h) - f(x)}{h} = kf'(x)$$

69. Let $P(x) = x^n$, Prove $P'(x) = nx^{n-1}$

$$P'(x) = \lim_{h \to 0} \frac{P(x+h) - P(x)}{h} = \lim_{h \to 0} \frac{(x+h)^n - x^n}{h}$$

$$= \lim_{h \to 0} \frac{[x^n + nx^{n-1}h + \frac{n(n+1)}{2}x^{n-2}h^2 + \cdots + nxh^{n-1} + h^n] - x^n}{h}$$

$$= \lim_{h \to 0} [nx^{n-1} + \frac{n(n+1)}{2}x^{n-2}h + \cdots + nxh^{n-2} + h^{n-1}] = nx^{n-1}$$

2.6 Differentiation Techniques, Part II

1. $f'(x) = 5x^2(x^2 - 6)' + (x^2 - 6)(5x^2)'$

$= 5x^2(2x) + (x^2 - 6)(10x)$

$= 10x^3 + 10x^3 - 60x$

$= 20x^3 - 60x$

3. $f'(x) = (x + 1)(x - 2)' + (x - 2)(x + 1)'$

$= (x + 1)(1) + (x - 2)(1)$

$= 2x - 1$

5. $g'(x) = (3x^2 + 5)(2x^2 - 5)' + (2x^2 - 5)(3x^2 + 5)'$

$= (3x^2 + 5)(4x) + (2x^2 - 5)(6x)$

$= 12x^3 + 20x + 12x^2 - 30x$

$= 24x^3 - 10x$

7. $g'(x) = 5x^4(2x^2 - 5x + 1)' + (2x^2 - 5x + 1)(5x^4)'$

$= 5x^4(4x - 5) + (2x^2 - 5x + 1)(20x^3)$

$= 20x^5 - 25x^4 + 40x^5 - 100x^4 + 20x^3$

$= 60x^5 - 125x^4 + 20x^3$

9. Use the quotient rule:

$$y' = \frac{(x - 3)x' - x(x - 3)'}{(x - 3)^2} = \frac{(x - 3)(1) - (x)(1)}{(x - 3)^2} = \frac{-3}{(x - 3)^2}$$

11. Use the quotient rule:

$$y' = \frac{(x - 3)(x + 5)' - (x + 5)(x - 3)'}{(x - 3)^2} = \frac{(x - 3) - (x + 5)(1)}{(x - 3)^2}$$

$$= \frac{x - 3 - x - 5}{(x - 3)^2} = \frac{-8}{(x - 3)^2}$$

13. Use the quotient rule:

$$y' = \frac{(x^2 - 5)(x^2 + 3)' - (x^2 + 3)(x^2 - 5)'}{(x^2 - 5)^2} = \frac{(x^2 - 5)(2x) - (x^2 + 3)(2x)}{(x^2 - 5)^2}$$

$$= \frac{2x^3 - 10x - 2x^3 - 6x}{(x^2 - 5)^2} = \frac{2x^3 - 10x - 2x^3 - 6x}{(x^2 - 5)^2} = \frac{-16x}{(x^2 - 5)^2}$$

15. Simplify first: $y = \dfrac{x^2 + x - 6}{x^2 - 4} = \dfrac{(x + 3)(x - 2)}{(x - 2)(x + 2)} = \dfrac{x + 3}{x + 2}$.

Now use the quotient rule:

$$y' = \frac{(x + 2)(x + 3)' - (x + 3)(x + 2)'}{(x + 2)^2} = \frac{(x + 2)(1) - (x + 3)(1)}{(x + 2)^2} = \frac{-1}{(x + 2)^2}$$

17. Use the product rule:

$y' = (2x^2 - 1)(x^3 + 2x^2 - 3)' + (x^3 + 2x^2 - 3)(2x^2 - 1)'$

$= (2x^2 - 1)(3x^2 + 4x) + (x^3 + 2x^2 - 3)(4x)$

$= 6x^4 + 8x^3 - 3x^2 - 4x + 4x^4 + 8x^3 - 12x$

$= 10x^4 + 16x^3 - 3x^2 - 16x$

19. Use the product rule:

$$y' = (4x^2 + x)(x^3 - 3x^2 + 13)' + (x^3 - 3x^2 + 13)(4x^2 + x)'$$
$$= (4x^2 + x)(3x^2 - 6x) + (x^3 - 3x^2 + 13)(8x + 1)$$
$$= 12x^4 - 21x^3 - 6x^2 + 8x^4 - 23x^3 - 3x^2 + 104x + 13$$
$$= 20x^4 - 44x^3 - 9x^2 + 104x + 13$$

21. Use the product rule:

$$y' = (3x^3 - 2x + 5)(2x^4 + 5x - 9)' + (2x^4 + 5x - 9)(3x^3 - 2x + 5)'$$
$$= (3x^3 - 2x + 5)(8x^3 + 5) + (2x^4 + 5x - 9)(9x^2 - 2)$$
$$= 24x^6 - 16x^4 + 55x^3 - 10x + 25 + 18x^6 - 4x^4 + 45x^3 - 81x^2 - 10x + 18$$
$$= 42x^6 - 20x^4 + 100x^3 - 81x^2 - 20x + 43$$

23. Multiply first (preferred): $y = 25x^{-1/3} + 15x^{5/3}$ so $y' = -\frac{25}{3}x^{-4/3} + 25x^{2/3}$

Or use product rule: $y' = (5x^{2/3})(5x^{-1} + 3x)' + (5x^{-1} + 3x)(5x^{2/3})'$

$$= (5x^{2/3})(-5x^{-2} + 3) + (5x^{-1} + 3x)\left(\tfrac{10}{3}x^{-1/3}\right)$$
$$= 25x^{-4/3} + 15x^{2/3} + \tfrac{50}{3}x^{-4/3} + 10x^{2/3}$$
$$= -\tfrac{25}{3}x^{-4/3} + 25x^{2/3}$$

25. Multiply first (preferred): $f(x) = 12x^{5/2} - 30x^{1/2}$ so $f'(x) = 30x^{3/2} - 15x^{-1/2}$

Or use the product rule: $f'(x) = (6x^{1/2})(2x^2 - 5)' + (2x^2 - 5)(6x^{1/2})'$

$$= (6x^{1/2})(4x) + (2x^2 - 5)(3x^{-1/2})$$
$$= 24x^{3/2} + 6x^{3/2} - 15x^{-1/2}$$
$$= 30x^{3/2} - 15x^{-1/2}$$

27. Use the quotient rule: $g'(x) = \dfrac{(5x^2 - 11x + 3)(2x)' - (2x)(5x^2 - 11x + 3)'}{(5x^2 - 11x + 3)^2}$

$$= \frac{(5x^2 - 11x + 3)(2) - (2x)(10x - 11)}{(5x^2 - 11x + 3)^2}$$
$$= \frac{(10x^2 - 22x + 6) - (20x^2 - 22x)}{(5x^2 - 11x + 3)^2}$$
$$= \frac{-10x^2 + 6}{(5x^2 - 11x + 3)^2}$$

29. The slope of the tangent line is given by $f'(x)$:

$$f'(x) = \frac{(x - 5)(5x^2 + 5x)' - (5x^2 + 5x)(x - 5)'}{(x - 5)^2}$$

$$= \frac{(x - 5)(10x + 5) - (5x^2 + 5x)(1)}{(x - 5)^2}$$
$$= \frac{10x^2 - 45x - 25 - 5x^2 - 5x}{(x - 5)^2}$$
$$= \frac{5x^2 - 50x - 25}{(x - 5)^2}$$

In particular, the slope of the tangent line at $x = 1$ is

$$f'(1) = \frac{5(1)^2 - 50(1) - 25}{(1 - 5)^2} = -\frac{35}{8}$$

31. The slope of the tangent line is given by

$$f'(x) = (x^{1/2})'(x^2 + 3) + x^{1/2}(x^2 + 3)'$$
$$= \tfrac{1}{2}x^{-1/2}(x^2 + 3) + x^{1/2}(2x)$$
$$= \tfrac{5}{2}x^{3/2} + \tfrac{3}{2}x^{-1/2}$$

In particular, the slope of the tangent line at $x = 4$ is

$$f'(x) = \tfrac{5}{2}(4)^{3/2} + \tfrac{3}{2}(4)^{-1/2}$$
$$= \tfrac{5}{2}(8) + \tfrac{3}{2}(\tfrac{1}{2})$$
$$= \tfrac{83}{4}$$

33. The slope of the tangent line is given by $f'(x)$

$$f'(x) = \frac{(x^2 + 4x + 4)(x^2 - 5x + 1)' - (x^2 - 5x + 1)(x^2 + 4x + 4)'}{(x^2 + 4x + 4)^2}$$
$$= \frac{(x^2 + 4x + 4)(2x - 5) - (x^2 - 5x + 1)(2x + 4)}{(x^2 + 4x + 4)^2}$$
$$= \frac{9x^2 + 6x - 24}{(x^2 + 4x + 4)^2}$$
$$= \frac{3[3x^2 + 2x - 8]}{(x^2 + 4x + 4)^2}$$
$$= \frac{3(3x - 4)(x + 2)}{[(x + 2)(x + 2)]^2}$$
$$= \frac{3(3x - 4)}{(x + 2)^3}$$

In particular, at $x = 0$ the slope of the tangent line is $f'(0) = \dfrac{3(-4)}{2^3} = -\dfrac{3}{2}$

35. The slope of the tangent line is $f'(x) = \dfrac{5x^2 - 50x - 25}{(x - 5)^2}$. (See Problem 29.)

In particular, at $x = 0$ $f'(0) = -1$. The point of tangency is at $(0, f(0))$ which is

(0, 0). Using the point-slope equation of a line: $y - y_1 = m(x - x_1)$; $y - 0 = -1(x - 0)$; $y = -x$ or $x + y = 0$

37. The slope of the tangent line at $x = 4$ is $f'(4) = \frac{83}{4}$. (See Problem 31.) The point of tangency is $(4, f(4))$ which is (4, 38). Using the point-slope equation of a line:

$$y - y_1 = m(x - x_1)$$
$$y - 38 = 83(x - 4) \quad \textit{Multiply both sides by 4}$$
$$4y - 152 = 83x - 332$$
$$83x - 4y - 180 = 0$$

39. The slope of the tangent line is $f'(x) = \dfrac{3(3x - 4)}{(x + 2)^3}$. (See Problem 33.) The slope at $x = 1$ is $f'(1) = -\frac{1}{9}$. The point of tangency is $(1, f(1))$ which is $(1, -\frac{1}{3})$. Using the point-slope equation of a line:

$$y - y_1 = m(x - x_1)$$
$$y + \tfrac{1}{3} = -\tfrac{1}{9}(x - 1)$$
$$9y + 3 = -1(x - 1) \quad \textit{Multiply both sides by 9}$$
$$9y + 3 = -x + 1$$
$$x + 9y + 2 = 0$$

41. Write $D(x) = 100{,}000(x^2 + 15x + 25)^{-1}$. The rate of change of D (demand) with respect to x (price) is: $D'(x) = -100{,}000(x^2 + 15x + 25)^{-2}(2x + 15)$

43. From Problem 41, $D'(x) = -100{,}000(x^2 + 15x + 25)^{-2}$; so $D'(15)$

≈ -19.944598 when the price is \$15, the demand is decreasing by about 20 items per unit change in price. The slope of the graph at $x = 15$ is -20.

45. $N(x) = 1500 - 575x^{-1}$ so $N'(x) = 575x^{-2}$

47. From Problem 45, $N'(x) = 575x^{-2}$ so $N'(10) = 575(10)^{-2} = 5.75$ the slope of the graph at $x = 10$ is 5.75. The number of items is increasing at the point $x = 10$ since N' is positive.

49. $$f'(t) = \frac{(.01 + .005t)t' - t(.01 + .005t)'}{(.01 + .005t)^2}$$

$$= \frac{.01 + .005t - t(.005)}{(.01 + .005t)^2}$$

$$= .01(.01 + .005t)^{-2}$$

51. $N'(t) = \dfrac{(2t^2 + 1)(10t)' - 10t(2t^2 + 1)'}{(2t^2 + 1)^2}$

$$= \frac{10(2t^2 + 1) - 10t(4t)}{(2t^2 + 1)^2}$$

$$= \frac{-20t^2 + 10}{(2t^2 + 1)^2}$$

2.7 The Chain Rule

The "work" in this problem set probably shows more detail than will your work. Try to do some of these steps in your head.

1. $f'(x) = 3(3x + 2)^2 \frac{d}{dx}(3x + 2)$

$$= 3(3x + 2)^2(3)$$

$$= 9(3x + 2)^2$$

3. $f'(x) = 4(5x - 1)^3 \frac{d}{dx}(5x - 1)$

$$= 4(5 - 1)^3(5)$$

$$= 20(5x - 1)^3$$

5. $y' = 3(2x^2 + x)^2 \frac{d}{dx}(2x^2 + x)$

$$= 3(2x^2 + x)^2(4x + 1) \text{ or } 3x^2(2x + 1)^2(4x + 1)$$

7. $y' = 2(2x^2 - 3x + 2)\frac{d}{dx}(2x^2 - 3x + 2)$

$$= 2(2x^2 - 3x + 2)(4x + 1)$$

9. $g'(x) = 4(x^3 + 5x)^3 \frac{d}{dx}(x^3 + 5x)$

$$= 4(x^3 + 5x)^3(3x^2 + 5) \text{ or } 4x^3(x^2 + 5)^3(3x^2 + 5)$$

11. $m'(x) = -2(2x^2 - 5x)^{-3}\frac{d}{dx}(2x^2 - 5x)$

$$= -2(2x^2 - 5x)^{-3}(4x - 5) \text{ or } -2x^{-3}(2x - 5)^{-3}(4x - 5)$$

13. $t'(x) = -1(x^4 + 3x^3)^{-2}\frac{d}{dx}(x^4 + 3x^3)$

$$= -(x^4 + 3x^3)^{-2}(4x^3 + 9x^2)$$

$$= -x^{-6}(x + 3)^{-2}x^2(4x + 9)$$

$$= -x^{-4}(x + 3)^{-2}(4x + 9)$$

15. $y' = 15(4x^3 + 3x^2)^2(4x^3 + 3x^2)'$

$= 15(4x^3 + 3x^2)^2(12x^2 + 6x)$

$= 15x^4(4x + 3)^2(6x)(2x + 1)$

$= 90x^5(4x + 3)^2(2x + 1)$

17. $y' = \frac{1}{4}(x^2 - 3x)^{-3/4}(x^2 - 3x)'$

$= \frac{1}{4}(x^2 - 3x)^{-3/4}(2x - 3)$ or $\dfrac{2x - 3}{4(x^2 - 3x)^{3/4}}$

19. $y = (x^2 + 16)^{1/2};$

$y' = \frac{1}{2}(x^2 + 16)^{1/2}(x^2 + 16)'$

$= \frac{1}{2}(x^2 + 16)^{-1/2}(2x)$

$= x(x^2 + 16)^{-1/2}$ or $\dfrac{x\sqrt{x^2 + 16}}{x^2 + 16}$

21. $y = 5(x^3 + 8)^{1/2};$

$y' = \frac{5}{2}(x^3 + 8)^{-1/2}(x^3 + 8)'$

$= \frac{5}{2}(x^3 + 8)^{-1/2}(3x^2)$

$= \frac{15}{2}x^2(x^3 + 8)^{-1/2}$

23. $y = x(3x + 1)^{1/2}$ Using product rule:

$y' = x\left[(3x + 1)^{1/2}\right]' + x'(3x + 1)^{1/2}$

$= x\left[\frac{1}{2}(3x + 1)^{-1/2}\frac{d}{dx}(3x + 1)\right] + (1)(3x + 1)^{1/2}$

$= x(\frac{1}{2})(3x + 1)^{-1/2}(3) + (3x + 1)^{1/2}$

$= \frac{3}{2}x(3x + 1)^{-1/2} + (3x + 1)^{-1/2}$

$= \frac{1}{2}(3x + 1)^{-1/2}[3x + 2(3x + 1)]$ *Factor out* $\frac{1}{2}(3x + 1)^{-1/2}$

$= \frac{1}{2}(3x + 1)^{-1/2}(9x + 2)$

25. $y' = \frac{1}{3}(2x + 5)^{-2/3}(2)$

$= \frac{2}{3}(2x + 5)^{2/3}$

27. $y = (5x+3)^{-1}$ so $\quad y' = -(5x+3)^{-2}(5)$
$$= -5(5x+3)^{-2}$$

29. $y = (x^2+3)^{-2}$ so $\quad y' = -2(x^2+3)^{-3}(2x)$
$$= -4x(x^2+3)^{-3}$$

31. $f'(x) = 2(3x^2+2)(6x)$ point: if $x = 1$, $f(1) = 5^2 = 60$
$$= 12x(3x^2+2) \qquad y - 25 = 60(x-1)$$
at $x = 1$ $\quad m = 60$ $\qquad 60x - y - 35 = 0$

33. $f(x) = 8(5-3x)^{-1}$ point: if $x = -1$, $f(-1) = 8(5+3)^{-1} = 1$
$$f(x) = -8(5-3x)^{-2}(-3) \qquad y - 1 = \tfrac{3}{8}(x+1)$$
$$= 24(5-3x)^{-2} \qquad 8y - 8 = 3x + 3$$
at $x = -1$ $\quad m = \frac{24}{64} = \frac{3}{8}$ $\qquad 3x - 8y + 11 = 0$

35. $f'(x) = x(\tfrac{1}{2})(x-4)^{-1/2}(1) + (x-4)^{1/2}$ $\qquad f(5) = 5$
$$f'(5) = \tfrac{5}{2} + 1 \qquad y - 5 = \tfrac{7}{2}(x-5)$$
$$= \tfrac{7}{2} \qquad 7x - 2y - 25 = 0$$

37. $f'(x) = 2x(x-3)^2 + x^2(2)(x-3)(1)$
$$= 2x(x-3)[(x-3)+x]$$
$$= 2x(x-3)(2x-3)$$
$f'(x) = 0$ if $x = 0, 3, \frac{3}{2}$

39. $f(x) = x(2x+3)^{-2}$
$$f'(x) = x(-2)(2x+3)^{-3}(2) + (2x+3)^{-2}$$
$$= (2x+3)^{-3}[-4x + (2x+3)]$$
$$= (2x+3)^{-3}(3-2x)$$
$f'(x) = 0$ if $x = \frac{3}{2}$

41. $f'(x) = \frac{1}{2}(x^2 - 5x + 20)^{-1/2}(2x-5)$
$f'(x) = 0$ if $x = \frac{5}{2}$

43. a. $f'(x) = (2x+1)^2[(3x+2)^3]' + (3x+2)^3\,[(2x+1)^2]'$
$$= (2x+1)^2[3(3x+2)^2(3x+2)'] + (3x+2)^3\,[2(2x+1)(2x+1)']$$
$$= (2x+1)^2[3(3x+2)^2 3] + (3x+2)^3\,[2(2x+1)(2)]$$

b. Factor out $(2x+1)(3x+2)^2$:
$$f'(x) = (2x+1)(3x+2)^2[9(2x+1) + 4(3x+2)]$$
$$= (2x+1)(3x+2)^2(30x+17)$$

45. a. $y' = (5x+1)^2[(4x+3)^{-1}]' + (4x+3)^{-1}[(5x+1)^2]'$

$= (5x+1)^2[-1(4x+3)^{-2}(4x+3)'] + (4x+3)^{-1}[2(5x+1)(5x+1)']$

$= (5x+1)^2[-1(4x+3)^{-2}(4)] + (4x+3)^{-1}[2(5x+1)(5x+1)']$

$= (5x+1)^2[-1(4x+3)^{-2}(4)] + (4x+3)^{-1}[2(5x+1)(5)]$

b. Factor out $2(5x+1)(4x+3)^{-2}$

$y' = 2(5x+1)(4x+3)^{-2}[-2(5x+1) + 5(4x+3)]$

$= 2(5x+1)(4x+3)^{-2}(10x+13)$

47. a. $g'(x) = \dfrac{(5x+3)[(2x-5)^2]' - (2x-5)^2[5x+3]'}{(5x+3)^2}$

$= \dfrac{(5x+3)[2(2x-5)(2)] - (2x-5)^2(5)}{(5x+3)^2}$

b. Factor out $2x - 5$:

$g'(x) = \dfrac{(2x-5)[4(5x+3) - 5(2x-5)]}{(5x+3)^2}$

$= \dfrac{(2x-5)(10x+37)}{(5x+3)^2}$

49. a. $t'(x) = \dfrac{(2x-5)^2[(x+5)^4]' - (x+5)^4[(2x-5)^2]'}{(2x-5)^4}$

$= \dfrac{(2x-5)^2[4(x+5)^3] - (x+5)^4 2(2x-5)(2)}{(2x-5)^4}$

b. Factor out $4(2x-5)(x+5)^3$:

$= \dfrac{4(2x-5)(x+5)^3[(2x-5) - (x+5)]}{(2x-5)^4}$

$= \dfrac{4(2x-5)(x+5)^3(x-10)}{(2x-5)^4}$

$= \dfrac{4(x+5)^3(x-10)}{(2x-5)^3}$

51. a. $f(x) = 3x(x^2+1)^{-1/2}$

$f'(x) = 3x(-\frac{1}{2})(x^2+1)^{-3/2}(2x) + (x^2+1)^{-1/2}(3)$

b. Factor out $3(x^2+1)^{-3/2}$:

$f'(x) = 3(x^2+1)^{-3/2}[(x^2+1) - x^2]$

$= 3(x^2+1)^{-3/2}$

53. The rate of change at any t is:

$N'(t) = 10^9[(50 - 2t)^3]' = 10^9[3(50 - 2t)^2(-2)] = 10^9(-6)(50 - 2t)^2$.

At time $t = 10$: $N'(10) = 10^9(-6)[50-2(10)]^2 = -5400 \cdot 10^9$.

The negative indicates the number is decreasing at the rate of 5,400,000,000,000 bacteria per minute.

55. $\frac{dA}{dr} = 100{,}000(1 + r)^9$

57. $\frac{dP}{dr} = -100{,}000(1 + r)^{-11}$

59. a. When $t = 0$ $N = 8{,}000 - \frac{4{,}000}{\sqrt{1}} = 4{,}000$; The present enrollment is 4,000.

b. When $t = 10$ $N = 8{,}000 - \frac{4{,}000}{\sqrt{1 + .2(10)}} \approx 5{,}691$ The enrollment in 10 years will be about 5,691. **c.** Let $N(t) = 8{,}000 - 4{,}000(1 + .2t)^{-1/2}$
$= 2000(1 + .2t)^{-3/2}(.2) = 400(1 + .2t)^{-3/2}$ **d.** $N'(0) = 400(1 + 0)^{-3/2}$
$= 400$; annual enrollment is increasing by about 400 students per year.

e. $N'(10) = 400[1 + .2(10)]^{-3/2} \approx 77$; annual enrollment is increasing by about 77 students per year.

61. $\frac{f}{g} = fg^{-1}$. Using the product rule,

$$\begin{aligned}(fg^{-1})' &= f'(g^{-1}) + f(g^{-1})' \\ &= f'g^{-1} + f(-g^{-2}g') \\ &= \frac{f'}{g} - \frac{fg'}{g^2} \\ &= \frac{fg'}{g^2} - \frac{fg'}{g^2} \\ &= \frac{gf' - fg'}{g^2}\end{aligned}$$

2.8 Review

Objective Questions

1. As we approach $x = 3$ from either the left or right we see the function approaches 10. It does not matter that the function is not defined at $x = 3$.

3. As we approach $x = 5$ from the left, the function approaches 6, but as we approach $x = 5$ from the right, the function approaches 4; therefore, the limit does not exist.

5.

x	2.9	2.99	2.999	3.1	3.01	3.001	3.0001
$f(x)$	155	1595	15995	-165	-1605	-16005	-160005

Limit does not exist.

7. $\lim_{|x| \mapsto \infty} \dfrac{5x + 1}{3 - x} = \lim_{|x| \mapsto \infty} \dfrac{5 + \frac{1}{x}}{\frac{3}{x} - 1} = -5$

9. $\lim_{x \to 1} \dfrac{x^2 + 5x - 6}{x - 1} = \lim_{x \to 1} \dfrac{(x - 1)(x + 6)}{x - 1} = \lim_{x \to 1} (x - 6) = 7$

11. $\lim_{x \to -1} \dfrac{x^2 + 7x + 6}{x - 1} = \dfrac{(-1)^2 + 7(-1) + 6}{(-1)(-1)} = \frac{0}{1} = 0$

13. Suspicious points at $x = -3$ and $x = 4$; discontinuous at $x = 4$ (hole).

15. Suspicious points at $x = 7$; discontinuous at $x = 7$ (hole).

17. Discontinuous (bacteria are nonnegative integers)

19. Continuous

21. *Without the restriction*, the function would be undefined at $x = 8$ (division by zero); but since we are limited to x values only between -5 and 5, $f(x)$ is continuous throughout the interval.

23. Suspicious point at $x = 8$:

$f(8) = 4;\ \lim_{x \to 8} \dfrac{x^2 - 15x + 56}{x - 8} = \lim_{x \to 8} \dfrac{(x - 8)(x - 7)}{x - 8} = \lim_{x \to 8} (x - 7) = 1;$

discontinuous at $x = 8$

25. Find $\dfrac{f(x + h) - f(x)}{h}$ where $x = 1$ and $h = 2$; $\dfrac{f(3) - f(1)}{2} = \dfrac{39 - 7}{2} = 16$

27. Find $\dfrac{f(x + h) - f(x)}{h}$ where $x = 0$ and $h = 6$; $\dfrac{f(6) - f(0)}{6} = \dfrac{\sqrt{18} - 0}{6} = \dfrac{3\sqrt{2}}{6}$

$= \dfrac{\sqrt{2}}{2}$

29. $\dfrac{dy}{dx} = 6x + 4$; at $x = 1$, $\left.\dfrac{dy}{dx}\right|_{x = 1} = 10$

31. $\frac{dy}{dx} = \frac{1}{2}(3x)^{-1/2}(3) = \frac{3}{2}(3x)^{-1/2}$ $\left.\frac{dy}{dx}\right|_{x=3} = \frac{1}{2}$

33. For a function $f(x)$, the derivative is $f'(x) = \lim_{h\to 0} \frac{f(x+h) - f(x)}{h}$ provided this limit exists.

35. $$\lim_{h\to 0} \frac{\frac{1}{x+h-5} - \frac{1}{x-5}}{h} = \lim_{h\to 0} \frac{x-5-x-h+5}{(x+h-5)(x-5)} \cdot \frac{1}{h}$$
$$= \lim_{h\to 0} \frac{-1}{(x+h-5)(x-5)}$$
$$= \frac{-1}{(x-5)^2}$$

37. $y' = 14x^{13}$ **39.** $y' = -\frac{7}{9}x^{-16/9}$ **41.** $y' = 0$ **43.** $y' = 6x^2 - 10x$

45.
$$\begin{aligned} y' &= (1-3x)(2+9x)' + (2+9x)(1-3x)' \\ &= (1-3x)(9) + (2+9x)(-3) \\ &= 3 - 54x \end{aligned}$$

47.
$$\begin{aligned} y' &= x^{1/2}(-1)(x-1)^{-2}(1) + (x-1)^{-1}(\tfrac{1}{2}x^{-1/2}) \\ &= \tfrac{1}{2}x^{-1/2}(x-1)^{-2}[-2x + (x-1)] \\ &= \tfrac{1}{2}x^{-1/2}(x-1)^{-2}(-x-1) \\ &= -\tfrac{1}{2}x^{-1/2}(x-1)^{-2}(x+1) \end{aligned}$$

49. $y = (3x^2+1)^{-1}$; $y' = -(3x^2+1)^{-2}(6x) = \frac{-6x}{(3x^2+1)^2}$

51.
$$\begin{aligned} y' &= \frac{(5x-3)(\frac{1}{2})(x^{-1/2}) - x^{1/2}(5)}{(5x-3)^2} \\ &= \frac{\frac{1}{2}x^{-1/2}[(x^{-1/2}) - x^{1/2}(5)]}{(5x-3)^2} \\ &= \frac{-5x-3}{2x^{1/2}(5x-3)^2} \end{aligned}$$

53. $f'(x) = 4(5x+9)^3(5) = 20(5x+9)^3$

55. $y' = \frac{1}{5}(5x^2 - 3x)^{-4/5}(10x - 3)$

Sample Test

1. 4 **3.** $\lim_{x\to 2} \frac{6x+1}{3-x} = 13$

5. Look for jumps, holes, or "poles" (vertical asymptotes). There is a hole at $x = 0$, a jump at $x = 3$, and a pole at $x = 10$.

7. Let $f(x) = y = 5 + x - 2x^2$. The average rate of change is $\dfrac{f(3) - f(1)}{3 - 1}$

$$= \frac{f(3) - f(1)}{3 - 1} = \frac{(-10) - (4)}{3 - 1} = -7$$

9. The derivative of $f(x)$ is $f'(x) = \lim\limits_{h \to 0} \dfrac{f(x + h) - f(x)}{h}$ provided this limit exists.

11. The derivative of *any* constant is zero: $y' = 0$.

13. $y' = 5(2)x - 2(-1)x^{-2} = 10x + 2x^{-2}$

15. $y' = \dfrac{(4x - 5)(15)' - (15)(4x - 5)'}{(4x - 5)^2} = \dfrac{0 - (15)(4)}{(4x - 5)^2} = \dfrac{-60}{(4x - 5)^2}$

17.
$$y' = \frac{(1 - 2x)[(x - 5)^3]' - (x - 5)^3(1 - 2x)'}{(1 - 2x)^2}$$

$$= \frac{(1 - 2x)[3(x - 5)^2] - (x - 5)^3(-2)}{(1 - 2x)^2}$$

$$= \frac{(x - 5)^2[3(1 - 2x) + 2(x - 5)]}{(1 - 2x)^2}$$

$$= \frac{(x - 5)^2[-4x - 7]}{(1 - 2x)^2}$$

$$= \frac{-(x - 5)^2(4x + 7)}{(1 - 2x)^2}$$

19. $\dfrac{P(25) - P(20)}{25 - 20} = \dfrac{\frac{1000 - 25^2}{100 - 25} - \frac{1000 - 20^2}{100 - 20}}{25 - 20} = \dfrac{5 - 7.5}{5} = -.5;$

The profit is decreasing by \$.50 per unit.

CHAPTER 3

Additional Derivative Topics

3.1 Implicit Differentiation

1. explicitly: $y = -5x^2 + 12$

$\frac{dy}{dx} = -10x$

implicitly: $\frac{dy}{dx} + 10x + 0 = 0$

$\frac{dy}{dx} = -10x$

3. explicitly: $y = \frac{5}{x} = 5x^{-1}$

$\frac{dy}{dx} = -5x^{-2}$

implicitly: $x\frac{dy}{dx} + y(1) = 0$

$\frac{dy}{dx} = \frac{-y}{x}$

5. explicitly: $3xy = x^2 - 50$

$y = \frac{1}{3}x - \frac{50}{3}x^{-1}$

$\frac{dy}{dx} = \frac{1}{3} + \frac{50}{3}x^{-2}$

implicitly: $2x - 3x\frac{dy}{dx} - 3y = 0$

$3x\frac{dy}{dx} = 2x - 3y$

$\frac{dy}{dx} = \frac{2}{3} - \frac{y}{x}$

7. explicitly: $y^2 = 4 - x^2$

$y = \sqrt{4 - x^2}, \quad y > 0$

$\frac{dy}{dx} = \frac{1}{2}(4 - x^2)^{-1/2}(-2x)$

$= -x(4 - x^2)^{-1/2}$

implicitly: $2x + 2y\frac{dy}{dx} = 0$

$\frac{dy}{dx} = -\frac{x}{y}, \quad y > 0$

9. explicitly: $5y^3 = 2x^4 - 7$

$y^3 = \frac{2}{5}x^4 - \frac{7}{5}$

$y = 3\sqrt[3]{\frac{2}{5}x^4 - \frac{7}{5}}$

$\frac{dy}{dx} = \frac{1}{3}(\frac{2}{5}x^4 - \frac{7}{5})^{-2/3}(\frac{8}{5}x^3)$

$= \frac{8}{15}x^3(\frac{2}{5}x^4 - \frac{7}{5})^{-2/3}$

implicitly: $8x^3 - 15y^2\frac{dy}{dx} = 0$

$\frac{dy}{dx} = \frac{8x^3}{15y^2}$

11.
$$2x - y - x\frac{dy}{dx} + 2y\frac{dy}{dx} = 0$$
$$-x\frac{dy}{dx} + 2y\frac{dy}{dx} = -2x + y$$
$$(x - 2y)\frac{dy}{dx} = 2x - y$$
$$\frac{dy}{dx} = \frac{2x - y}{x - 2y}$$

13.
$$2(2x + 1)(2) + 2(3y - 5)(3\frac{dy}{dx}) = 0$$
$$4(2x + 1) + 6(3y - 5)\frac{dy}{dx} = 0$$
$$6(3y - 5)\frac{dy}{dx} = -4(2x + 1)$$
$$\frac{dy}{dx} = \frac{-2(2x + 1)}{3(3y - 5)}$$

15.
$$\frac{x^2 - y^2}{2x + y^2} = 10$$
$$x^2 - y^2 = 10(2x + y^2)$$
$$D_x(x^2 - y^2) = D_x[10(2x + y^2)]$$
$$2x - 2y\frac{dy}{dx} = 20 + 20y\frac{dy}{dx}$$
$$20y\frac{dy}{dx} + 2y\frac{dy}{dx} = 2x - 20$$
$$(20y + 2y)\frac{dy}{dx} = 2x - 20$$
$$\frac{dy}{dx} = \frac{x - 10}{11y}$$

17.
$$(3x^2y^3)' - (3xy^2)' + (5xy)' = (2)'$$
$$3[x^2(y^3)' + (x^2)'y^3] - 3[x(y^2)' + x'y^2] + 5(xy' + x'y) = (2)'$$
$$3[x^2(3y^2y') + 2xy^3] - 3[x(2yy') + y^2] + 5(xy' + y) = 0$$
$$9x^2y^2y' + 6xy^3 - 6xyy' - 3y^2 + 5xy' + 5y = 0$$
$$9x^2y^2y' - 6xyy' + 5xy' = -6xy^3 + 3y^2 - 5y$$
$$y'(9x^2y^2 - 6xy + 5x) = -6xy^3 + 3y^2 - 5y$$
$$y' = \frac{-6xy^3 + 3y^2 - 5y}{9x^2y^2 - 6xy + 5x}$$

19.
$$(x^3)' + (2x^2)' + (xy)' - (4)' = (0)'$$
$$3x^2 + 4x + (xy' + x'y) - 0 = 0$$
$$3x^2 + 4x + xy' + y = 0$$
$$xy' = -3x^2 - 4x - y$$

$$y' = \frac{-3x^2 - 4x - y}{x}$$

21. First, the *slope* of the tangent line is given by y':

$$(x^2)' + (y^2)' - (4)' = (0)'$$
$$2x + 2yy' - 0 = 0$$
$$y' = -\frac{2x}{2y} = -\frac{x}{y}.$$

At (0, 2), $y' = -\frac{0}{2} = 0$. The slope is 0 and the line passes through (0, 2):

$$y - 2 = 0(x - 0)$$
$$y - 2 = 0$$

23. First, the *slope* of the tangent line is given by y':

$$(2x^4)' - (5y^3)' - (7)' = (0)'$$
$$8x^3 - 15y^2y' - 0 = 0$$
$$8x^3 = 15y^2y'$$
$$\frac{8x^3}{15y^2} = y'$$

At $(-1, -1)$: $y' = -\frac{8}{15}$

The slope is $-\frac{8}{15}$ and the line passes through $(-1, -1)$: $y + 1 = -\frac{8}{15}(x + 1)$

Multiply by 15: $15y + 15 = -8(x + 1)$

$$15y + 15 = -8x - 8$$
$$8x + 15y + 23 = 0$$

25. First, the *slope* of the tangent line is given by y':

$$(x^2)' + (xy)' + (y^2)' - (7)' = (0)'$$
$$2x + (xy' + x'y) + 2yy' - 0 = 0$$
$$2x + xy' + y = 2yy' = 0$$
$$xy' + 2yy' = -2x - 7$$
$$y'(x + 2y) = -2x - y$$
$$y' = \frac{-2x - y}{x + 2y}$$

At(1, −3): $y' = \frac{1}{-5}$

The slope is $-\frac{1}{5}$ and the line passes through $(1, -3)$: $y + 3 = -\frac{1}{5}(x - 1)$

$$5y + 15 = -(x - 1)$$
$$x + 5y + 14 = 0$$

27. First, the *slope* of the tangent line is given by y':

$$[(x - 2)^2]' + [(y - 1)^2]' = (9)'$$
$$2(x - 2) + 2(y - 1)(y - 1)' = (9)'$$
$$2x - 4 + 2(y - 1)y' = 0$$
$$2(y - 1)y' = 4 - 2x$$
$$y' = \frac{4 - 2x}{2(y - 1)}$$
$$= \frac{2 - x}{y - 1}$$

At $x = 2$ the slope is 0. To find y when $x = 2$, substitute into the equation of the curve:

$$(2 - 2)^2 + (y - 1)^2 = 9$$
$$(y - 1)^2 = 9$$
$$y - 1 = \pm 3$$
$$y = 4 \text{ or } -2$$

Since the slope is 0, the line passing through (2, 4) has equation: $y - 4 = 0$. The line passing through $(2, -2)$ has equation $y + 2 = 0$.

29. The *slope* of the tangent line is given by y':

$$(x^2)' - (2xy)' + (y^2)' = (0)'$$
$$2x - 2(xy' + x'y) + 2yy' = 0$$
$$x - (xy' + y) = yy' = 0 \qquad \textit{Divide by 2}$$
$$x - xy' - y + yy' = 0$$
$$yy' - xy' = y - x$$
$$y'(y - x) = y - x$$
$$y' = \frac{y - x}{y - x} = 1$$

At $x = -1$ (or anywhere else) the slope is 1. To find y when $x = 1$, substitute into the equation of the curve:

$$(-1)^2 - 2(-1)y + y^2 = 0$$
$$1 + 2y + y^2 = 0$$
$$(y + 1)^2 = 0$$
$$y = -1$$

The line with slope 1, passing through $(-1, -1)$ has the equation$y + 1 = 1(x + 1)$

$$x - y = 0$$

31. The slope of the tangent line is given by y':

$$\frac{(x+1)^2}{25} + \frac{(y-1)^2}{4} = (1)'$$

$$\frac{1}{25}[2(x+1)] + \frac{1}{4}[2(y-1)y'] = 0 \qquad \textit{Multiply by 100}$$

$$8(x+1) + 50(y-1)y' = 0$$

$$50(y-1)y' = -8(x+1)$$

$$y' = \frac{-8(x+1)}{50(y-1)}$$

$$= \frac{-4(x+1)}{25(y-1)}$$

At $x = -1$ the slope is 0.

To find y when $x = -1$, substitute into the equation of the curve:

$$\frac{(-1+1)^2}{25} + \frac{(y-1)^2}{4} = 1$$

$$0 + \frac{(y-1)^2}{4} = 1$$

$$(y-1)^2 = 4$$

$$y - 1 = \pm 2$$

$$y = 3 \text{ or } -1.$$

The line with slope 0, passing through $(-1, 3)$ is $y - 3 = 0$. The line with slope 0, passing through $(-1, -1)$ is $y + 1 = 0$.

33.

$$2yy' = 12x^2$$

$$y' = \frac{6x^2}{y}$$

at $(1, -2)$: $$y' = \frac{6}{-2} = -3$$

35.

$$6x^2 + 6y^2y' = 9y + 9xy'$$

$$(2y^2 - 3x)y' = 3y - 2x^2$$

$$y' = \frac{3y - 2x^2}{2y^2 - 3x}$$

at $(1, 2)$: $$y' = \frac{6-2}{8-6} = \frac{4}{2} = 2$$

37.

$$24(x^2 + y^2)(2x + 2yy') = 625y + 625xy'$$

$$24(2x^3 + 2x^2yy' + 2xy^2 + 2y^3y') = 625y + 625xy'$$

$$48x^2yy' + 48y^3y' - 625xy' = 625y - 48x^3 - 48xy^2$$

$$y' = \frac{625y - 48x^3 - 48x^2y}{48x^2y + 48y^3 - 625x}$$

at (3, 4):

$$y' = \frac{625(4) - 48(3)^3\,48(3)^2(4)}{48(3)^2(4) + 48(4)^3 - 625(3)}$$

$$= \frac{-524}{2925} \approx -.2$$

39.

$$x' = (p^2)' - (5p)' + 500'$$
$$1 = 2pp' - 5p' + 0$$
$$1 = p'(2p - 5)$$
$$\frac{1}{2p - 5} = p'$$

3.2 Differentials

1. $\frac{dy}{dx} = 15x^2$, so $dy = 15x^2\ dx$

3. $\frac{dy}{dx} = -10x^{-2}$, so $dy = -10x^{-2}dx$

5. $\frac{dy}{dx} = 300x^2 - 50$, so $dy = (300x^2 - 50)dx$

7. $y = 3(x - 1)^{1/2}$; $\frac{dy}{dx} = \frac{3}{2}(x - 1)^{-1/2}$, so $dy = \frac{3}{2}(x - 1)^{-1/2}dx$

9. $\frac{dy}{dx} = 5$, so $dy = 5\ dx$

11.

$$\frac{dy}{dx} = (5x - 3)^2(x^2 - 3)' + [(5x - 3)^2]'(x^2 - 3)$$
$$= (5x - 3)^2(2x) + 2(5x - 3)(5)(x^2 - 3)$$
$$= (5x - 3)[(5x - 3)(2x) + 10(x^2 - 3)]$$
$$= (5x - 3)(20x^2 - 6x - 30)$$
$$= 2(5x - 3)(10x^2 - 3x - 15)$$

So $dy = 2(5x - 3)(10x^2 - 3x - 15)dx$

13.

$$\frac{dy}{dx} = \frac{(x - 2)(3x + 1)' - (3x + 1)(x - 2)'}{(x - 2)^2} = \frac{(x - 2)(3) - (3x + 1)}{(x - 2)^2}$$
$$= \frac{-7}{(x - 2)^2}$$

So $dy = -7(x - 2)^{-2}dx$

15. $y = x^2 - x + 3$; $\frac{dy}{dx} = 2x - 1$, so $dy = (2x - 1)dx$

17. First expand: $y = \left(5 - \frac{1}{x^2}\right)\left(1 + \frac{1}{x}\right)$

$= (5 - x^{-2})(1 + x^{-1}) = 5 + 5x^{-1} - x^{-2} - x^{-3}$

$\frac{dy}{dx} = -5x^{-2} + 2x^{-3} + 3x^{-4}$, so $dy = (-5x^{-2} + 2x^{-3} + 3x^{-4})dx$

19. $\frac{dy}{dx} = \frac{(x^2 + 3x - 2)(x^2 - 5x + 1)' - (x^2 - 5x + 1)(x^2 + 3x - 2)'}{(x^2 + 3x - 2)^2}$

$= \frac{(x^2 + 3x - 2)(2x - 5) - (x^2 - 5x + 1)(2x + 3)}{(x^2 + 3x - 2)^2} = \frac{8x^2 - 6x + 7}{(x^2 + 3x - 2)^2}$

So $dy = \frac{8x^2 - 6x + 7}{(x^2 + 3x - 2)^2}\, dx$

21. $dy = f'(x)dx = (2x - 2)dx$. Substituting $x = 10$ and $dx = \Delta x = .1$,

$dy = [2(10) - 2](.1) = 1.8$

$\Delta y = f(x + \Delta x) - f(x)$

$= (x + \Delta x)^2 - 2(x + \Delta x) + 5 - (x^2 - 2x + 5)$

$= 2x\Delta x + \Delta x^2 - 2\Delta x$. Substituting $x = 10$, $\Delta x = .1$,

$\Delta y = 2(10)(.1) + (.1)^2 - 2(.1) = 1.81$

23. $dy = f'(x)dx = \frac{5}{2}(5x)^{-1/2}dx$. Substituting $x = 5$ and $dx = \Delta x = .15$,

$dy = \frac{5}{2}[5(5)]^{-1/2}(.15) = \frac{5}{2}(\frac{1}{5})(.15) = .075$

$\Delta y = f(x + \Delta x) - f(x) = \sqrt{5x(x + \Delta x)} - \sqrt{5x}$

Substituting $x = 5$ and $\Delta x = .15$,

$\Delta y = \sqrt{5(5.15)} - \sqrt{5(5)} \approx .0744457825$

25. $dy = \frac{(5x + 2)(2x - 3)' - (2x - 3)(5x + 2)'}{(5x + 2)^2}\, dx$

$= \frac{(5x + 2)(2) - (2x - 3)(5)}{(5x + 2)^2}\, dx$

$= \frac{19}{(5x + 2)^2}\, dx$ For $x = 100$, $dx = \Delta x = 3$: $\Delta y \approx dy$

$= \frac{19}{[5(100) + 2]^2}(3) = \frac{57}{502^2} \approx .0002261869$

27. $\frac{dy}{dx} = 20\left(3 - \frac{1}{x^2}\right)' = 20\left(\frac{2}{x^3}\right) = \frac{40}{x^3}$ So $dy = \frac{40}{x^3}\, dx$

At $x = 10$, $dx = \Delta x = .02$:

$$\Delta y \approx \frac{40}{10^3}(.02) = .0008$$

29. $$dy = 450\left[\frac{\sqrt{x - 50}\,(x + 300)' - (x + 300)\,\sqrt{x - 50}'}{\sqrt{x - 50}^2}\right]dx$$

$$= 450\left[\frac{\sqrt{(x - 50)}(1) - (x + 300)\left[\frac{1}{2}(x - 50)^{-1/2}\right]}{x - 50}\right]dx$$

$$= 450\left[\frac{\sqrt{x - 50} - \frac{x + 300}{2\sqrt{x - 50}}}{x - 50}\right] \cdot \frac{2\sqrt{x - 50}}{2\sqrt{x - 50}}\,dx$$

$$= 450\left[\frac{2(x - 50) - (x + 300)}{2(x - 50)^{3/2}}\right]dx$$

At $x = 1000$, $dx = \Delta x = 10$: $\Delta y \approx dy = 450\left[\frac{2(950) - (1300)}{2(950)^{3/2}}\right](10) = 46.105036$

31. First, $c = 100$ and $\Delta c = 110 - 100 = 10$. $\frac{ds}{dc} = 500 - 2c$,

so $ds = (500 - 2c)dc$ $\Delta s \approx ds = (500 - 2c)dc = (500 - 2(100))(10) = 3{,}000$

so the sales will increase by about 3,000 units.

33. First, $x = 3$ and $\Delta x = 3.5 - 3 = .5$

$$\frac{dA}{dx} = \frac{(100 + x^2)(3x)' - (3x)(100 + x^2)'}{(100 + x^2)^2} = \frac{(100 + x^2)(3) - (3x)(2x)}{(100 + x^2)^2} = \frac{300 - 3x^2}{(100 + x^2)^2}$$

So $\Delta A \approx dA = \frac{300 - 3x^2}{(100 + x^2)^2}dx = \frac{300 - 3(3)^2}{(100 + 3x^2)^2}(.5) = .011489$ The alcohol will

increase by about .01 percent.

35. The area of a circle is given by $A = \pi r^2$. $\frac{dA}{dr} = \pi r$, so $dA = 2\pi r\, dr$. At $r = 2$ and $\Delta r = 2.1 - 2 = .1$, $\Delta A \approx dA = 2\pi(2)(.1) = .4\pi \approx 1.256637$ square miles.

37. $dR = 60dx$; For $x = 100$ and $\Delta x = 110 - 100 = 10$: $dR = 60(10) = 600$, so the change in revenue is $\Delta R \approx dR = \$600$. $P = R(x) - C(x)$ $= 60x - (400 + 30x) = 30x - 400$ $dP = (30)dx$, so the change in profit is $\Delta P \approx dP = 30(10) = \$300$.

39. $\frac{dN}{dx} = \frac{1}{5}x - \frac{3}{50{,}000}x^2$; $dN = (\frac{1}{5}x - \frac{3}{50{,}000}x^2)dx$. The change in N for $x = 1{,}000$

and $x = 1{,}100 - 1{,}000 = 100$ is: $\Delta N \approx dN = \left[\frac{1}{5}(1{,}000) - \frac{3}{50{,}000}(1{,}000)^2\right](100)$ $= 14{,}000$ votes.

3.3 Business Models Using Differentiation

1. $C'(x) = 6 - 2x + 3x^2$; $C'(1) = 6 - 2(1) + 3(1)^2 = 7$
3. $C'(x) = 6 - 2x + 3x^2$; $C'(4) = 6 - 2(4) + 3(4)^2 = 46$
5. $R'(x) = 9 - .002x$; $R'(0) = 9$
7. $R'(x) = 9 - .002x$; $R'(50) = 9 - .002(50) = 8.9$
9. $P'(x) = 3x^2 - 16x + 2$; $P'(2) = 3(2)^2 - 16(2) + 2 = -18$
11. $P'(x) = 3x^2 - 16x + 2$; $P'(4) = 3(4)^2 - 16(4) + 2 = -14$
13. $\overline{C}(x) = 200x^{-1} + 6 - x + x^2$; $\overline{C}'(x) = -200x^{-2} - 1 + 2x$
15. $\overline{C}(x) = 5{,}000x^{-1} + .4x$; $\overline{C}'(x) = -5{,}000x^{-2} + .4$
17. $\overline{R}(x) = 50 - .5x$; $\overline{R}'(x) = -.5$
19. $\overline{P}(x) = x^2 - 8x + 2 + 50x^{-1}$; $\overline{P}'(x) = 2x - 8 - 50x^{-2}$
21. $\overline{P}(x) = x^2 - 50x + 5 + 200x^{-1}$; $\overline{P}'(x) = 2x - 50 - 200x^{-2}$

For Problems 22-27; Let p = price and x = no. of items sold in a year and consider (x, p).

23. From Problem 22, $p = -.001x + 30$

$$R(x) = xp = x(-.001x + 30) = -.001x^2 + 30x$$

25. $R'(x) = -.002x + 30$;

This is approximately the change of revenue relative to a unit change in x.

27.
$$\begin{aligned} P(x) &= R(x) - C(x) \\ &= (-.001x^2 + 30x) - (15{,}000 + 8.5x) \\ &= -.001x^2 + 21.5x - 15{,}000 \\ P'(x) &= -.002x + 21.5 \end{aligned}$$

29. $\overline{C}(x) = \dfrac{50{,}000 + 50x}{x} = 50{,}000x^{-1} + 50$ so $\overline{C}'(x) = 50{,}000x^{-2}$

31. $R(x) = xp = x(200 - .04x) = 200x - .04x^2$, so $R'(x) = 200 - .08x$

33. See Problem 31; $R'(1{,}000) = 200 - .08(1{,}000) = 120$; at a production level of 1,000 items, the revenue is increasing at a rate of \$120 per item. $R'(2{,}500) = 0$; at a production level of 2,500 items the revenue does not change per unit change in the number of items.

35. $P(x) = (200x - .04x^2) - (50{,}000 + 50x) = 150x - .04x^2 - 50{,}000$

$P'(x) = 150 - .08x$, so $P'(1{,}000) = 70$ which means that at a production level of 1,000 items, the profit is increasing at \$70 per item. Similarly, $P'(2{,}500) = -50$ which means the profit is decreasing at \$50 per item.

37. The graph of $C(x) = 50{,}000 + 50x$ is a straight line passing through $[0, C(0)] = (0, 50{,}000)$ and $[1{,}000, C(1{,}000)] = (1000, 100000)$. The graph of $R(x) = 200x - .04x^2$ is a parabola, some of the ordered pairs are

$$x = 0,\ R(0) = 0$$

$$x = 1{,}000,\ R(1{,}000) = 200(1{,}000) - .04(1{,}000)^2 = 160{,}000$$

$$x = 2{,}000,\ R(2{,}000) = 200(2{,}000) - .04(2{,}000)^2 = 240{,}000$$

You should calculate several other ordered pairs, The graphs are below:

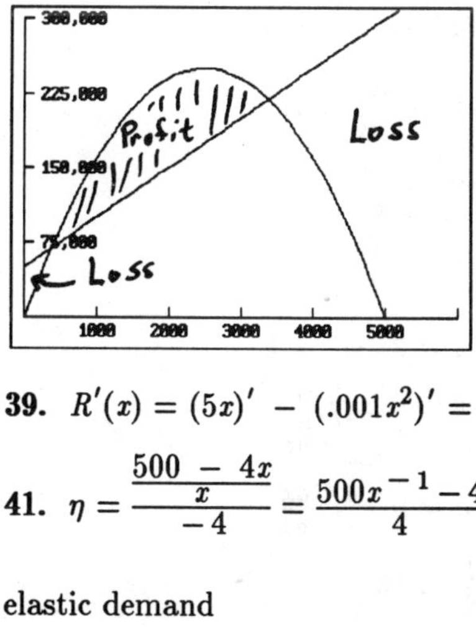

39. $R'(x) = (5x)' - (.001x^2)' = 5 - .002x$

41. $\eta = \dfrac{\frac{500 - 4x}{x}}{-4} = \dfrac{500x^{-1} - 4}{4} = 125x^{-1} - 1$ at $x = 50$, $\eta = 1.5$;

elastic demand

43. $\eta = \dfrac{\frac{100 - .5x^2}{x}}{-x} = -100x^{-2} + .5$ at $x = 20$, $\eta = .25$; inelastic demand

3.4 Related Rates

1. Differentiate both sides with respect to t:

$$x^2 + y^2 = 25$$
$$2x\frac{dx}{dt} + 2y\frac{dy}{dt} = 0$$
$$x\frac{dx}{dt} + y\frac{dy}{dt} = 0$$
$$3(4) + y\frac{dy}{dt} = 0$$
$$y\frac{dy}{dt} = -12$$
$$\frac{dy}{dt} = \frac{-12}{y}$$

We need to know y when $x = 3$:

$$x^2 + y^2 = 25$$
$$3^2 + y^2 = 25$$
$$y^2 = 16$$
$$y = 4$$

So, $\frac{dy}{dt} = \frac{-12}{y} = \frac{-12}{4} = -3$

3. Differentiate both sides with respect to t:

$$5x^2 - y = 100$$
$$10x\frac{dx}{dt} - \frac{dy}{dt} = 0$$
$$10(10)(10) - \frac{dy}{dt} = 0$$
$$\frac{dy}{dt} = 1{,}000$$

5. Differentiate both sides with respect to t:

$$y = 2x^{1/2} - 9$$
$$\frac{dy}{dt} = x^{-1/2}\frac{dx}{dt}$$
$$5 = 9^{-1/2}\frac{dx}{dt}$$
$$\frac{dx}{dt} = 15$$

7.

$$xy = 10$$
$$x\frac{dy}{dt} + y\frac{dx}{dt} = 0 \qquad \text{Product Rule}$$
$$5\frac{dy}{dt} + y(-2) = 0$$

$$\frac{dx}{dt} = \frac{2y}{5} \qquad \text{but when } x = 5,\ y = 2$$

$$\frac{dy}{dt} = \frac{4}{5}$$

9. $$x^2 + xy - y^2 = 20$$

$$2x\frac{dx}{dt} + \left[x\frac{dy}{dt} + y\frac{dx}{dt}\right] - 2y\frac{dy}{dt} = 0$$

$$2(4)\frac{dx}{dt} + \left[4(5) + (2)\frac{dx}{dt}\right] - 2(2)(5) = 0 \quad \text{If } x = 4, \text{ then } 16 + 4y - y^2 = 20 \Rightarrow y = 2$$

$$8\frac{dx}{dt} + 20 + 2\frac{dx}{dt} - 20 = 0$$

$$10\frac{dx}{dt} = 0$$

$$\frac{dx}{dt} = 0$$

11. Differentiate with respect to t: $\frac{dP}{dt} = 125\frac{dx}{dt} - \frac{1}{100}x\frac{dx}{dt}$. We are given $\frac{dx}{dt} = -1$

(This means that production is decreasing 1 unit per week) and that $x = 200$:

$\frac{dp}{dt} = 125(-1) - \frac{1}{100}(200)(-1) = -123.$

So profit is decreasing at a rate of \$123 per week.

13. Let x represent the distance the product has moved along the conveyor belt.

By the Pythagorean Theorem, $d^2 = 8^2 + x^2$, $d^2 = 64 + x^2$

Differentiate both sides with respect to time (t): $2d\frac{dd}{dt} = 2x\frac{dx}{dt}$

when $x = 9$, $d^2 = 8^2 + 9^2$, so $d = \sqrt{145}$. Also, $\frac{dx}{dt} = 3$ (ft./sec.): $2d\frac{dd}{dt} = 2x\frac{dx}{dt}$,

$2(\sqrt{145})\frac{dd}{dt} = 2(9)(3)$. Solving for $\frac{dd}{dt}$; $\frac{dd}{dt} = \frac{2(9)(3)}{2\sqrt{145}} = \frac{27}{\sqrt{145}} \approx 2.24$

So the distance is changing at about 2.24 feet per second.

15. If the other details remain the same, we only need to change $\frac{dV}{dt} = 500$:

$\frac{dV}{dt} = 400\pi\frac{dh}{dt}$, $500 = 400\pi\frac{dh}{dt}$, $\frac{500}{400\pi} = \frac{dh}{dt}$, $\frac{dh}{dt} = \frac{5}{4\pi} \approx .398$ (feet per minute).

17. The general situation is the same as in Example 5. The specific situation is:

$\frac{dV}{dt} = 1000(.01)(-.0015) = -.015$

19. The general situation is the same as in Example 5. The specific situation is:

$\frac{dV}{dt} = 1000(.02)(-.003) = -.06$

21. The general situation: $A = \pi r^2$ Then $\frac{dA}{dt} = 2\pi r\frac{dr}{dt}$ The specific situation: $r = 4$ and $\frac{dr}{dt} = 1.5$; $\frac{dA}{dt} = 2\pi(4)(1.5) = 12\pi \approx 37.7$ The area is increasing at a rate of about 37.7 square miles per year.

23. Start with $A = \pi r^2$ and differentiate both sides with respect to t: $\frac{dA}{dt} = 2\pi r\frac{dr}{dt}$. The radius is changing at 1 foot per second, so $\frac{dr}{dt} = 1$: $\frac{dA}{dt}\, 2\pi(10)(1) = 20\pi$. So the area is growing at a rate of about 62.8 square feet per second.

25. Differentiate both sides with respect to t (time): $d = \frac{-5x}{x + 1000} + 70$

(Quotient Rule): $\frac{dd}{dt} = \frac{(x + 1000)(-5\frac{dx}{dt}) - (-5x)(\frac{dx}{dt})}{(x + 1000)^2} + 0$

$= \frac{5\frac{dx}{dt}(-(x + 1000) + x)}{(x + 1000)^2} = \frac{(-5000)\frac{dx}{dt}}{(x + 1000)^2}$ We are given that $x = 2000$ and

$\frac{dx}{dt} = -2000$: $\frac{dd}{dt} = \frac{-5000(-200)}{(2000 + 1000)^2} = \frac{1{,}000{,}000}{9{,}000{,}000} \approx .11$ So the price is increasing at \$.11 per day.

27. Differentiate both sides of $V = \frac{4}{3}\pi r^3$ with respect to t (time): $\frac{dV}{dt} = 4\pi r^2\frac{dr}{dt}$

We are given that $\frac{dr}{dt} = .3$ and $r = 4$: $\frac{dV}{dt} = 4\pi(4)^2(.3) \approx 60.32(cm^3$ per minute).

29. First draw a picture:

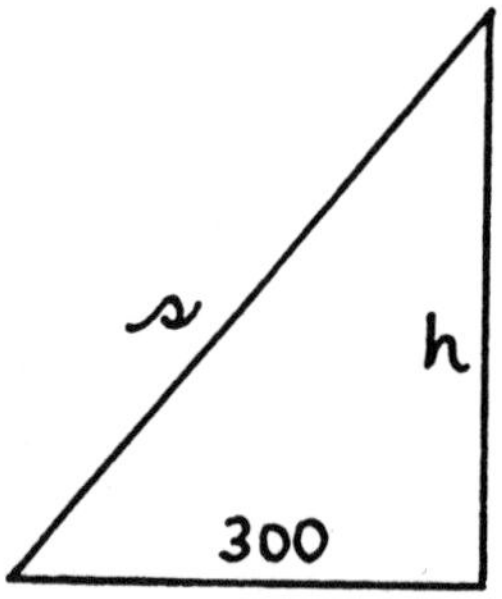

By the Pythagorean Theorem, $s^2 = 300^2 + h^2$,

$s^2 = 90{,}000 + h^2$

Differentiating with respect to t(time):

$2s\frac{ds}{dt} = 2h\frac{dh}{dt}$

Isolate $\frac{ds}{dt}$: $\frac{dx}{dt} = \frac{2h\frac{dh}{dt}}{2s} = \frac{h\frac{dh}{dt}}{s}$

We know $\frac{dh}{dt} = 10$(feet per second) and that $h = 400$. We need to know s:

$s^2 = 300^2 + h^2 = 300^2 = 400^2$, s^2, 250,000, $s = 500$. So $\frac{ds}{dt} = \frac{h(\frac{dh}{dt})}{s} = \frac{(400)(10)}{500} = 8$ (feet per second)

3.5 Review

Objective Answers

1.

$$x^5 + 2y^2 + y + 10 = 0$$
$$5x^4 + 4y\frac{dy}{dx} + \frac{dy}{dx} + 0 = 0$$
$$(4y + 1)\frac{dy}{dx} = -5x^4$$
$$\frac{dy}{dt} = \frac{-5x^4}{4y + 1}$$

3.

$$(x - 3)^2 - (y + 1)^2 = 1$$
$$D_x[(x - 3)^2 - (y + 1)^2] = D_x(1)$$
$$2(x - 3) - 2(y + 1)\frac{dy}{dt} = 0$$
$$2(y + 1)\frac{dy}{dt} = 2(x - 3)$$
$$\frac{dy}{dt} = \frac{x - 3}{y + 1}$$

5.

$$2y\frac{dy}{dx} = 24x^2$$
$$\frac{dy}{dx} = \frac{24x^2}{2y} = \frac{12x^2}{y}$$
at (2, 8), $m = \frac{dy}{dx} = \frac{12(2)^2}{8} = 6$

7.

$$\frac{dy}{dx} = 10x(6x - 5)^3 + 5x^2(3)(6x - 5)^2(6)$$
$$= 2x(6x - 5)^2[5(6x - 5) + 45x]$$
$$= 2x(6x - 5)^2(75x - 25)$$
$$= 50x(6x - 5)^2(3x - 1)$$
at $x = 1$, $m = \frac{dy}{dx} = 50(6 - 5)^2(3 - 1)$
$$= 100$$

9. From Problem 5, $m = 6$ at (2, 8):

$$y - y_1 = m(x - x_1)$$
$$y - 8 = 6(x - 2)$$
$$6x - y - 4 = 0$$

11. From Problem 7, $m = 100$:

Also if $x = 1$, then $y = 5$

$$y - y = m(x - x_1)$$
$$y - 5 = 100(x - 1)$$
$$100x + y + 95 = 0$$

13. $dy = 6(\frac{1}{2})(3x^2 + 5)^{-1/2}(6x)\,dx$

$= 18x(3x^2 + 5)^{-1/2}dx$

15. $$\frac{dy}{dx} = \frac{(3x^2 + x - 4)(2x^2 + x - 1)' - (2x^2 + x - 1)(3x^2 + x - 4)'}{(3x^2 + x - 4)^2}$$

$$= \frac{(3x^2 + x - 4)(4x + 1) - (2x^2 + x - 1)(6x + 1)}{(3x^2 + x - 4)^2}$$

$$= \frac{-x^2 - 10x - 3}{(3x^2 + x - 4)^2}. \text{ So } dy = \frac{-x^2 - 10x - 3}{(3x^2 + x - 4)^2}dx$$

17.

$dy = f'(x)dx$	$\Delta y = f(x + \Delta x) - f(x)$
$= 10x\,dx$	$= f(10.1) - f(10)$
$= 10(10)(.1)$	$= 510.05 - 500$
$= 10$	$= 10.05$

19.

$dy = f'(x)dx$	$\Delta y = f(x + \Delta x) = f(x)$
$= \frac{1}{2}(6x)^{-1/2}(6)\,dx$	$= f(6.2) - f(6)$
$= 3(6x)^{-1/2}dx$	$\approx .0991802$
$= 3(36)^{-1/2}(.2)$	
$= .1$	

21. $\Delta y \approx dy = 15x^{-4}dx = 15(10)^4(.03) = .000045$

23. $y = \frac{x^2 - 1}{x + 2}$ $\quad \Delta y \approx dy = f'(x)dx$

$$= \frac{(x + 2)(2x) - (x^2 - 1)(1)}{(x + 2)^2}\Delta x$$

$$= \frac{22(40) - (400 - 1)}{22^2}$$

$$= \frac{481}{484}$$

$$\approx .9938$$

25. a. $\frac{P(15) - P(10)}{5} \approx 1.17571$, about \$1.18.

b. $$P'(x) = \frac{(10 - x^2)(-3x^2) - (200 - x^3)(-2x)}{(10 - x^2)^2}$$

$$= \frac{x^4 - 30x^2 + 400x}{(10 - x^2)^2}$$

27. $C'(x) = 20x^3(2)(5x - 100)(5) + 60x^2(5x - 100)^2$

$= 20x^2(5x - 100)[10x + 3(5x - 100)]$

$= 20x^2(5x - 100)(25x - 300)$

$= 2{,}500x^2(x - 20)(x - 12)$

29. $8x\frac{dx}{dt} + 18y\frac{dy}{dt} = 0$

$\frac{dy}{dt} = \frac{-8x\frac{dx}{dt}}{18y} = \frac{-4x\frac{dx}{dt}}{9y}$ If $x = 2$ then $16 + 9y^2 = 36$

$= \frac{-4(\cancel{2})(\cancel{3})}{\cancel{9}(\frac{\cancel{2}\sqrt{5}}{\cancel{3}})}$ $y = \pm\frac{2\sqrt{5}}{3}$

$= \frac{-4}{\sqrt{5}}$

$= -\frac{4}{5}\sqrt{5}$

31. $2x\frac{dx}{dt} + 2x\frac{dy}{dt} + 2y\frac{dx}{dt} + 2y\frac{dy}{dt} = 0$

$\frac{dy}{dt} = \frac{(-2x - 2y)\frac{dx}{dt}}{2x + 2y} = \frac{-dx}{dt} = -5$

Sample Test

1. $\frac{dy}{dx} - 20x + 6 - 0 = 0$; $\frac{dy}{dx} = 20x - 6$

3. $4x^3 + 5x\frac{dy}{dt} + 5y - 6xy - 3x^2 = 0\ \frac{dy}{dt} + 9y^2 + 18xy\frac{dy}{dt}\ 0$

$(5x - 3x^2 + 18xy)\frac{dy}{dt} = -4x^3 - 5y + 6xy - 9y^2$

$\frac{dy}{dt} = \frac{4x^3 - 6xy + 9y^2 + 5y}{3x^2 - 18xy - 5x}$

5. $9x^2 + 16y^2 = 145$ To find y, use $x = 3$ in.

$9x^2 + 16y^2 = 145$

$18x\frac{dx}{dt} + 32y\frac{dy}{dt} = 0$ $9(3)^2 + 16y^2 = 145$

$16y^2 = 64$

$32y\frac{dy}{dt} = -18x\frac{dx}{dt}$ $y = \pm 2$

$\frac{dy}{dt} = \frac{-18x\frac{dx}{dt}}{32y} = \frac{-9x\frac{dx}{dt}}{16y}$ $\frac{dy}{dt} = \frac{-9x\frac{dx}{dt}}{16y} = \frac{-9(3)(4)}{16(\pm 2)} = \pm\frac{27}{8}$

7.
$$y = 25x^2$$
$$\frac{dy}{dt} = 50x\frac{dx}{dt}$$
$$(-4) = 50(3)\frac{dx}{dt}$$
$$\frac{dx}{dt} = \frac{-4}{150} = \frac{-2}{75}$$

9. Let x be the number of \$50 price raises

Rent = 500 + 50x; Number = 100 − 2x

$$R(x) = \text{Revenue} = \text{Rent} \cdot \text{Number}$$
$$= (500 + 50x)(100 - 2x)$$
$$= 50{,}000 + 4{,}000 - 100x^2$$

$$R'(x) = 4{,}000 - 200x$$

The marginal revenue is $4{,}000 - 200x$.

CHAPTER 4

Applications and Differentiation

4.1 First Derivatives and Graphs

1. $f'(x) = 16x$; $16x = 0$ when $x = 0$.

The vertex is at $(0, f(0)) = (0, 0)$.

A few points are:

$x = \pm 1, f(\pm 1) = 8(1)^2 = 8$

$x = -+1.5, f(\pm 1.5) = 8(\pm 1.5)^2 = 18$

You should calculate a few more points.

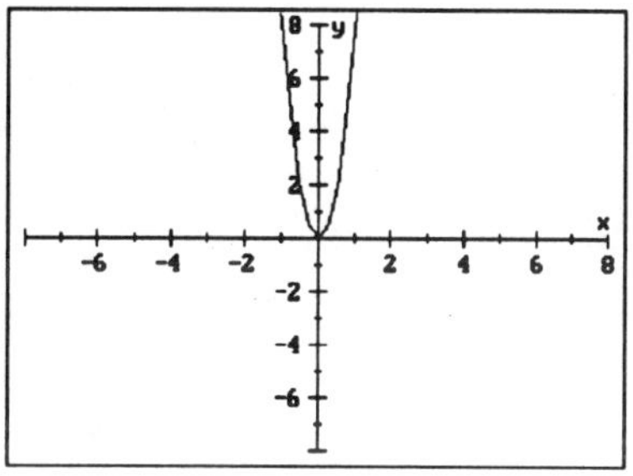

3. $f'(x) = -40x$; $-40x = 0$ when $x = 0$.

The vertex is at $(0, f(0)) = (0, 0)$.

A few points are:

$x = \pm 1, f(\pm 1) = -20(\pm 1)^2 = -20$

$x = \pm 1.5, f(\pm 1.5) = -20(\pm 1.5)^2 = -45$

You should calculate a few more points.

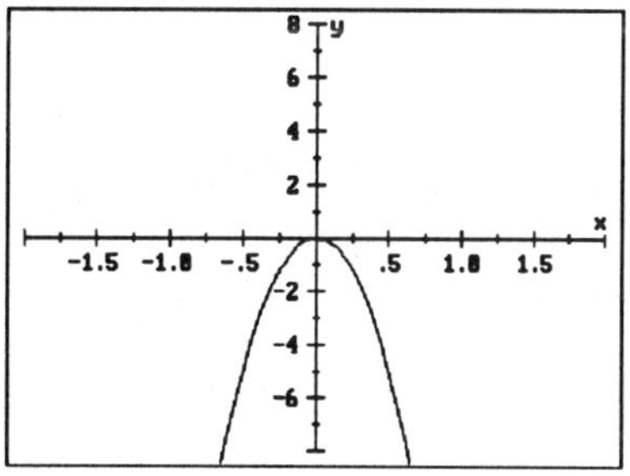

5. $(5y)' + (15x^2)' = 0'$, $5y' + 30x = 0$,

$5y' = -30x$, $y' = -6x = 0$ when $x = 0$

the vertex and a few other points are:

$5y + 15x^2 = 0$, $y = -3x^2$

$x = 0$, $y = -3(0)^2 = 0$

$x = \pm 1$, $y = -3(\pm 1)^2 = -3$

$x = \pm 2$, $y = -3(\pm 2)^2 = -12$

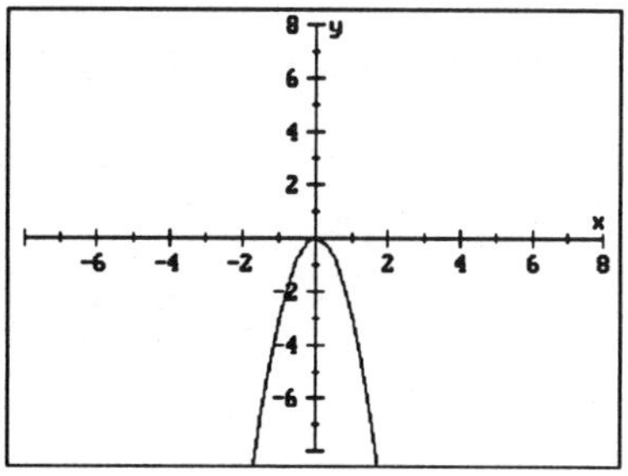

7. $f'(x) = 10x - 20;$

$10x - 20 = 0$ when $x = 2.$

The vertex is at $(2, f(2)) = (2, -18).$

A few points are:
$x = 1, f(1) = 5(1)^2 - 20(1) + 2 = -13$
$x = 0, f(0) = 5(0)^2 - 20(0) + 2 = 2$
$x = 3, f(3) = 5(0)^2 - 20(3) + 2 = -13$

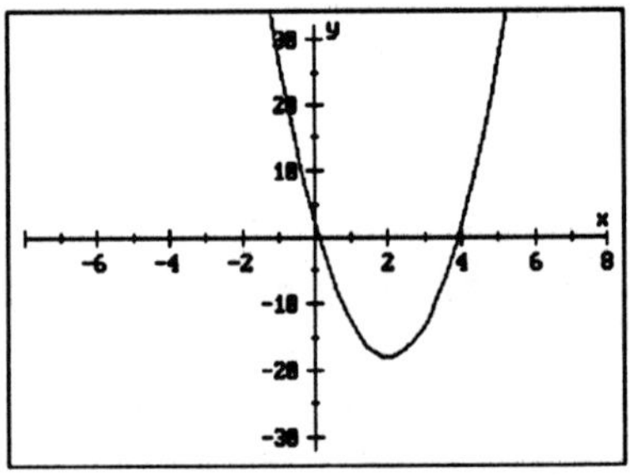

9. $(x^2)' + (4y)' - (3x)' + 1' = 0'$

$2x + 4y' - 3 = 0$

$4y' = 3 - 2x$

$y' = \frac{3 - 2x}{4}; \frac{3 - 2x}{4} = 0$ when $x = \frac{3}{2}$

The vertex and a few other points are:

$$x = 1,\ y = \frac{3(1) - 1 - (1)^2}{4} = \frac{1}{2}$$

$$x = 0,\ y = \frac{3(0) - 1 - (0)^2}{4} = -\frac{1}{4}$$

$$x = \tfrac{3}{2},\ y = \frac{3(\frac{3}{2}) - 1 - (\frac{3}{2})^2}{4} = \frac{5}{16}$$

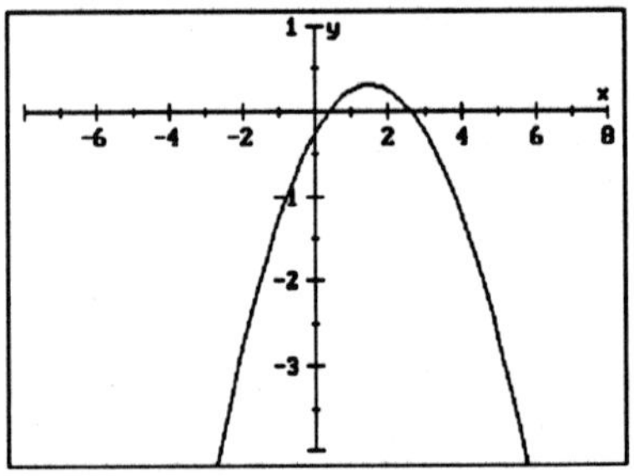

11. $(9x^2)' + (6x)' + (18y)' - 23' = 0'$

$18x + 6 + 18y' = 0$

$y' = \frac{-18x - 6}{18} = \frac{-3x - 1}{3};$

$\frac{-3x - 1}{3} = 0$ when $x = -\frac{1}{3}$

The vertex and a few other points are:

$$x = -\tfrac{1}{3},\ y = \frac{-9(-\frac{1}{3})^2 - 6(-\frac{1}{3}) + 23}{18} = \frac{4}{3}$$

$$x = 0,\ y = \frac{-9(0)^2 - 6(0) + 23}{18} = \frac{23}{18}$$

$$x = -1,\ y = \frac{-9(-1)^2 - 6(-1) + 23}{18} = \frac{10}{9}$$

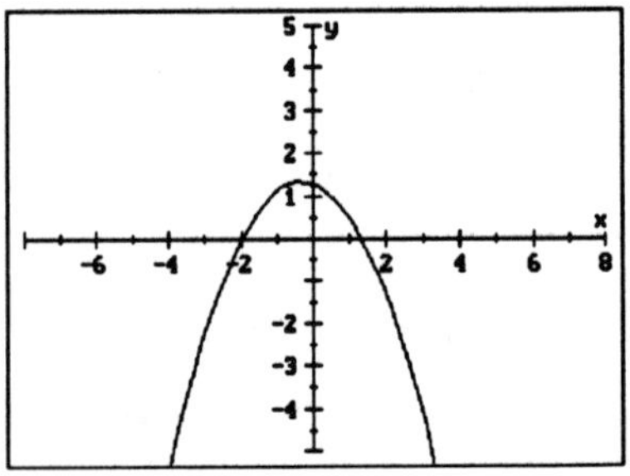

13. $y' = 10x - 3$; $10x - 3 = 0$ when $x = \frac{3}{10}$

The vertex is:

$x = \frac{3}{10}$, $y = 5(\frac{3}{10})^2 - 3(\frac{3}{10}) + 1 = \frac{11}{10}$

You should calculate a few more points.

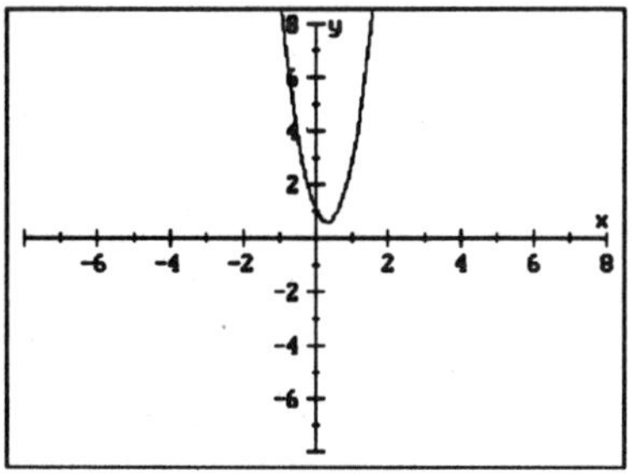

15. $y' = 14x + 2$; $14x + 2 = 0$

when $x = -\frac{1}{7}$

The vertex is: $x = -\frac{1}{7}$,

$y = 7(-\frac{1}{7})^2 + 2(-\frac{1}{7}) - 3 = -\frac{22}{7}$

You should calculate a few more points.

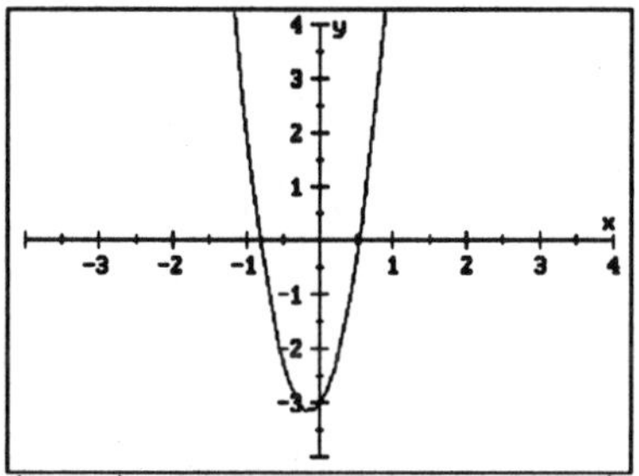

17. Increasing on $(-\infty, 2)$ and decreasing on $(2, \infty)$. horizontal tangent at $(2, 4)$.

19. Decreasing on $(-\infty, 0)$ and on $(5, \infty)$; increasing on $(0, 3)$; constant on $(3, 5)$. horizontal tangent at $(0, 3)$ and for all $3 < x < 5$.

21. Increasing on $(-\infty, 2)$ and $(2, \infty)$; no horizontal tangents.

23.

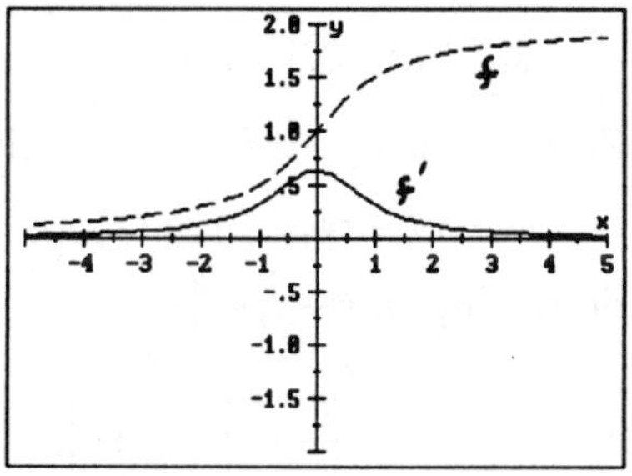

25.

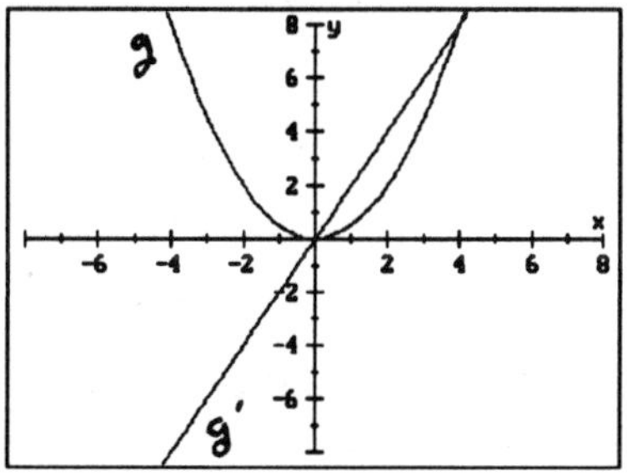

27. $f'(x) = 10 - 2x$

a. Always defined

$10 - 2x = 0$

$x = 5$

critical value is 5

b.

$10 - 2x$:	$+ + +$	$- - -$
	5	
Derivative:	pos	neg
Curve:	rising	falling

$f(x)$ is increasing on $(-\infty, 5)$

and decreasing on $(5, \infty)$

29. $f'(x) = 2x - 12$

a. Always defined

$$2x - 12 = 0$$

$$x = 6$$

critical value is 6

b.

	$x < 6$	$x > 6$
$2x - 12$:	− − −	+ + +
	6	
Derivative:	neg	pos
Curve:	falling	rising

$f(x)$ is increasing on $(6, \infty)$ and decreasing on $(-\infty, 6)$

31. $g'(x) = 6x^2 - 6x - 36$

a. Always defined

$$6x^2 - 6x - 36 = 0$$

$$x^2 - x - 6 = 0$$

$$(x - 3)(x + 2) = 0$$

$$x = 3, -2$$

critical values: $3, -2$

b.

		-2		3	
$x - 3$:	− − −		− − −		+ + +
$x + 2$:	− − −		+ + +		+ + +
Derivative:	pos		neg		pos
Curve:	rising		falling		rising

$g(x)$ is increasing on $(-\infty, -2)$ and $(3, \infty)$ and decreasing on $(-2, 3)$

33. $y' = 3x^2 + 10x + 8$

a. Always defined

$$3x^2 + 10x + 8 = 0$$

$$(3x + 4)(x + 2) = 0$$

$$x = -\tfrac{4}{3}, -2$$

critical values: $-\frac{4}{3}, -2$

b.

		-2		$-\frac{4}{3}$	
$3x + 4$:	− − −		− − −		+ + +
$x + 2$:	− − −		+ + +		+ + +
Derivative:	pos		neg		pos
Curve:	rising		falling		rising

y is increasing on $(-\infty, -2)$ and $(-\frac{4}{3}, \infty)$ and decreasing on $(-2, -\frac{4}{3})$

35. $y' = 12x^3 + 6x^2 - 18x$

a. Always defined

$$12x^3 + 6x^2 - 18x = 0$$

$$2x^3 + x^2 - 3x = 0$$

$$x(2x^2 + x - 3) = 0$$

$$x(2x + 3)(x - 1) = 0$$

$$x = 0, -\tfrac{3}{2}, 1$$

critical values: $0, -\frac{3}{2}, 1$

b.

		$-\frac{3}{2}$		0		1	
x:	− −		− −		+ +		+ +
$2x + 3$:	− −		+ +		+ +		+ +
$x - 1$:	− −		− −		− −		+ +
Derivative:	neg		pos		neg		pos
Curve:	falling		rising		falling		rising

y is increasing on $(-\frac{3}{2}, 0)$ and $(1, \infty)$ and decreasing on $(-\infty, -\frac{3}{2})$ and $(0, 1)$

37. $f'(x) = 15x^4 - 75x^2 - 540$

a. Always defined

$15x^4 - 75x^2 - 540 = 0$

$x^4 - 5x^2 - 36 = 0$

$(x^2 - 9)(x^2 + 4) = 0$

$x = 3, -3$

critical values: $3, -3$

b.

		-3		3	
$x - 3$:	$- - -$		$+ + +$		$+ + +$
$x + 3$:	$- - -$		$+ + +$		$+ + +$
$x^2 + 4$:	$+ + +$		$+ + +$		$+ + +$
Derivative:	pos		neg		pos
Curve:	rising		falling		rising

$f(x)$ is increasing on $(-\infty, -3)$ and $(3, \infty)$ and decreasing on $(-3, 3)$

39. $y' = \dfrac{(x - 2)(1) - (x + 1)(1)}{(x - 2)^2}$

$= \dfrac{-3}{(x - 2)^2}$

a. No critical values ($x = 2$ is not in the domain)

b. $y' < 0$ for all values ($x \neq 2$)

y is increasing on $(-\infty, 2)$ and $(2, \infty)$

41. $f'(x) = -x^{-2} - 2x^{-3}$

a. Not defined at $x = 0$

$-\dfrac{1}{x^2} - \dfrac{2}{x^3} = 0$

$-x - 2 = 0$

$x = -2$

critical values: $0, -2$

b.

		-2		0	
$-(x + 2)$:	$+ + +$		$+ + +$		$- - -$
x^3:	$- - -$		$+ + +$		$+ + +$
Derivative:	neg		pos		neg
Curve:	falling		rising		falling

$f(x)$ is increasing on $(-2, 0)$ and and decreasing elsewhere

43. $S(x)$ is increasing when $S'(x) > 0$: $S'(x) = 5000 - 50x - 3x^2 > 0$

$3x^2 + 50x - 5000 < 0$

$(3x - 100)(x + 50) < 0$

This has a solution when $-50 < x < \frac{100}{3}$.

Since x must be positive, when $x < \frac{100}{3} = 33.\overline{33}$ (thousands of dollars). So $S(x)$ is increasing when the amount spent on advertising is under $33,333.

45. $f'(x) = \frac{(x^2 + 50)(10) - 10x(2x)}{(x^2 + 50)^2}$

$= \frac{10x^2 + 500 - 20x^2}{(x^2 + 50)^2}$

$= \frac{500 - 10x^2}{(x^2 + 50)^2}$

Critical values:

$\frac{500 - 10x^2}{(x^2 + 50)^2} = 0$

$500 - 10x^2 = 0$

$x^2 - 50 = 0$

$x = \pm\sqrt{50} \approx$ 7 mo. or 7 mo. 2 days

Begin campaign April 1.

(You could check first derivative test to make sure there is a maximum at $x = \sqrt{50}$; there is.)

4.2 Second Derivatives and Graphs

1. $f'(x) = 10x^4 - 12x^3 + 3x^2 - 10x + 19$; $f''(x) = 40x^3 - 36x^2 + 6x - 10$;

$f'''(x) = 120x^2 - 72x + 6$

3. $y = \sqrt{5}x^{1/2}$; $y' = \frac{\sqrt{5}}{2}x^{-1/2}$; $y'' = -\frac{\sqrt{5}}{4}x^{-3/2}$; $y''' = \frac{3\sqrt{5}}{8}x^{-5/2}$ $\frac{d^4y}{dx^4} =$

$y^4 = -\frac{15\sqrt{5}}{16}x^{-7/2}$ or $= \frac{-15\sqrt{5x}}{16x^4}$

5. $g'(x) = 12x^2 + 2x^{-2}$; $g''(x) = 24x - 4x^{-3}$; $g'''(x) = 24 + 12x^{-4}$;

$g^{(4)}(x) = -48x^{-5}$

7. $y' = 5x^{2/3}$; $y'' = \frac{10}{3}x^{-1/3}$; $\frac{d^3y}{dx^3} = y''' = \frac{-10}{9}x^{-4/3}$ or $= \frac{-10\sqrt[3]{x^2}}{9x^2}$

9. $y = 3(x - 1)^{-1}$; $y' = -3(x - 1)^{-2}$; $y'' = 6(x - 1)^{-3}$; $\frac{d^3y}{dx^3} =$

$y''' = -18(x - 1)^{-4}$

11. $y' = \frac{(x + 4)(x^2 - 1)' - (x^2 - 1)(x + 4)'}{(x + 4)^2}$

$= \frac{(x + 4)(2x) - (x^2 - 1)(1)}{(x + 4)^2}$

$$= \frac{2x^2 + 8x - x^2 + 1}{(x+4)^2}$$

$$= \frac{x^2 + 8x + 1}{(x+4)^2}$$

$$\frac{d^2y}{dx^2} = y'' = \frac{(x+4)^2(x^2+8x+1)' - (x^2+8x+1)[(x+4)^2]}{(x+4)^4}$$

$$= \frac{(x+4)^2(2x+8) - (x^2+8x+1)[2(x+4)]}{(x+4)^4}$$

$$= \frac{(x+4)^2 2(x+4) - (x^2+8x+1)2(x+4)}{(x+4)^4}$$

$$= \frac{2(x+4)[(x+4)^2 - (x^2+8x+1)]}{(x+4)^4}$$

$$= \frac{2[x^2 + 8x + 16 - x^2 - 8x - 1]}{(x+4)^3}$$

$$= \frac{2[15]}{(x+4)^3}$$

$$= \frac{30}{(x+4)^3}$$

13. $f'(x) = 4(2x+1)^3(2) = 8(2x+1)^3$, which is defined for all x, so set $f'(x) = 0$: $8(2x+1)^3 = 0$, $2x+1 = 0$, $x = -\frac{1}{2}$. Trying the second derivative test, $f''(x) = 24(2x+1)^2(2) = 48(2x+1)^2$; $f'(-\frac{1}{2}) = 0$, so the test fails (i.e. no information). Using the first derivative test we will try $c^- = -1$ and $c^+ = 0$: $f'(-1) = 8[2(-1)+1]^3 = -8 < 0$, so curve is decreasing at $x = -1$. $f'(0) = 8[2(0)+1]^3 = 8 > 0$, so curve is increasing at $x = 0$. Therefore, a relative minimum is 0 at $x = -\frac{1}{2}$

15. $f(x) = \frac{x^2+9}{x} = \frac{x^2}{x} + \frac{9}{x} = x + 9x^{-1}$; $f'(x) = 1 - 9x^{-2}$, which is defined for all $x \neq 0$. Since $f(0)$ is also undefined, the only critical values result from solving $f'(x) = 0$:

$$1 - 9x^{-2} = 0;\ 1 = 9x^{-2};\ x^2 = 9;\ x = \pm 3$$

Using the second derivative test, $f''(x) = 18x^{-3}$: $f''(3) = \frac{18}{3^3} = \frac{2}{3} > 0$, Curve is concave up at $x = 3$ So there is a relative minimum 6 at $x = 3$. $f''(-3) = \frac{18}{(-3)^3} = -\frac{2}{3} < 0$. Curve is concave down at $x = -3$ so there is a relative maximum -6 at $x = -3$.

17. $f'(x) = \frac{1}{2}(x^2+1)^{-1/2}(2x)$ To find critical values, set $f'(x) = 0$. The only critical value is $x = 0$.

$f''(x) = (x^2 + 1)^{-1/2} + (-\frac{1}{2}) + (x^2 + 1)^{-3/2}(2x) > 0$ when $x = 0$.

There is a relative minimum of 1 at $x = 0$.

19. $f'(x) = 16x - 4x^3$ which is defined for all x, so the only critical values result from solving

$f'(x) = 0$: $16x - 4x^3 = 0$ *Do not divide both sides by x*

$$4x - x^3 = 0$$

$$x(4 - x^2) = 0$$

$x(2 - x)(2 + x) = 0$; so $x = 0$, $x = 2$, or

$x = -2$ are critical values.

Using the second derivative test, $f''(x) = 16 - 12x^2$:

$f''(0) = 16 > 0$ Curve is concave up, relative minimum at $x = 0$; this is point (0, 0). $f''(-2) = -32 < 0$ Curve is concave down, relative maximum at $x = -2$; point $(-2, 16)$. $f''(2) = -32 < 0$ Curve is concave down, relative maximum at $x = 2$; point (2, 16).

21. $f'(x) = 3x^2 + 10x - 8$ which is defined for all x so the only critical values result from solving

$f'(x) = 0$: $3x^2 + 10x - 8 = 0$, $(3x - 2)(x + 4) = 0$; so $x = \frac{2}{3}$ or $x = -4$.

Using the second derivative test, $f''(x) = 6x + 10$: $f''(\frac{2}{3}) = 14 > 0$; Curve is concave up, relative minimum at $x = \frac{2}{3}$; this is point $(\frac{2}{3}, \frac{194}{27})$.

$f''(-4) = -14 < 0$; Curve is concave down, relative maximum at $x = -4$; point $(-4, 58)$.

23. $y' = 12x^2 - 54x - 30$. The only critical values result from solving $y' = 0$:

$$12x^2 - 54x - 30 = 0$$

$$2x^2 - 9x - 5 = 0$$

$(2x + 1)(x - 5) = 0$, so $x = -\frac{1}{2}$ or 5.

Using the second derivative test, $y'' = 24x - 54$:

y'' at $x = -\frac{1}{2}$: $24(-\frac{1}{2}) - 54 = -66 < 0$, relative maximum at $(-\frac{1}{2}, \frac{7}{4})$.

y'' at $x = 5$: $24(5) - 54 = 66 < 0$, relative minimum at $(5, -331)$.

25. $g(x) = x + 2x^{-1}$; $g'(x) = 1 - 2x^{-2}$. The only critical value results from solving $g'(x) = 0$:

$x^2 - 2 = 0$

$x = \pm\sqrt{2}$

Using the second derivative test, $g''(x) = 4x^{-3}$:

$g''(\sqrt{2}) = 2 > 0$; Curve concave up, relative minimum at $x = \sqrt{2}$; this is the point $(\sqrt{2}, 2\sqrt{2})$. $g''(-1) = -2 < 0$; Curve concave down, relative maximum at $x = -\sqrt{2}$; point $(-\sqrt{2}, 2\sqrt{2})$.

27. $g'(x) = \dfrac{(x - 1)(x^2)' - x^2(x - 1)'}{(x - 1)^2} = \dfrac{(x - 1)(2x) - x^2}{(x - 1)^2} = \dfrac{x^2 - 2x}{(x - 1)^2}$

The only critical values will result from setting $g'(x) = 0$:

$\dfrac{x^2 - 2x}{(x - 1)^2} = 0$

$x^2 - 2x = 0$

$x(x - 2) = 0$; so $x = 0$ or 2.

Using the first derivative test, $g'(-1) = (-1)(-1 - 2) = 3 > 0$ Curve is increasing left of $x = 0$.

$g'(1) = 1(1 - 2) = -1 < 0$; Curve is decreasing for $0 < x < 2$.

$g'(3) = 3(3 - 2) = 3 > 0$; Curve is increasing right of $x = 2$.

Therefore, a relative maximum at (0, 0) and a relative minimum at (2, 4).

29. $f'(x) = \frac{2}{3}(x + 1)^{-1/3} = \dfrac{2}{3\sqrt[3]{x + 1}}$ which is never zero but undefined at $x = -1$ and since $f(-1)$ is defined, $x = -1$ is a critical value. Using the first derivative test: $f'(-2) = -\frac{2}{3} < 0$; Curve is decreasing left of $x = -1$.

$f'(0) = \frac{2}{3} > 0$; Curve is increasing right of $x = -1$.

Therefore, a relative minimum occurs at $(-1, 0)$.

31. a. $f'(x) = 12 - 2x = 0$ when $x = 6$ **b.** increasing when $f'(x) > 0$: $12 - 2x > 0$ if $x < 6$ or on $(-\infty, 6)$; decreasing when $x > 6$ or on $(6, \infty)$.

c. $f''(x) = -2$ which is always negative; curve in concave down for all x.

33. a. $g'(x) = 3x^2 - 14x - 5 = 0$ when $(3x + 1)(x - 5) = 0$; $x = -\frac{1}{3}$ or 5

b. Look at number line to find that curve is increasing left of $x = -\frac{1}{3}$ or $(-\infty, -\frac{1}{3})$; decreasing for $-\frac{1}{3} < x < 5$ or on $(-\frac{1}{3}, 5)$ and increasing right of $x = 5$ or $(5, \infty)$. **c.** $g''(x) = 6x - 14 < 0$ when $x < \frac{7}{3}$ or $(-\infty, \frac{7}{3})$. Curve is concave up for $x > \frac{7}{3}$ or on $(\frac{7}{3}, \infty)$.

35. a. $y' = 3x^2 + 22x - 45 = 0$ when $(3x - 5)(x + 9) = 0$; $x = \frac{5}{3}$ or -9

b. Look at number line to find that curve increasing left of -9 namely $(-\infty, -9)$; decreasing for $(-9, \frac{5}{3})$ and increasing right of $\frac{5}{3}$ namely $(\frac{5}{3}, \infty)$.

c. $y'' = 6x + 22 > 0$ when $x > -\frac{11}{3}$; curve is concave up on $(-\frac{11}{3}, \infty)$ and $y'' = 6x + 22 < 0$ when $x < -\frac{11}{3}$; curve is concave down on $(-\infty, -\frac{11}{3})$.

37. a. Differentiate implicitly (with respect to x): $36x^2 - 10x - 4 - 2y' = 0$

$$-2y' = -36x^2 + = 10x + 4$$

$$y' = 18x^2 - 5x - 2$$

$$\text{Set } y' = 0\text{: } 18x^2 - 5x - 2 = 0$$

$$(9x + 2)(2x - 1) = 0$$

$$x = -\frac{2}{9} \text{ or } \frac{1}{2}$$

b. Look at number line to find that curve is increasing left of $-\frac{2}{9}$ or on $(-\infty, -\frac{2}{9})$; decreasing on $(-\frac{2}{9}, \frac{1}{2})$; increasing right of $\frac{1}{2}$ or on $(\frac{1}{2}, \infty)$.

c. $y'' = 36x - 5 > 0$ when $x > \frac{5}{36}$ so concave up on $(\frac{5}{36}, \infty)$.
$y'' = 36x - 5 < 0$ when $x < \frac{5}{36}$. Curve is concave down on $(-\infty, \frac{5}{36})$.

39. $f(x) = (2 - x)^3 - 8$

$f'(x) = 3(2 - x)^2(-1)$

$= -3(2 - x)^2$

a. critical value $x = 2$

b. Graph decreasing on $(-\infty, 2)$ and $(2, \infty)$

c. concavity

$f''(x) = -6(2 - x)(-1)$

$= 12 - 6x$

$12 - 6x = 0$

$x = 2$

concave up on $(-\infty, 2)$ down on $(2, \infty)$

41. $S'(x) = -3x^2 + 90x + 60$

$S''(x) = -6x + 90$ ← we want to know when this rate changes from positive to negative

$-6x + 90 = 0$

$x = 15$

The point of diminishing returns is at $x = 15$. This represents $15,000.

43. The rate of change is increasing when the second derivative is positive. From Problem 41, this is time when $x < 15$ or on the interval $[0, 15)$.

45. Maximum *rate* of change is maximum value of $S'(x)$ critical values for

$-6x + 90 = 0$

$x = 15$ This is the point of diminishing returns with sales of

$S(15) = 4{,}000 - 15^3 = 45(15)^2 + 60(15)$

$= 11{,}650$

47. *Function* | *1st Derivative* | *2nd Derivative*

$S(x) = 4{,}000 - x^3 + 45x^2 + 60x$ $\quad S'(x) = -3x^2 + 90x + 60$ $\quad S''(x) = -6x + 90$

critical values: $x = \dfrac{30 \pm \sqrt{980}}{2}$

$S\left(\dfrac{300 + \sqrt{980}}{2}\right) \approx 19{,}320$ $\quad x = \dfrac{30 + \sqrt{980}}{2} \approx 30.65$ $\quad S''\left(\dfrac{30 + \sqrt{980}}{2}\right) \approx -93.9 < 0$

(30.7, 19,320) (other value not in domain) concavity:

relative maximum $\quad -6x + 90 = 0$

$x = 15$

$S(15) \approx 11{,}560$ $\quad x < 15$ concave up

Endpoints: $\quad x > 15$ concave down

$S(10) = 8100$

$S(35) = 11{,}650$

For $S'(x)$ graph

$y = -3x^2 + 90x + 60$

$y' = -6x + 90$; vertex at $x = 15$

Graph of S

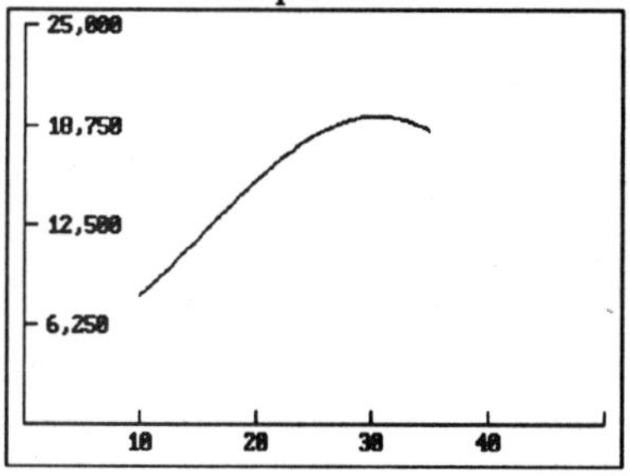

Graph of S'

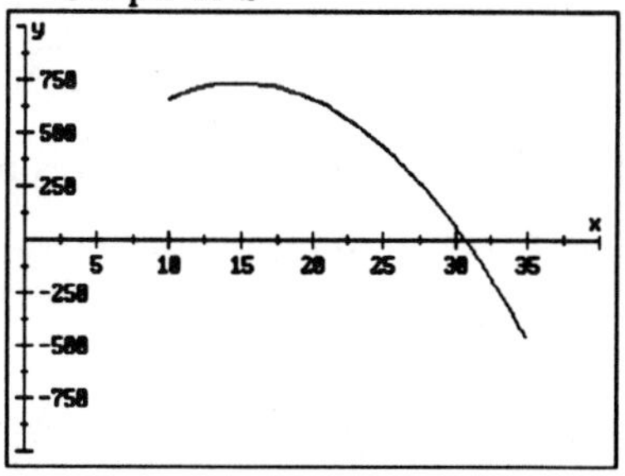

49. $R(x) = px$

$= 1{,}156x - \frac{1}{12}x^3$

$R'(x) = 1{,}156 - \frac{1}{4}x^2$ $\quad R''(x) = -\frac{1}{2}x$

critical values

$1{,}156 - \frac{1}{4}x^2 = 0$

$$x^2 = 4{,}624$$

$$x = \pm 68 \qquad R''(68) < 0 \text{ relative maximum at } x = 68$$

$$\text{reject } x = -68$$

51. Revenue = (price)(number of items)

a. $R(x) = xp(x) = \left[5 - \left(\frac{x}{100}\right)^2\right](x) = 5x - \frac{1}{10{,}000}x^3$ **b.** $R'(x) = 5 - \frac{3}{10{,}000}x^2$

c. Look at derivative of marginal revenue:

$R''(x) = -\frac{3}{5{,}000}x$ at $x = 100$: $R''(100) = -\frac{3}{50} < 0$ so $R'(x)$, the marginal revenue, is decreasing.

4.3 Curve Sketching-- Relative Maximums and Minimums

1. $f'(x) > 0$ means the curve must be increasing,

and $f''(x) > 0$ means the curve must be concave up.

Ones possibility is shown.

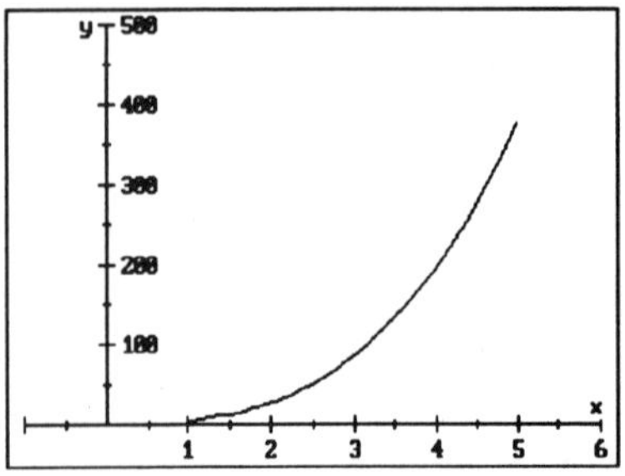

3. $f'(x) < 0$ means the curves must be decreasing,

and $f''(x) > 0$ means the curve must be concave up.

One possibility is shown.

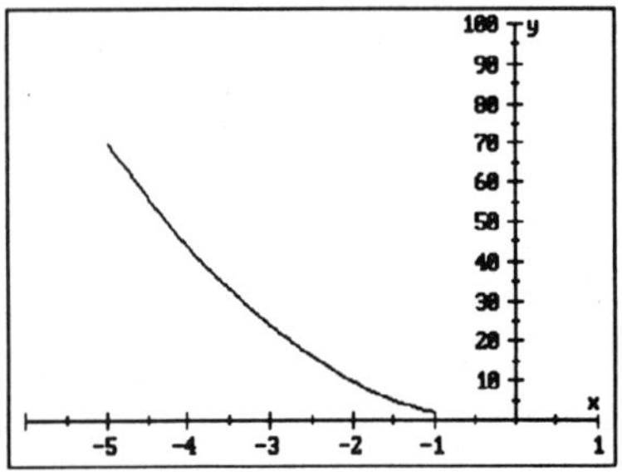

5. $f'(-2) = 0$ and $f'(2) =$ means the curve must have a horizontal tangent at $x = -2$ and 2 (curve is not in-creasing or decreasing).

$f''(x) > 0$ if $x < 0$ means for negative values of x the curve is concave up;

$f''(x) < 0$ if $x > 0$ means for positive values of x the curve is concave down. One possibility is shown.

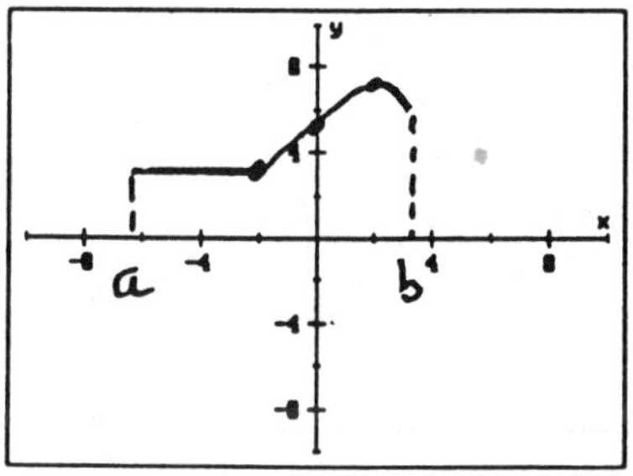

7. $y' = 6 - 2x$ which is positive for $x < 3$ and negative for $x > 3$; hence the curve is increasing for $x < 3$ and decreasing for $x > 3$. The only critical point is $x = 3$. $y'' = -2 \le 0$ for all x; hence the curve is concave down for all x. Some points are:

(vertex)

$x = 3,\ y = 6(3) - (3)^2 = 9$

$x = 5,\ y = 6(5) - 5^2 = 5$

$x = 1,\ y = 6(1) - 1^2 = 5$

You should calculate a few more points.

The curve, a parabola, is shown at right.

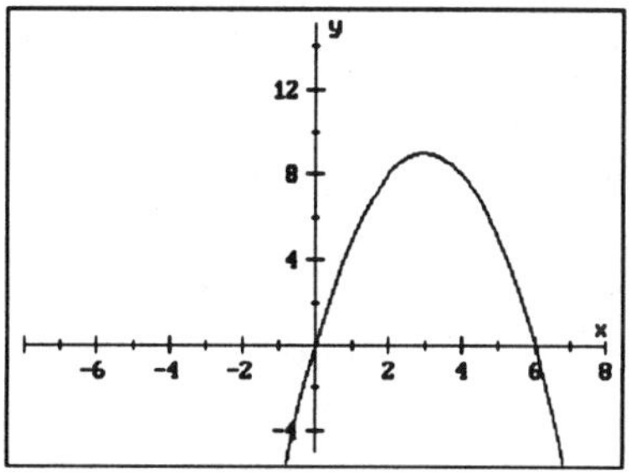

9. $y' = 4x - 3$ which is positive for $x > \frac{3}{4}$; hence the curve is decreasing for $x < \frac{3}{4}$ and increasing for $x > \frac{3}{4}$. The only critical point is $x = \frac{3}{4}$. $y'' = 4 > 0$ for all x; hence the curve is concave up for all x.

Some points are: (vertex) $x = \frac{3}{4}$,

$y = 2(\frac{3}{4})^2 - 3(\frac{3}{4}) + 5 = 3\frac{7}{8}$

$x = 1,\ y = 2(1)^2 - 3(1) + 5 = 4$

$x = 0,\ y = 5$

The curve, a parabola, is shown.

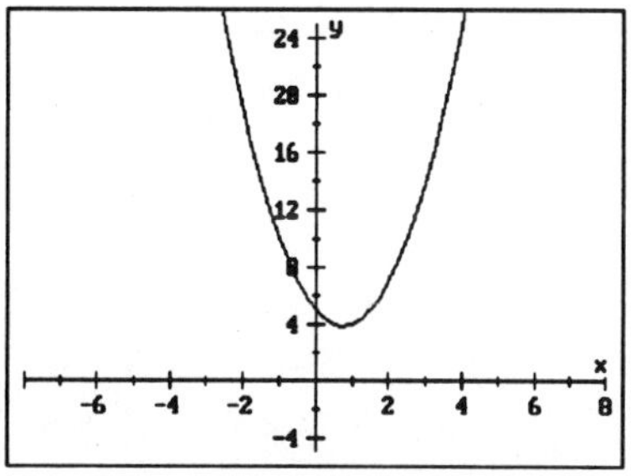

11. $y' = 3x^2 + 6x - 9$

$= 3(x+3)(x-1)$

$x + 3$:	$- \ -$	$- \ -$	$+ \ +$
$x + 2$:	$- \ -$	$- \ -$	$+ \ +$
	-3	1	

critical values $x = 3, -3$

Derivative:	pos	neg	pos
Curve:	rising	falling	rising

$y'' = 6x + 6 = 0$

concavity:

$6x + 6 = 0$

$x = -1$

$x < -1$ concave down

$x > -1$ concave up

concave down ↑ $x = -1$ concave up

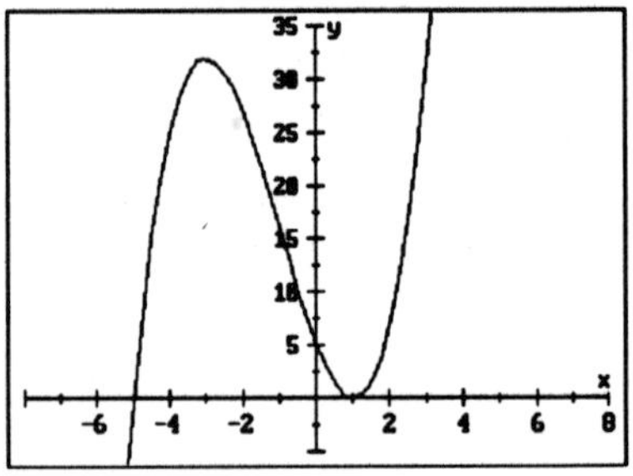

13. $y' = 3x^2 - 48$

$= 3(x^2 - 16)$

$x - 4$:	$- \ -$	$- \ -$	$+ \ +$
$x + 4$:	$- \ -$	$+ \ +$	$+ \ +$
	-4	4	

critical values $x = \pm 4$

Derivative:	pos	neg	pos
Curve:	rising	falling	rising

$y'' = 6x$

concavity:

$6x = 0$

$x = 0$

$x < 0$ concave down

$x > 0$ concave up

concave down ↑ $x = 0$ concave up

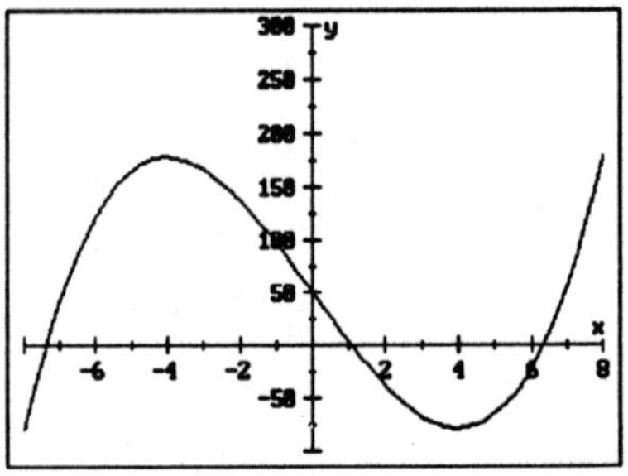

15. $f'(x) = 3x^2$ critical value $x = 0$

never negative so concave is increasing

for all nonzero x horizontal all

tangent at $x = 0$

$f''(x) = 6x$

concave down for $x < 0$

concave up for $x > 0$

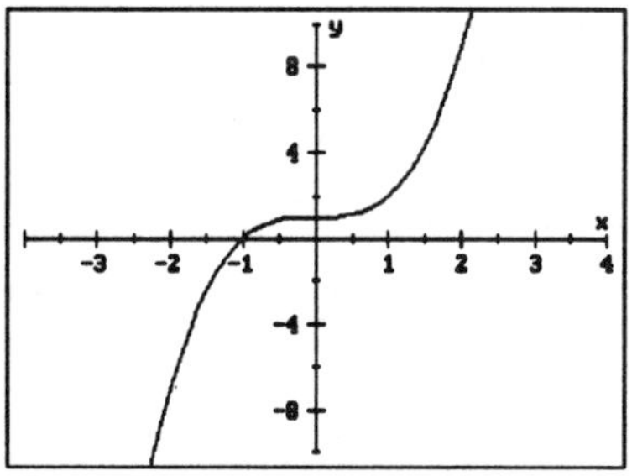

17. $g'(x) = 4x^3$

critical value $x = 0$

negative when $x < 0$ (decreasing)

positive when $x > 0$ (increasing)

horizontal tangents at $x = 0$

$g''(x) = 12x^2$ never negative

concave up for all non-zero x

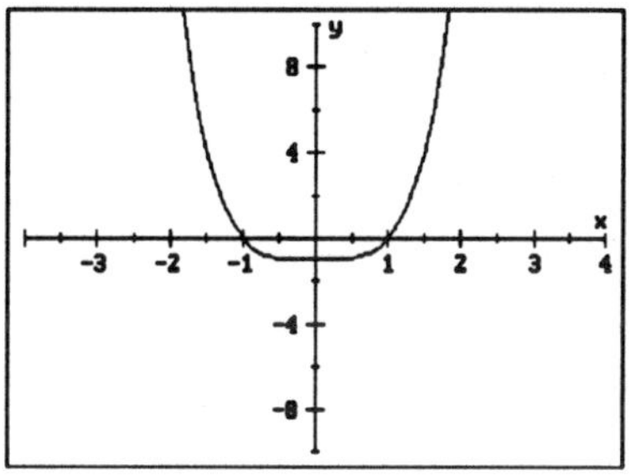

19. $f'(x) = 4x^3 - 4x$

$4x^3 - 4x = 4x(x^2 - 1)$

$= 4x(x - 1)(x + 1)$

x:	$- -$	$- -$	$+ +$	$+ +$
$x - 1$:	$- -$	$- -$	$- -$	$+ +$
$x + 1$:	$- -$	$+ +$	$+ +$	$+ +$
	-1	0	1	
Derivative:	neg	pos	neg	pos
Curve:	falling	rising	falling	rising
	concave up		concave up	

critical values , -1, 0

$f''(x) = 12x^2 - 4$

concave up for all non-zero x

$f''(0) < 0$ relative maximum at $x = 0$

$f''(1) > 0$ relative minimum at $x = 1$

$f''(-1) > 0$ relative minimum at $x = -1$

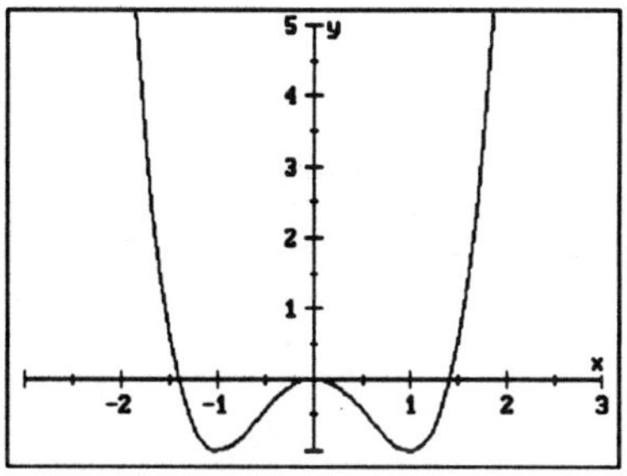

21.

$$g'(x) = 1 - 4x^{-2}$$
$$1 - 4x^{-2} = 0$$
$$x^2 - 4 = 0$$
$$(x - 2)(x + 2) =$$
$$x = 2, -2$$

critical values $x = \pm 2$

$$g''(x) = 8x^{-3}$$

concave down for $x < 0$

concave up for $x > 0$

		-2		2	
$x - 2$:	$- \; -$		$- \; -$		$+ \; +$
$x + 2$:	$- \; -$		$+ \; +$		$+ \; +$
Derivative:	pos		neg		pos
Curve:	rising		falling		rising

concave down ↑ concave up

$x = 0$

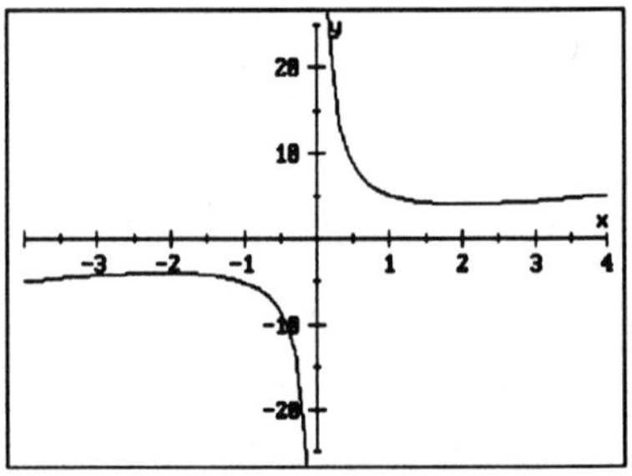

23.

$$\begin{aligned} f'(x) &= 12x^3 + 24x^2 - 12x - 24 \\ &= 12x^2(x + 2) - 12(x + 2) \\ &= (12x^2 - 12)(x + 2) \\ &= 12(x^2 - 1)(x + 2) \\ &= 12(x + 1)(x - 1)(x + 2) \end{aligned}$$

		-2		-1		1	
$x + 1$:	$- \; -$		$+ \; +$		$+ \; +$		$+ \; +$
$x - 1$:	$- \; -$		$- \; -$		$+ \; +$		$+ \; +$
$x + 2$:	$- \; -$		$- \; -$		$- \; -$		$+ \; +$
Derivative:	neg		pos		neg		pos
Curve:	falling		rising		falling		rising

A few points are:

$(-3, f(-3)) = (-3, 40)$;

$(-2, f(-2)) = (-2, 3)$;

$(-1, f(-1)) = (-1, 8)$.

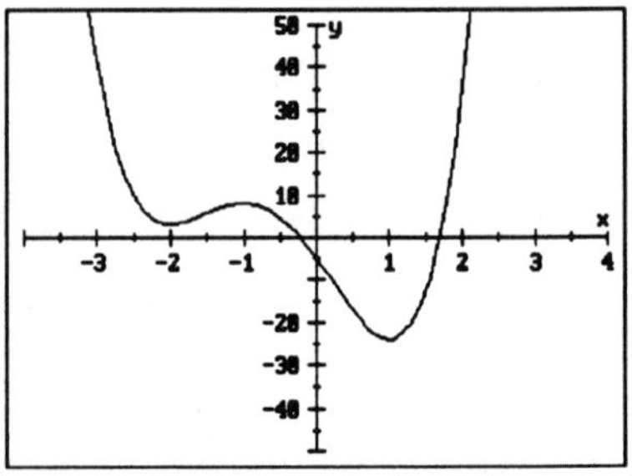

25. Domain excludes $x = 1$; Asymptote at $x = 1$

$y' = \dfrac{(x - 1)(x^2)' - (x^2)(x - 1)'}{(x - 1)^2} = \dfrac{(x - 1)(2x) - x^2}{(x - 1)^2} = \dfrac{x^2 - 2x}{(x - 1)^2}$, which is zero if and only if $x^2 - 2x = 0$; which occurs when $x = 0$ or $x = 2$. Since the denominator is always positive, y' is positive when $x^2 - 2x = x(x - 2)$ is positive: $x(x - 2) > 0$ when $x > 2$ and when $x < 0$. Hence the curve is increasing when $x > 2$ and when $x < 0$ and decreasing when $0 < x < 2$.

$$\begin{aligned} y'' &= \frac{(x - 1)^2(x^2 - 2x)' - (x^2 - 2x)[(x - 1)^2]'}{(x - 1)^4} \\ &= \frac{(x - 1)^2(2x - 2) - (x^2 - 2x)[2(x - 1)]}{(x - 1)^4} \\ &= \frac{(x - 1)^2 2(x - 1) - 2(x^2 - 2x)(x - 1)}{(x - 1)^4} \\ &= \frac{(x - 1)[2(x - 1)^2 - 2(x^2 - 2x)]}{(x - 1)^4} \\ &= \frac{[2(x^2 - 2x + 1) - 2x^2 + 4x]}{(x - 1)^3} \\ &= \frac{2}{(x - 1)^3} \end{aligned}$$

which is negative if $x < 1$ and positive when $x > 1$, so the curve is concave down for $x < 1$, concave up for $x > 1$. The function is undefined at $x = 1$, and the graph is shown.

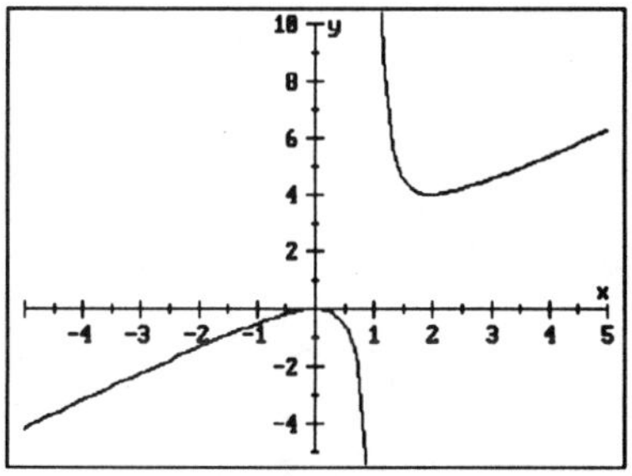

27. Domain excludes $x = 1$; Vertical asymptote $x = 1$. For the horizontal asymptotes $y = \lim_{|x| \mapsto \infty} \frac{x + 1}{x - 1} = 1$; It might be helpful to divide: $y = \frac{x + 1}{x - 1}$:

$$\begin{array}{r} 1 \\ x - 1 \,\overline{)\, x + 1} \\ \underline{x - 1} \\ 2 \end{array} \qquad \text{so } y = \frac{x + 1}{x - 1} = 1 + \frac{2}{x - 1} = 1 + 2(x - 1)^{-1}$$

$y' = -2(x - 1)^{-2} = \frac{-2}{(x - 1)^2}$ which is negative for all $x \neq 1$ hence the curve is always decreasing. $y'' = 4(x - 1)^{-3}$ which is negative if $x < 1$, positive when $x > 1$; hence the curve is concave down for $x < 1$, concave up for $x > 1$. Also, the function is undefined at $x = 1$ and $\lim\limits_{|x| \mapsto \infty} \dfrac{x + 1}{x - 1} = 1$ which give the horizontal and vertical asymptotes.

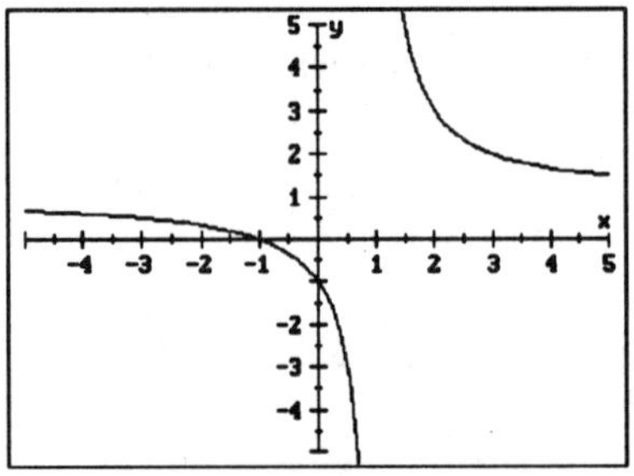

29. $y = \dfrac{x - 5}{x + 5}$ $\qquad y' = \dfrac{(x + 5)(1) - (x - 5)(1)}{(x + 5)^2}$

Domain excludes $x = -5$ $\qquad = \dfrac{10}{(x + 5)^2}$

Critical value $x = 5$

Horizontal asymptote $x = -5$

Vertical asymptotes:

$y = \lim\limits_{|x| \mapsto \infty} \dfrac{x - 5}{x + 5} = 1$

$y'' = -20(x + 5)^{-3}$

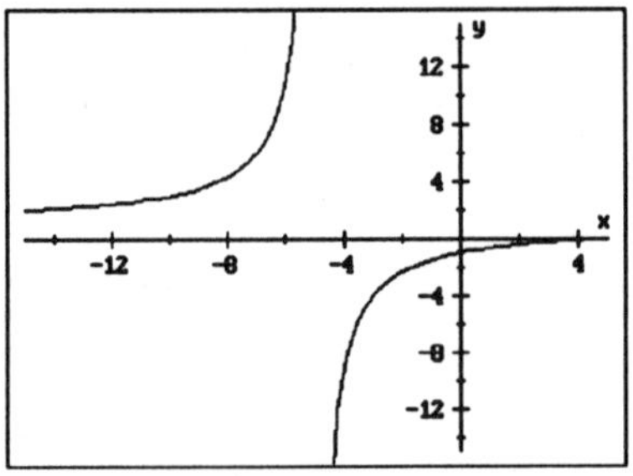

This is undefined at $x = -5$.

Second derivative test fails. Graph is rising everywhere except at $x = -5$; since derivative is nonnegative

31. $y' = 2x^{-1/3} = \dfrac{2}{\sqrt[3]{x}}$ which is undefined at $x = 0$.

Since y *is* defined at $x = 0$ a vertical tangent line occurs at $x = 0$. Also, if $x < 0$ y' is negative, and if $x > 0$ y' is positive; hence the curve is decreasing for $x < 0$ and increasing for $x > 0$.

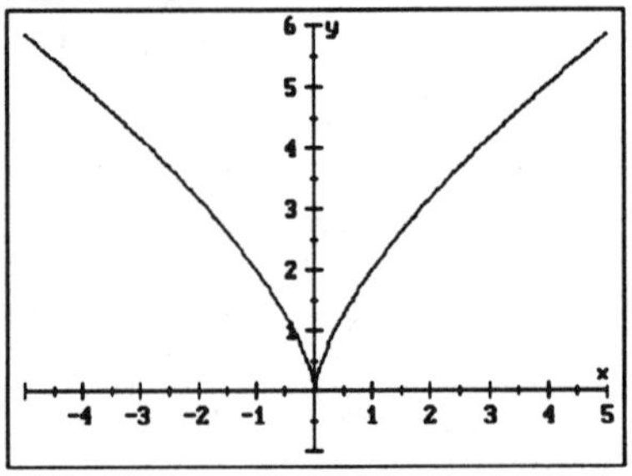

$y'' = -\frac{2}{3}x^{-4/3} = \dfrac{-2}{3\sqrt[3]{x^4}}$ which is negative for all $x \neq 0$; hence the curve is concave down for all $x \neq 0$.

33. $f'(x) = 3x^{-1/2} - 6x^{1/2}$, set equal to zero:

$$3x^{-1/2} - 6x^{1/2} = 0$$
$$3 - 6x = 0$$

$\frac{1}{2} = x$

If $x < \frac{1}{2}$ (but $x > 0$) $f'(x)$ is positive. If $x > \frac{1}{2}$ $f'(x)$ is negative; hence the curve is increasing on $(0, \frac{1}{2})$ and decreasing for $(\frac{1}{2}, \infty)$

$$f''(x) = -\tfrac{3}{2}x^{-3/2} - 3x^{-1/2}$$

$$= \frac{-3}{2\sqrt{x^3}} - \frac{3}{\sqrt{x}}$$

$$= \frac{-3}{2x\sqrt{x}} - \frac{6x}{2x\sqrt{x}} = \frac{-3 - 6x}{2x\sqrt{x}},$$

which is always negative (since $x \geq 0$).

Hence, the curve is concave down.

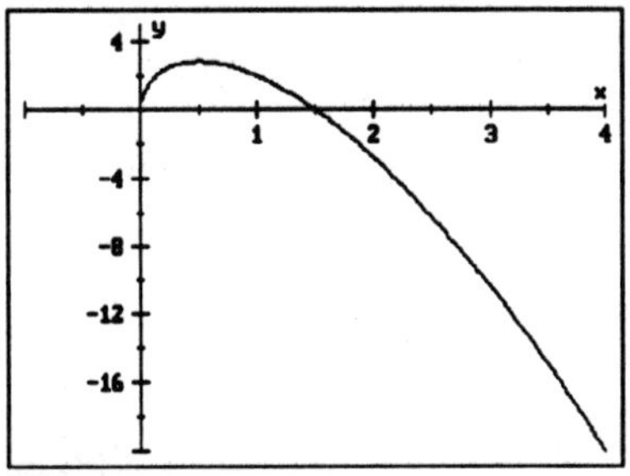

35. $$C'(p) = \frac{(100 - p)(20{,}000p)' - (20{,}000p)(100 - p)'}{(100 - p)^2}$$

$$= \frac{(100 - p)(20{,}000) - (20{,}000p)(-1)}{(100 - p)^2}$$

$$= \frac{20{,}000[(100 - p) + p]}{(100 - p)^2}$$

$$= \frac{2{,}000{,}000}{(100 - p)^2}$$

$$= 2{,}000{,}000(100 - p)^2$$

which is always positive, so the curve is increasing. $C''(p) = 4{,}000{,}000(100 - p)^{-3}$ which is positive for $0 < p < 100$; hence the curve is concave up.

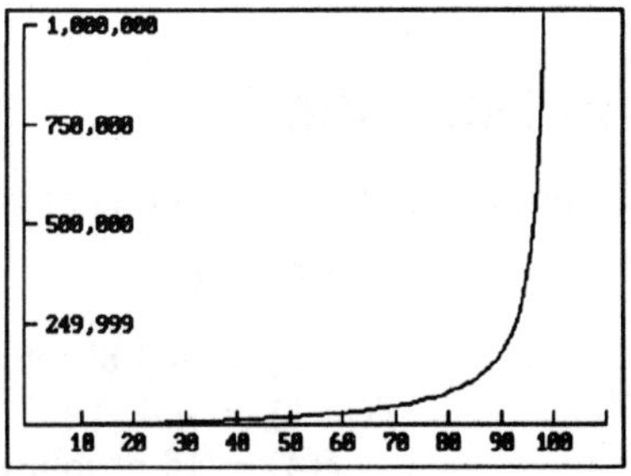

37. $C'(x) = 2x$ is the marginal cost; average cost function

$\overline{C}(x) = \frac{C(x)}{x} = 5000x^{-1} + \frac{1}{2}x$ $\overline{C}'(x) = -5000x^{-2} + \frac{1}{2}$ $\overline{C}(x) = 10{,}000x^{-3}$

critical values: $\overline{C}''(100) > 0$ relative minimum

$$\frac{-5000}{x^2} + \frac{1}{2} = 0$$

$$-10{,}000 + x^2 = 0$$

$$x^2 = 10{,}000$$

$$x = \pm 100$$

↑
reject negative value

at $x = 100$

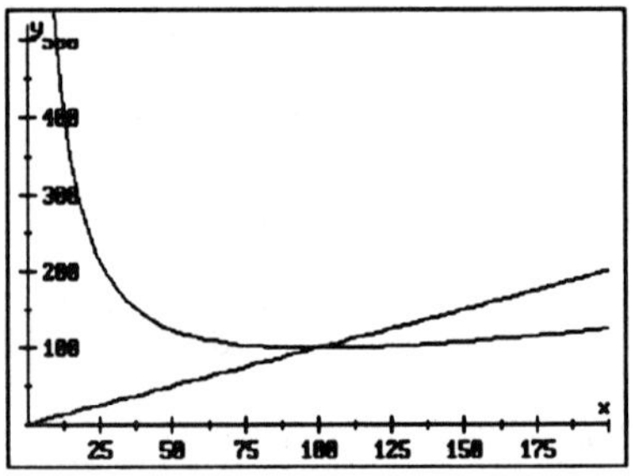

39. $C(t) = \dfrac{10t}{8 + t^3}$

Domain $t > 0$

$$C'(t) = \frac{(8 + t^3)\,10 - 10t(3t^2)}{(8 + t^3)^2}$$

$$= \frac{80 - 20t^3}{(8 + t^3)^2}$$

$$= \frac{20(4 - t^3)}{(8 + t^3)^2}$$

critical values

$$4 - t^3 = 0$$

$$t = \sqrt[3]{4}$$

$$C''(t) = \frac{(8 + t^3)^2(-3t^2) - 20(4 - t^3)(2)(8 + t^3)(3t^2)}{(8 + t^3)^2}$$

$C''(\sqrt[3]{4}) < 0$ Relative maximum at $t = \sqrt[3]{4}$

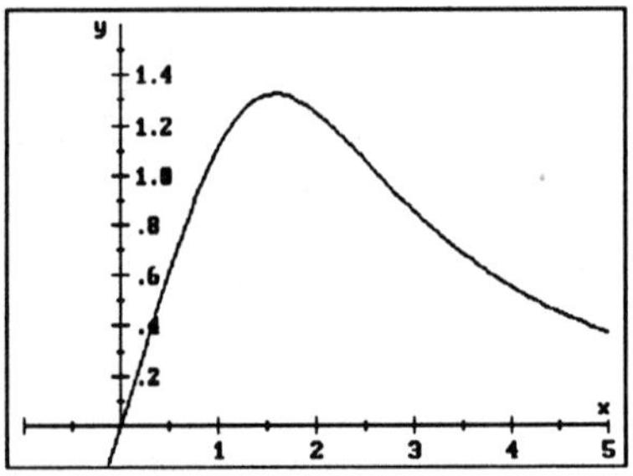

4.4 Absolute Maximum and Minimum

1. $f'(x) = 3x^2 - 6x$ check critical values and endpoints for $f(x) = x^3 - 3x^2$:

critical values

$$3x^2 - 6x = 0$$
$$3x(x - 2) = 0$$
$$x = 0,\ 2$$

$f(0) = 0$

$f(2) = -4$ absolute minimum

$f(-1) = -4$

$f(4) = 16$ absolute maximum

Maximum $= 16$ at $x = 4$ and minimum $= -4$ occurs at $x = 2$ and $x = -1$

3. $f'(x) = 15x^4 - 150x^2 + 135 = 15(x^4 - 10x^2 + 9) = 15(x^2 - 9)(x^2 - 1)$

$= 15(x + 3)(x - 3)(x + 1)(x - 1)$ Compute $f(x)$ for all critical values on $[-5, 0]$, and also at the endpoints: $f(-5) = -3{,}050$; $f(-3) = 966$; $f(-1) = 662$;

$f(0) = 750$. Maximum $= 966$ occurs at $x = -3$; minimum $= -3{,}050$ occurs at $x = -5$.

5. $g'(x) = 3x^2 - 4x - 15 = (3x + 5)(x - 3)$. Compute $g(x)$ for all critical values on $[0, 5]$, and also the endpoints: $g(0) = 42$, $g(3) = 6$, $g(5) = 42$. Maximum $= 42$ occurs at $x = 0$ and $x = 5$; minimum $= 6$ occurs at $x = 3$.

7.
$$\begin{aligned} h'(x) &= 15x^4 - 75x^2 + 60 \\ &= 15(x^4 - 5x^2 + 4) \\ &= 15(x^2 - 1)(x^2 - 4) \\ &= 15(x - 1)(x + 1)(x - 2)(x + 2). \end{aligned}$$

Compute $h(x)$ for all critical values on $[0, 3]$, and also the endpoints: $h(0) = -200$, $h(1) = -162$, $h(2) = -184$, $h(3) = 34$. Maximum $= 34$ occurs at $x = 3$; minimum $= -200$ occurs at $x = 0$.

9. $f'(x) = \frac{1}{3}x^{-3/2} = \dfrac{1}{3\sqrt[3]{x^2}}$ which is undefined at $x = 0$. Since $f'(x) \neq 0$, $x = 0$ is the only critical value. Calculate $f(x)$ at $x = 0$ and the endpoints:

$f(-1) = 0$; $f(0) = 1$; $f(8) = 3$.

Maximum $= 3$ occurs at $x = 8$; minimum $= 0$ occurs at $x = -1$.

11. $f'(x) = \frac{1}{2}x^{-1/2} + 1$ which is undefined at $x = 0$. Set $f'(x) = 0$ to find other critical values: Solve $\frac{1}{2}x^{-1/2} + 1 = 0$ which has no solution since $\sqrt{x}$ is non-negative. Calculate $f(x)$ at $x = 0$ and the endpoints: $f(0) = 0$; $f(4) = 6$.

Maximum $= 6$ occurs at $x = 4$; minimum $= 0$ occurs at $x = 0$.

13.
$$\begin{aligned} f'(x) = (x + 1)^2(x - 3)' + (x - 3)[(x + 1)^2]' &= (x + 1)^2 + (x - 3)^2(x + 1) \\ &= (x + 1)[(x + 1) + 2(x - 3)] \\ &= (x + 1)[3x - 5] \end{aligned}$$

$f'(x) = 0$ when $x = -1$ or $\frac{5}{3}$. Since $x = -1$ is not in the interval, calculate $f(x)$ at $\frac{5}{3}$ and the endpoints: $f(0) = -3$; $f(\frac{5}{3}) = -\frac{256}{27}$; $f(5) = 72$. Maximum $= 72$ occurs at $x = 5$; minimum $= -\frac{256}{27}$ occurs at $x = \frac{5}{3}$.

15. $f'(x) = -2(x - 3)^{-2} = \dfrac{-2}{(x - 3)^2}$ (which is undefined at $x = 3$, but $f(x)$ is also undefined at $x = 3$, so it is not a critical value). A fraction is zero if and only if the numerator is zero, hence $f'(x) \neq 0$. Since we have no critical values, calculate $f(x)$ at the endpoints: $f(0) = -\frac{2}{3}$; $f(5) = 1$. These are *not* maximums or minimums, because the graph of $f(x)$ has a vertical asymptote at $x = 3$.

17. $f'(x) = \begin{cases} 1 & \text{if } x > 0 \\ -1 & \text{if } x < 0 \end{cases}$ and $f'(x)$ is undefined at $x = 0$ Calculate $f(x)$ at the only critical value, $x = 0$, and the endpoints: $f(-3) = 3$; $f(0) = 0$; $f(3) = 3$. Maximum $= 3$ occurs at $x = 3$ or $x = -3$; minimum $= 0$ occurs at $x = 0$.

19. $$f'(x) = \frac{(x+1)^2(x-1)' - (x-1)[(x+1)^2]'}{(x+1)^4}$$
$$= \frac{(x+1)^2 - (x-1)2(x+1)}{(x+1)^4}$$
$$= \frac{x+1 - 2(x-1)}{(x+1)^3}$$
$$= \frac{3-x}{(x+1)^3}$$ The critical values are (setting $f'(x) = 0$) at $x = 3$ and $x = -1$. Calculate $f(x)$ at $x = 0$ and 3: $f(0) = -1$; $f(3) = \frac{1}{8}$. Maximum $= \frac{1}{8}$ occurs at $x = 3$; minimum $= -1$ occurs at $x = 0$.

21. Set $R'(x) = C'(x)$: $35 - .06x = 5$
$$-.06x = -30$$
$$x = 500$$
$P(x) = R(x) - C(x)$ so
$$P(500) = R(500) - C(500)$$
$$= 10{,}000 - 8{,}500 = 1{,}500$$

23. Set $R'(x) = C'(x)$: $5{,}000 - .2x = 200$
$$-.2x = -4{,}800$$
$$x = 24{,}000$$
$P(x) = R(x) - C(x)$ so
$$P(24{,}000) = R(24{,}000) - C(24{,}000)$$
$$= 62{,}400{,}000 - 4{,}880{,}000$$
$$= 57{,}520{,}000$$

25. Set $R'(x) = C'(x)$; $5 + \dfrac{(x-64)(1) - (x)(1)}{(x-64)^2} = 1$
$$\frac{-64}{(x-64)^2} = -4$$

$$16 = (x - 64)^2$$
$$\pm 4 = x - 64$$
$$x = 68, 60$$

↑ reject, outside domain

$$P(x) = R(x) - C(x) \text{ so}$$
$$P(60) = R(60) - C(60)$$
$$= 285 - 70$$
$$= 215$$

27. Solution follows Example 2. $P(x) = -x^2 + 120x - 2{,}500$

Critical point at $x = 60$. Endpoints: $P(x) \geq 100$

$$150 - x \geq 100$$
$$50 \geq x$$
$$x \leq 50$$

Domain is [0, 50]. To find the absolute maximum, find $P(0) = -2{,}500$, $P(50) = 1{,}000$. Absolute maximum at endpoint: $x = 50$: (The maximum profit is 1,000 at a price of \$100) $P(50) = 100$.

29. $P(x) = \frac{60}{x} + \frac{240}{20 - x}$ $P'(x) = -60x^{-2} + 240(20 - x)^{-2}$ Find critical values:

$$\frac{-60}{x^2} + \frac{240}{(20 - x)^2} = 0$$

$$(3x - 20)(x + 20) = 0$$

$$x = \frac{20}{3}, \ -20 \text{ reject}$$

Check $P(1) = 87$ *ppm*

$P(19) = 243$ *ppm*

$P(\frac{20}{3}) = 27$ *ppm*

The minimum pollution is found at $6\frac{2}{3}$ miles from plant P_1.

31. Volume, V, = (length)(width)(height) = $(10 - 2x)(5 - 2x)\,x$
$= 4x^3 - 30x^2 + 50x$. Taking the derivative and setting equal to zero:

$$12x^2 - 60x + 50 = 0$$
$$6x^2 - 30x + 25 = 0$$

$$x = \frac{30 \pm \sqrt{30^2 - 4(6)(25)}}{2(6)} = \frac{30 \pm \sqrt{300}}{12} \approx 3.94 \text{ or } 1.057$$

The interval is [0, 2.5] so we calculate the volume at the endpoints and at $x = 1.06$: $V(0) = 0$, $V(1.06) = 24.06$, $V(2.5) = 9$. So the maximum volume is about 24.06

cubic inches which occurs when the height (x) is 1.06, width ($5 - 2x$) is 2.89, and the length ($10 - 2x$) is 7.89. For all practical purposes, the box has dimensions $1'' \times 3'' \times 8''$ with a volume of 24 cubic inches.

33. $C(x) = 25x + 500 - \frac{x^2}{20}$

a. $\overline{C}(x) = \frac{C(x)}{x} = -\frac{1}{20}x + 25 + 500x^{-1}$

b. $\overline{C}'(x) = -\frac{1}{20} - 500x^{-2}$

critical value $x = 100$

$x^2 + 100x + 10{,}000 = 0$

$\overline{C}''(x) = 1{,}000x^{-3} > 0$ for $x = 100$ so an absolute minimum at

$x = 100$ $\quad$ $C(100) = 2{,}500$

$-\frac{1}{20} - 500x^{-2} = 0$

$-x^2 - 10{,}000 = 0$

$x = \pm 100$ $\quad$ Reject -100 (not in domain)

	0		100	
$x - 100$:		$- - -$		$+ + +$
$x + 100$:		$+ + +$		$+ + +$
Derivative:		neg		pos
Curve		falling		rising

The average is decreasing on (0, 100) and increasing on (100, 200).

c. The minimal average cost is at $\overline{C}(100) = -\frac{1}{20}(100) + 25 + 500(10)^{-1} = \$25$

35. Revenue = (price)(no. of items)

$= (50 - x)x$

$= 50x - x^2$

$P(x) = R(x) - C(x)$

$= (50x - x^2) - (100 + 20x)$

$= -x^2 + 30x - 100$

If $P \geq 0$ then $50 - x \geq 0$

$50 \geq x$

$x \leq 50$

Domain [0, 50]

$P'(x) = -2x + 30$

critical values:

$-2x + 30 = 0$

$x = 15$

check $\quad P(0) = -100$

$P(15) = -(15)^2 + 30(15) - 100 = 125$

$P(50) = -(50)^2 + 30(50) - 100 = -1{,}100$

Maximum profit is \$125 when $x = 15$. The price is \$35.

37. From Problem 35, $R(x) = 50x - x^2$, $C(x) = 100 + 20x + 4x = 100 + 24x$

The maximum profit is found when

$$R'(x) = C'(x)$$
$$50 - 2x = 24$$
$$-2x = -26$$
$$x = 13$$

$$\begin{aligned} P(13) &= R(13) - C(13) \\ &= 481 - 412 \\ &= 69 \end{aligned}$$

The maximum profit is \$69 when $x = 13$. The price is \$37. Notice the \$4 tax raised the price \$2 and lowered the profit \$56.

39. $C'(x) = 5 - 5{,}000x^{-2}$

Critical values ($x = 0$ not in domain)

$$5 - 50{,}000x^{-2} = 0$$
$$5x^2 = 50{,}000$$
$$x^2 = 10{,}000$$
$$x = 100$$
$$\text{reject } x = -100$$

Test endpoints:

$$\begin{aligned} C(100) &= 5(100) + \frac{50{,}000}{100} \\ &= 1{,}000 \end{aligned}$$

$$\begin{aligned} C(125) &= 5(125) + \frac{50{,}000}{125} \\ &= 1{,}025 \end{aligned}$$

$$C(1) = 5 + \frac{50{,}000}{1} = 50{,}005$$

The minimum cost is \$1,000 when $x = 100$.

41. Maximize $A = \ell w$

$$\ell + 2w = 1{,}000$$
$$\ell = 1{,}000 - 2w$$
$$A = (1{,}000 - 2w)w$$
$$A = 1{,}000w - 2w^2$$

Domain [0, 500]

$$A' = 1{,}000 - 4w$$

Critical values

$$1{,}000 - 4w = 0$$
$$w = 250$$
$$\text{length} = 1{,}000 - 500 = 500$$

Endpoints

$A(0) = 0$; $A(500) = 0$;

$A(250) = 125{,}000$

The largest area occurs when the width is 250 and the length in 500 ft.

43. Let x represent the number of new people signed up.

The price of each ticket will be

$$2{,}000 \text{ if } x = 100$$
$$1{,}995 \text{ if } x = 101$$
$$1{,}990 \text{ if } x = 102$$
$$\vdots$$

In general, the price will always be $2{,}000 - 5x$ (if x new people are signed up)

$$\begin{aligned} \text{Revenue} = R(x) &= (\text{price})(\text{number}) \\ &= (2{,}000 - 5x)(100 + x) \\ &= 200{,}000 + 150x - 5x^2 \end{aligned}$$

$$\begin{aligned} \text{Cost} = C(x) &= \text{fixed costs} + \text{variable costs} \\ &= 125{,}000 + 500(100 + x) \\ &= 175{,}000 + 500x \end{aligned}$$

$$\begin{aligned} \text{Profit} = P(x) = R(x) - C(x) &= (200{,}000 + 1{,}500x - 5x^2) - (175{,}000 + 500x) \\ &= -5x^2 + 1{,}000x + 25{,}000 \end{aligned}$$

$P'(x) = -10x^2 + 1{,}000x$ so $P'(x) = 0$ when $x = 100$.

The interval for x is $[0, 50]$, so we evaluate P only at the endpoints: $P(0) = 25{,}000$ and $P(5) = 62{,}500$. Maximum is \$62,500 which occurs when $x = 50$. (That is, there are 150 people on the flight.) In this case the price is $2{,}000 - 5(50) = 1{,}750$.

45. Let x represent the number of grapevines per acre. The number of pounds will

$x = 50$, 150 pounds per grapevine

$x = 51$, 148 pounds per grapevine

$x = 52$, 146 pounds per grapevine, etc.

In general, if x vines are planted per acre, each vine will produce $250 - 2x$ pounds. The yield per acre will be $Y(x) = (250 - 2x)(x) = 250x - 2x^2$. $Y'(x) = 250 - 4x = 0$ when $x = 62.5$, but since there must be a whole number of vines, plant 62 vines per acre.

4.5 Review

Objective Questions

1. $y' = 4x - 8$

$4x - 8 = 0$

$x = 2$

$y'' = 4 > 0$

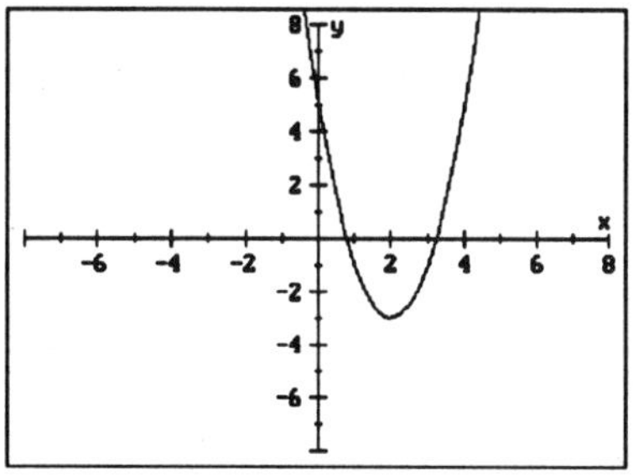

3. Differentiate implicitly: $2x + 4 + 2y' = 0$

Set $y' = 0$:

$2x + 4 = 0$

$x = -2$ which is where the vertex occurs.

Differentiate again implicitly:

$(2x)' + 4' + (2y')' = 0'$

$2 + 0 + 2y'' = 0$

$y'' = -1 < 0$

so the graph is concave down.

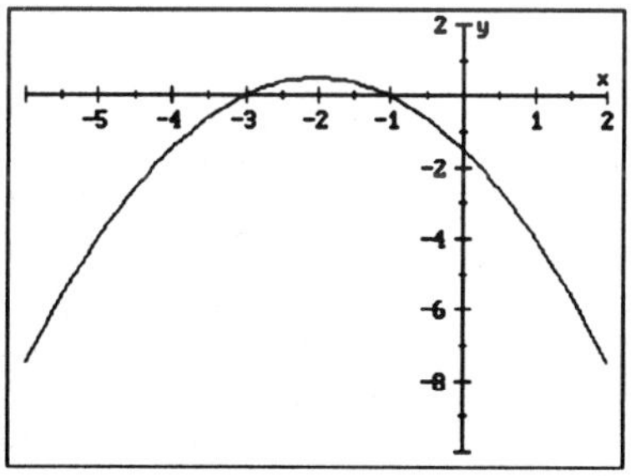

A few points are:

Let $x = -2$,

$(-2)^2 + 4(-2) + 2y + 3 = 0 \Rightarrow y = \frac{1}{2}$

Let $x = 0$, $(0)^2 + 4(0) + 2y + 3 = 0 \Rightarrow y = -\frac{3}{2}$

Let $x = 1$, $(1)^2 + 4(1) + 2y + 3 = 0 \Rightarrow y = -4$

5.

Function	*1st Derivative*	*2nd Derivative*
$f(x) = x^3 - 27x$	$f'(x) = 3x^2 - 27$	$f''(x) = 6x$
	critical values	
	$3x^2 - 27 = 0$	
	$x^2 = 9$	
	$x = \pm 3$	
	check $x = 3$	
$f(3) = -54$	Relative minimum	$f''(3) > 18$

Check $x = -3$

$f(-3) = 54$ Relative maximum $f''(-3) < 18$

concavity

$6x = 0$

$f(0)$ $f'(0^-) < 0$ $x = 0$

$f'(0^+) > 0$ not a point of inflection

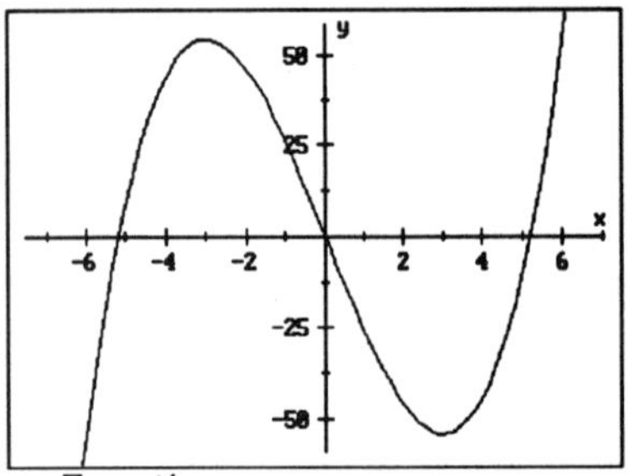

7. *Function* | *1st Derivative* | *2nd Derivative*

$g(x) = x^3 + 3x^2 - 9x + 5$ $g'(x) = 3x^2 + 6x - 9$ $g''(x) = 6x + 6$

critical values

$3x^2 + 6x - 9 = 0$

$x^2 + 2x - 3 = 0$

$(x + 3)(x - 1) = 0$

$x = 1, -3$

check $x = 1$

$g(1) = 1^3 + 3(1)^2 - 9(1) + 5 = 0$ Relative minimum $g''(1) = 12 > 0$

$g(3) = 3^3 + 3(3) - 9(3) + 5 = 14$ Check $x = -3$

Relative maximum $g''(-3) < 0$

concavity

$g(-1) = (-1)^3 + 3(-1) - 9(-1) + 5$ point of inflection $6x + 6 = 0$

$= 10$ $x = -1$

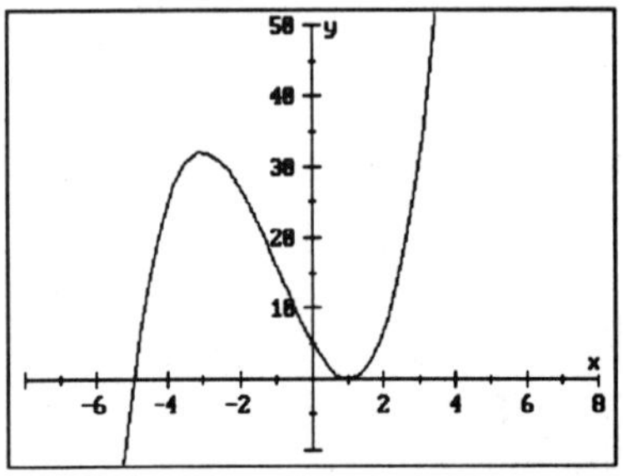

concave down on $(-\infty, -1)$

concave up on $(-1, \infty)$

9. $y' = 12x^3 - 3x^2 + 10x$; $y'' = 36x^2 - 6x + 10$; $y''' = 72x - 6$; $y^{(4)} = 72$; $y^{(n)} = 0$ for > 4

11. $f'(x) = -\frac{1}{2}x^{-3/2}$; $f''(x) = \frac{3}{4}x^{-5/2}$;
$f'''(x) = -\frac{15}{8}x^{-7/2}$; $f^{(4)}(x) = \frac{105}{6}x^{-9/2}$

13. $$f'(x) = 3x^2 - 2x - 5$$
$$= (3x - 5)(x + 1)$$

Critical values $x = \frac{5}{3}, -1$

$$f''(x) = 6x - 2$$

$f''(\frac{5}{3}) > 0$ relative minimum at $x = \frac{5}{3}$; $f(\frac{5}{3}) = \frac{-175}{27}$

$f''(-1) < 0$ relative maximum at $x = -1$; $f(-1) = 3$

15. Set the first derivative equal to zero to find critical values

$$t'(x) = 8x - 4x^3 = 0 \Rightarrow 2x - x^3 = 0 \Rightarrow x(x - x^2) = 0 \Rightarrow x = 0, \sqrt{2}, -\sqrt{2}$$
$$t''(x) = 8 - 12x^2$$
$$t''(0) = 8 - 12(0)^2 = 8 > 0$$
$$t''(\pm\sqrt{2})^2 = -16 < 0$$

So we have a relative maximum = 4, which occurs at $x = \pm\sqrt{2}$ and a relative minimum = 0, which occurs at $x = 0$.

17.

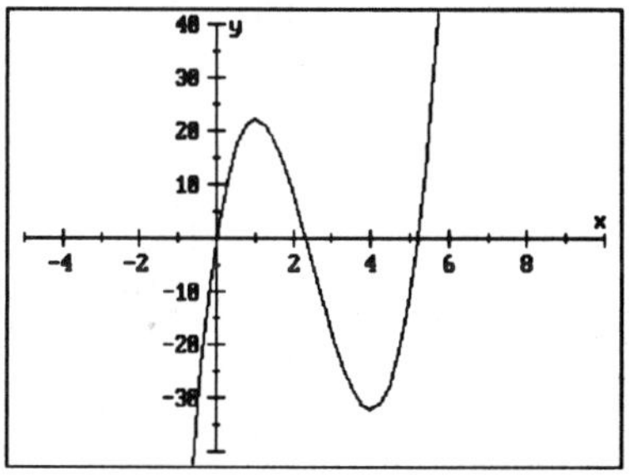

19.

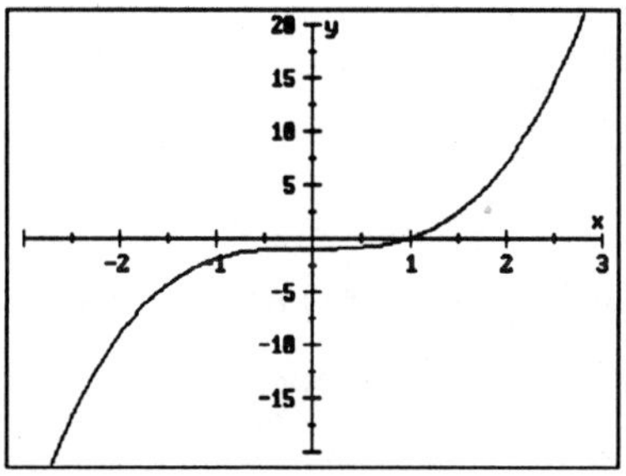

21. $$y' = 15x^4 - 255x^2 + 240$$

$$15x^4 - 255x^2 + 240 = 0$$
$$x^4 - 17x^2 + 16 = 0$$
$$(x^2 - 16)(x^2 - 1) = 0$$
$$(x - 4)(x + 4)(x - 1(x + 1) = 0$$

$f(4) = -1{,}408$ minimum

$f(-2) = 140$ maximum

$f(1) = 158$ maximum

$f(-1) = -158$

$x = 4, -4, 1, -1$ (reject -4; not in domain)

The maximum = 158 which occurs at $x = 1$; the minimum = $-1{,}408$, which occurs at $x = 4$.

23. Critical values

$$\frac{dy}{dx} = (x - 1)(3)(x+3)^2 + (1)(x+3)^3 \qquad 4x(x+3)^2 = 0$$

$$= (x+3)^2[(3x - 3) + (x+3)] \qquad x = 0$$

$$= (x+3)^2(4x) \qquad x = -3$$

$$= 4x(x+3)^2$$

at $x = -5$: $y = -6(-2)^3 = 48$

$x = -3$: $y = -4(0) = 0$

$x = 0$: $y = (-1)(3)^3 = -27$ Absolute minimum -27 at $x = 0$

$x = 5$: $y = 4(8)^3 = 2{,}048$ Absolute maximum 2,048 at $x = 5$

25. $R'(x) = 50 - .1x$ and $C'(x) = 2$; maximum profit when

$$50 - .1x = 2$$

$$x = 480$$

27. Since Profit, $P(x) = R(x) - C(x)$,

$$P(x) = -.005x^3 + .345x^2 + 3x - 4{,}000$$

$$P'(x) = -.015x^2 + .69x + 3.$$

Set $P(x) = 0$ and solve to find:

$x = -4$ or 50. Rejecting the negative values tells us to produce 50 items. Instead, you can solve $R'(x) = C'(x)$

29. $R(x) = xp = 25x - .000004x^2$

$R'(x) = 25 - .000012x^2$; This is the marginal revenue.

$R''(x) = -.000024 < 0$ so,

The marginal revenue is increasing for all values, in particular at $x = 100$

31. Let x = the number of \$50 per month increases.

$$R(x) = \text{Revenue} = (\text{no. rented})(\text{monthly rent})$$

$$= (100 - 2x)(500 + 50x)$$

Domain [0, 50]

Critical values:

$$R'(x) = (100 - 2x)(50) + (-2)(500 + 50x)$$

$$= 5{,}000 - 100x - 1{,}000 - 100x$$

$$= 4{,}000 - 200x$$

$$4{,}000 - 200x = 0$$

$x = 20$

check values:

$R(0) = 100(500) = 50{,}000$

$R(50) = 0(500 + 50 \cdot 50) = 0$

$R(20) = (100 - 40)(500 + 50 \cdot 20) = 90{,}000$

The rent should be \$1,500 per month (60 will be rented with 40 vacancies).

Sample Test

1. a. $y' = -9x^2 + 5x^4$; $y'' = -18x + 20x^3$; $y''' = -18 + 60x^2$; $y^{(4)} = 120x$; $y^{(5)} = 120$; $y^{(6)} = y^{(7)} = \cdots = 0$ **b.** $g' = 15x^4 + 3x^{-2}$; $g'' = 60x^3 - 6x^{-3}$; $g''' = 180x^2 + 18x^{-4}$; $g^{(4)} = 360x - 72x^{-5}$

3. $g'(x) = 3x^2 + 6x - 9 = 3(x - 1)(x + 3)$;

$g'(x) = 0$ when $x = 1$ or $x = -3$.

$g'(x) > 0$ when $x < -3$ or $x > 1$;

$g'(x) < 0$ when $-3 < x < 1$.

A few points are:

$(-4, g(-4)) = (-4, 25)$; $(-3, g(-3)) = (-3, 32)$; $(-2, g(-2)) = (-2, 27)$.

You should calculate more points; the graph is shown.

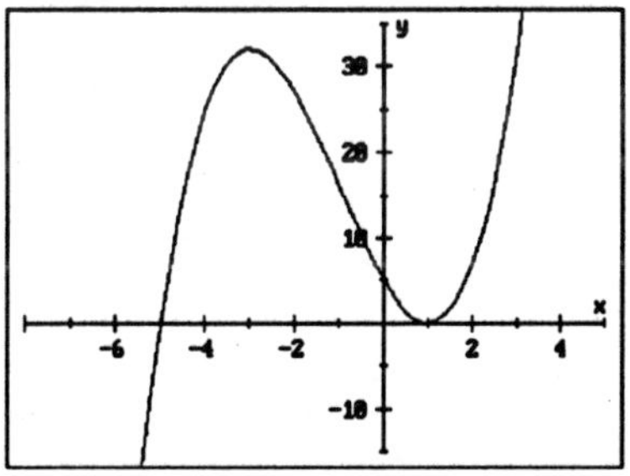

5. $y' = \dfrac{(x+1)(x-2)' - (x-2)(x+1)'}{(x+1)^2} = \dfrac{3}{(x+1)^2}$ which is always positive,

so the curve is always increasing. $y'' = \dfrac{-6}{(x+1)^3}$ which is negative if $x > -1$ and positive if $x < -1$, so the curve is concave down to the right of $x = -1$ and concave up to the left of $x = -1$. Also, there is a vertical asymptote at $x = -1$, and since $\lim_{|x| \to \infty} \dfrac{x-2}{x+1} = 1$, $y = 1$ is a horizontal asymptote. The graph is shown at right.

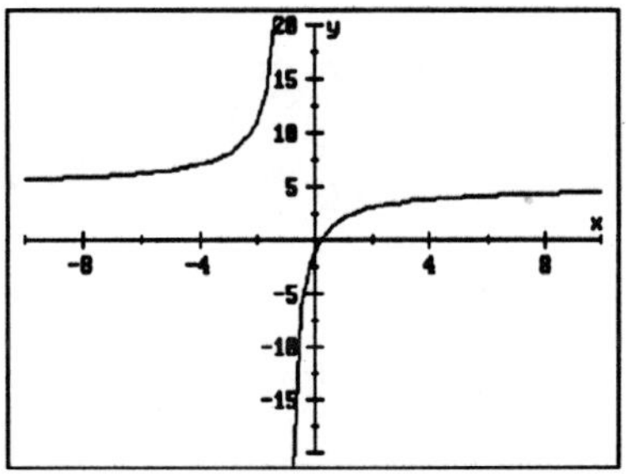

7. $g'(x) = 1 - 8x^{-2}$. Set g' equal to 0: $1 - 8x^{-2} = 0$

$$x^2 - 8 = 0$$

$$x = \pm\sqrt{8} = \pm 2\sqrt{2}.$$

$g''(x) = 16x^{-3}$; $g''(2\sqrt{2}) > 0$ and $g''(-2\sqrt{2}) < 0$. By the second derivative test, there is a relative maximum at $x = -2\sqrt{2}$ and a relative minimum at $x = 2\sqrt{2}$.

9. Revenue = (price)(number of items) $= \left[50 - \left(\frac{x}{1000}\right)^2 x\right](x)$.

$R(x) = 50x - \frac{x^3}{1000^2}$. Marginal revenue $= R'(x) = 50 - \frac{3}{1000^2}x^2$.

Take the derivative of marginal revenue to see if increasing or decreasing:

$R''(x) = -\frac{6}{1000^2}x$; $R''(100) = -\frac{6}{1000^2}(100) < 0$, so $R'(x)$, the marginal revenue, is decreasing when $x = 100$.

CUMULATIVE REVIEW CHAPTERS 2-4

1. As x approaches (but never equals) 3, the graph gets closer and closer to a y-value of 5. The limit is 5.

3. $\lim_{x \to 5} \frac{2x^2 - 7x - 15}{x - 5} = \lim_{x \to 5} \frac{(2x + 3)(x - 5)}{x - 5} = \lim_{x \to 5} (2x + 3) = 13$

5. $\lim_{x \to \infty} \frac{(3x + 1)(5x - 2)}{x^2} = \lim_{x \to \infty} \frac{15x^2 - x - 2}{x^2} \cdot \frac{\frac{1}{x^2}}{\frac{1}{x^2}}$

$= \lim_{x \to \infty} \frac{15 - \frac{1}{x} - \frac{2}{x^2}}{1} = \frac{15 - 0 - 0}{1} = 15$

7. We should be suspicious of $x = 8$, so check to see if $\lim_{x \to 8} f(x) = f(8)$:

$\lim_{x \to 8} f(x) = \lim_{x \to 8} \frac{x^2 - 15x + 56}{x - 8} = \lim_{x \to 8} \frac{(x - 8)(x - 7)}{x - 8} = \lim_{x \to 8} (x - 7) = 1$

But $f(8) = 7 \neq 1 = \lim_{x \to 8} f(x)$, so $f(x)$ is discontinuous at $x = 8$.

9. $y = 4 - x^{-1}$; $\frac{dy}{dx} = x^{-2} = \frac{1}{x^2}$. At $x = 1$, $\frac{dy}{dx} = \frac{1}{1^2} = 1$

11. $y = -24x - 15x^2$

13. $y' = 5(2 - 5x)^4(2 - 5x)' = 5(2 - 5x)^4(-5) = -25(2 - 5x)^4$

15. Differentiate implicitly with respect to x (remember to use the Product Rule):

$$3x^5y^4 = 10$$
$$x^5y^4 = \frac{10}{3}$$
$$(x^5y^4)' = \left(\frac{10}{3}\right)$$
$$x^5(4y^3y') + (5x^4)y^4 = 0$$
$$4x^5y^3y' = -5x^4y^4$$
$$y' = \frac{-5x^4y^4}{4x^5y^3} = \frac{-5y}{4x}$$

17. The slope is given by $f'(x) = 2 - 18x^2$.

At $x = 1$, $f'(1) = 2 - 18(1)^2 = -16$. Now use the point-slope form of the equation for a line:

$$y - y_1 = m(x - x_1)$$
$$y - (-4) = -16(x - 1)$$
$$y + 4 = -16(x - 1)$$
$$y + 4 = -16x + 16$$
$$16x + y - 12 = 0$$

19. $y' = 8x^3 - 3x^2 + 6x$; $y'' = 24x^2 - 6x + 6$; $y''' = 48x - 6$; $y^{(4)} = 48$; $y^{(5)} = y^{(6)} = \cdots = 0$.

21. First find the critical values by setting $y' = 0$:

$y = x^{1/2} - x$; $y' = \frac{1}{2}x^{-1/2} - 1$; $\frac{1}{2}x^{-1/2} - 1 = 0$

Multiply by $2x^{1/2}$: $\quad 1 - 2x^{1/2} = 0$

$$1 - 2\sqrt{x} = 0$$
$$\frac{1}{2} = \sqrt{x}$$
$$\frac{1}{4} = x.$$

Now calculate $f(x)$ here and at the endpoints: $f(0) = 0$, $f(9) = -6$, $f(\frac{1}{4}) = \frac{1}{4}$

The maximum is $\frac{1}{4}$, which occurs at $x = \frac{1}{4}$; the minimum is -6, which occurs at $x = 9$.

23. Revenue = (price)(number of items)

$$R(x) = (25 - \frac{x}{100})(x)$$
$$= 25x - \frac{1}{100}x^2$$

Marginal revenue = $R'(x) = 25 - \frac{1}{50}x$. To see if marginal revenue, $R'(x)$, is increasing or decreasing we examine the derivative of marginal revenue,

$R'(x)$: $R''(x) = -\frac{1}{50}$ which is always negative; hence

$R'(x)$ is always decreasing. Marginal revenue is decreasing at $x = 100$.

25. Let $x =$ width of print and $y =$ length of print. Then $xy = 20$ and the frame size is $(x + 4)(y + 8)$. Since $y = 20x^{-1}$ we want to minimize

$$A = (x + 4)(20x^{-1} + 8) = 52 + 8x + 80x^{-1}$$

$$A' = 8 - 80x^{-2} \text{ so } A' = 0 \text{ if } x^2 = 10$$

thus $x \approx 3.16$ in. and $y \approx 20(3.16)^{-1} \approx 6.32$ in. The frame should be 7.16 in. by 14.32 in. to the nearest inch the frame is 7 in. by 14 in.

CHAPTER 5

EXPONENTIAL AND LOGARITHMIC FUNCTIONS

5.1 Exponential Functions

1. $25^{1/2} = \sqrt{25} = 5$

3. $(-25)^{1/2} = \sqrt{-25}$ which is not a real number

5. $-27^{1/3} = -\sqrt[3]{27} = -3$

7. $7^{1/3} \cdot 7^{2/3} = 7^{1/3 + 2/3} = 7^1 = 7$

9. $1{,}000^{-1/3} = \dfrac{1}{1{,}000^{1/3}} = \dfrac{1}{\sqrt[3]{1{,}000}} = \dfrac{1}{10}$

11. $100^{-3/2} = \dfrac{1}{100^{3/2}} = \dfrac{1}{(\sqrt{100})^3} = \dfrac{1}{10^3} = \dfrac{1}{1{,}000}$

13. $\dfrac{2^3 \cdot 2^{-4}}{2^5 \cdot 2^{-2}} = 2^{3-4-5+2} = 2^{-4}$

15. $\dfrac{9^{1/2} \cdot 27^{2/3}}{3^2 \cdot 81} = \dfrac{3 \cdot 3^2}{3^2 \cdot 3^4} = 3^{-3}$

17. $\frac{1}{2} + \frac{1}{3} = \frac{5}{6}$ **19.** 20.085537 **21.** 1.0512711 **23.** 1.0460279

25. $x^{1/2}(x^{1/2} + x^{1/2}) = x^1 + x^1 = 2x$ **27.** $x^{2/3}(x^{-2/3} + x^{1/3}) = x^0 + x^1 = 1 + x$

29. $\left(x^{1/2} + y^{1/2}\right)^2 = \left(x^{1/2} + y^{1/2}\right)\left(x^{1/2} + y^{1/2}\right) = x + 2x^{1/2}y^{1/2} + y$

31. Some points are

x	3	2	1	0	-1	-2
y	$3^3 = 27$	$3^2 = 9$	3	$3^0 = 1$	$3^{-1} = \frac{1}{3}$	$3^{-2} = \frac{1}{9}$

The graph is shown at right:

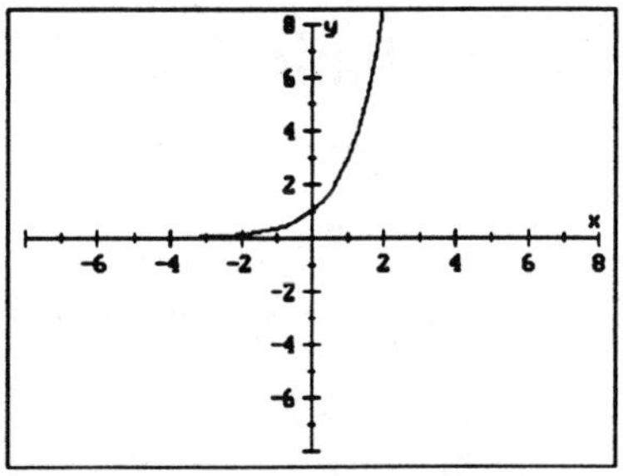

33. Some points are

x	3	2	1	0	-1	-2
y	$\left(\frac{1}{3}\right)^3 = 27$	$\left(\frac{1}{3}\right)^2 = 9$	$\frac{1}{3}$	$\left(\frac{1}{3}\right)^0 = 1$	$\left(\frac{1}{3}\right)^{-1} = 3$	$\left(\frac{1}{3}\right)^{-2} = 9$

The graph is shown:

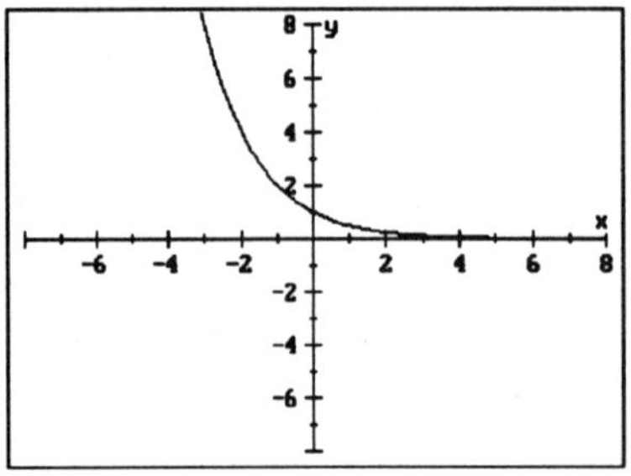

35.

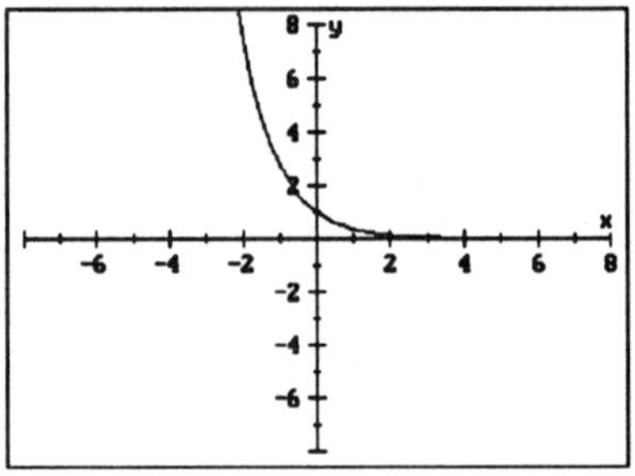

37.

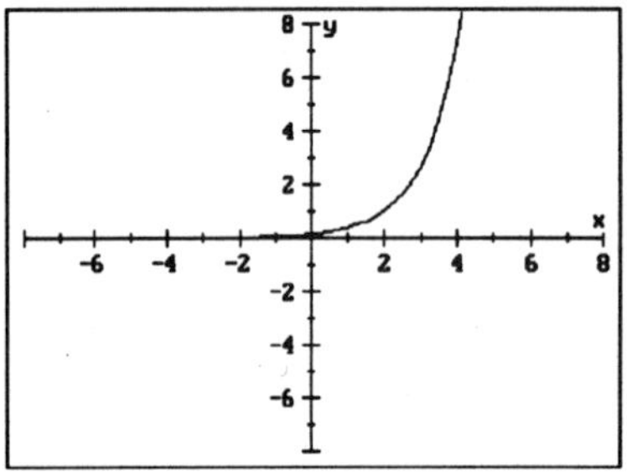

39.

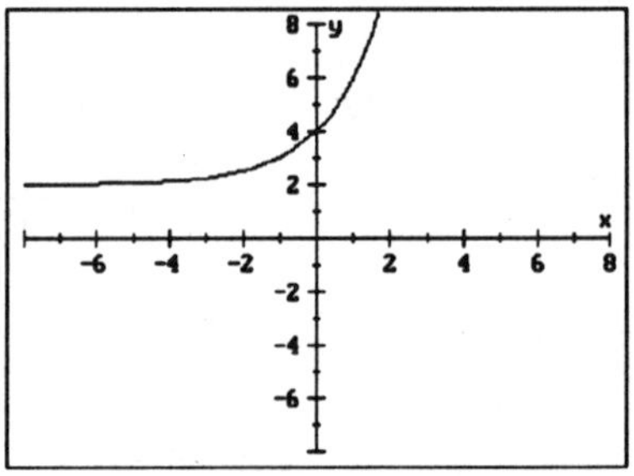

41. $A = Pe^{rt} = 1{,}000e^{(.16)(25)} = 1{,}000e^4 = 1{,}000(54.59815) = \$54{,}598.15$

43. $A = P(1 + i)^N = 8{,}500(1 + \frac{.18}{12})^{(12)(4)} = 8{,}500(1.015)^{48} = 8{,}500(2.043478)$

$= \$17{,}369.57$

45. $A = P(1 + i)^N = 9{,}400(1 + \frac{.14}{360})^{(1/2)(360)} = 9{,}400(1.000388889)^{180}$

$= 9{,}400(1.0724936) = \$10{,}081.44$

47. $P = 10{,}000(1 + \frac{.115}{2})^{-10} \approx \$5{,}717.37$

49. $P = 10{,}000e^{-.115(5)} \approx 5{,}627.05$

51. If $p < x < q$ and $0 < b < 1$ then $b^q < b^x < b^p$

5.2 Logarithmic Functions

1. Remember a logarithm is an *exponent;* since the exponent $= 6$ logarithm $= 6$; base $= 2$; $\log_2 64 = 6$

3. The exponent $= -2$, logarithm $= -2$, base $= \frac{1}{3}$, $\log_{1/3} 9 = -2$

5. The exponent $= c$, logarithm $= c$; base $= b$, $\log_b a = c$

7. logarithm $= \frac{1}{2}$, exponent $= \frac{1}{2}$, base $= 4$, $4^{1/2} = 2$

9. logarithm $= -2$, exponent $= -2$; base $= 10$, $10^{-2} = .01$ (log with no base is understood to be base 10.)

11. logarithm $= 2$, exponent $= 2$; base $= e$, $e^2 = e^2$ ($\ln x = \log_e x$)

13. Let $x = \log_b b^2$. In exponential form this is $b^x = b^2$, hence $\log_b b^2 = x = 2$.

15. Let $x = \log_\pi \sqrt{\pi}$. In exponential form this is $\pi^x = \sqrt{\pi} = \pi^{1/2}$, hence $x = \log_\pi \sqrt{\pi} = \frac{1}{2}$.

17. Let $x = \log 1{,}000$; In exponential form this is $10^x = 1{,}000$, hence $x = \log 1{,}000 = 3$.

19. Let $x = \log_{16} 1$. In exponential form this is $16^x = 1 = 16^0$, hence $x = \log_{16} 1 = 0$.

21. Using a calculator, .033423755

23. Using a calculator, 3.989449818

25. $-.493494967$ **27.** .81977983 **29.** .69314718

31. A few points are

x	$x \le 0$	$\frac{1}{3}$	1	$\sqrt{3}$	3	9
y	undefined	-1	0	$\frac{1}{2}$	1	2

The graph is at right.

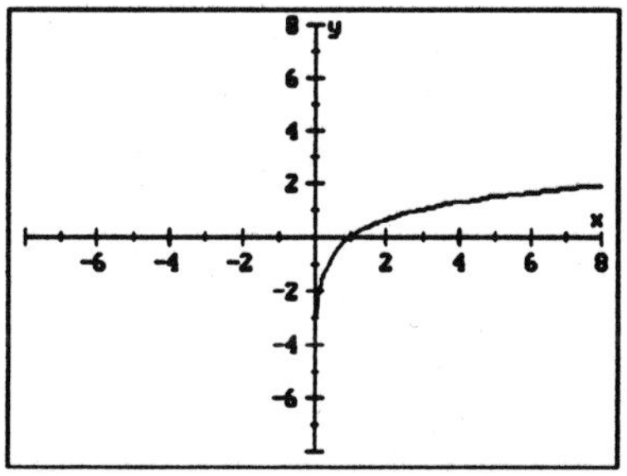

33. Using a calculator, a few points are:

x	$x \le 0$	$\frac{1}{3}$	1	$\sqrt{3}$	3	9
y	undefined	-1	0	$\frac{1}{2}$	1	2

The graph is at right.

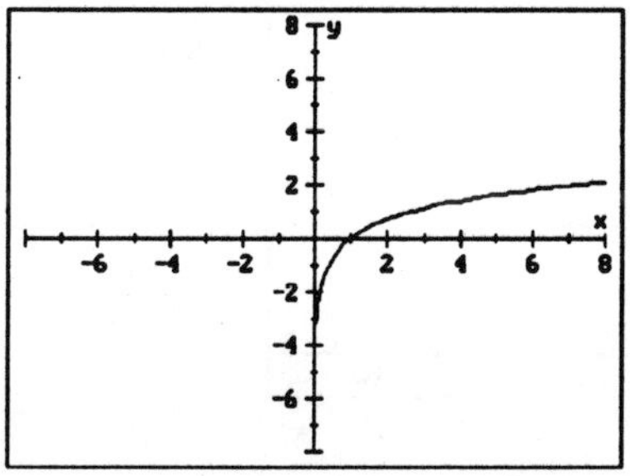

35.

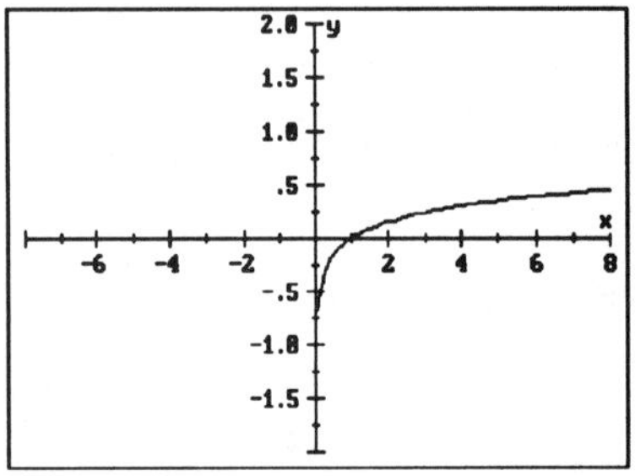

37.

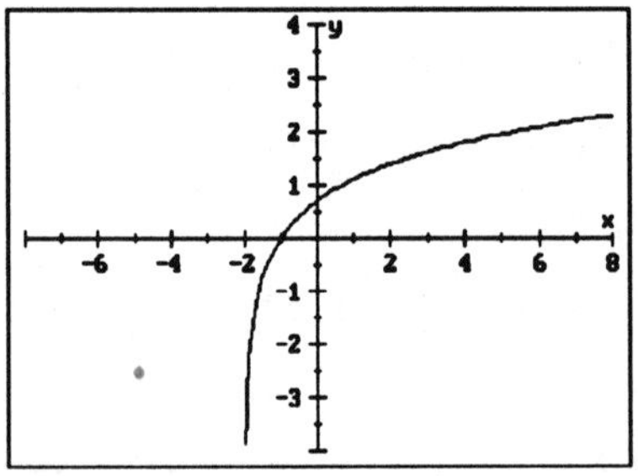

39.

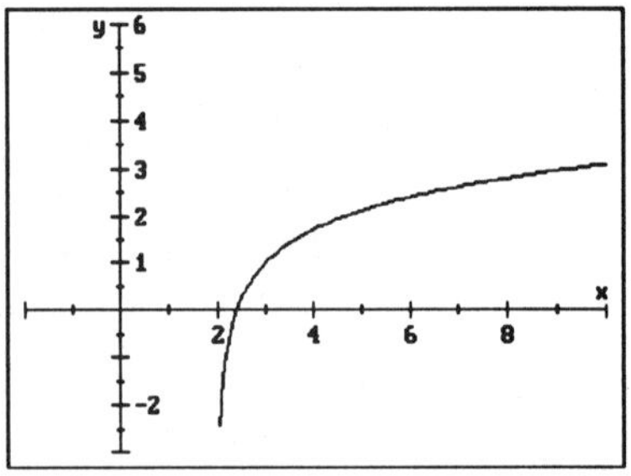

41.

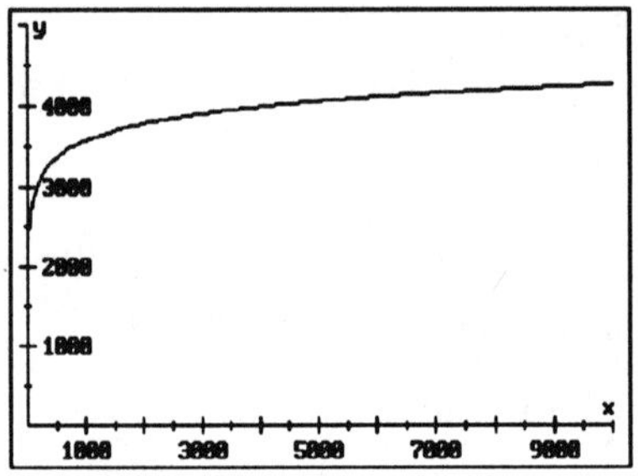

43. a. $t = -62.5 \ln(1 - \frac{30}{80}) \approx 29$; it would take 29 days
b. No, since $N = 80$ gives ln 0 which is not defined.

45.

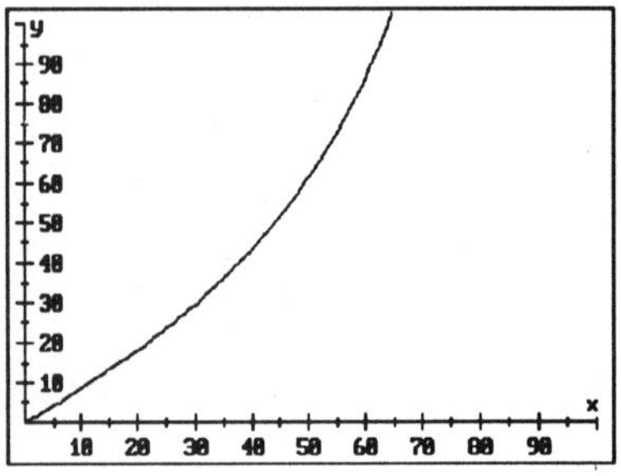

47. a. $M = \dfrac{\log 15^{15} - 11.8}{1.5} \approx \dfrac{17.64137 - 11.8}{1.5} \approx 3.9$

b. $M = \dfrac{\log 10^{25} - 11.8}{1.5} = \dfrac{25 - 11.8}{1.5} = 8.8$

49.

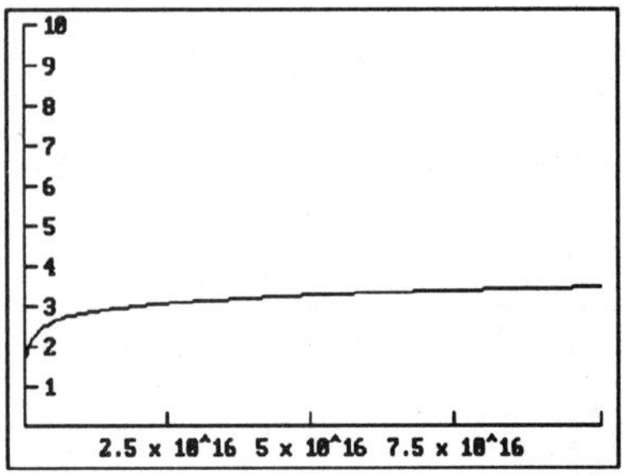

51. Since $b^x = b^x$, then $x = \log_b b^x$ by definition of logarithm.

53. Since $\log_b x = \log_b x$, then $b^{\log_b x} = x$ by definition of logarithm.

5.3 Logarithmic And Exponential Equations

1. The equation is equivalent to $5^x = 25$, so $x = 2$. Another way would be to use the change of base formula (with a calculator): $x = \log_5 25 = \dfrac{\log 25}{\log 5} \approx \dfrac{1.39794}{.69897}$ ≈ 2.

3. The equation is equivalent to $10^x = \frac{1}{10} = 10^{-1}$, so $x = -1$. Another way would be to use a calculator: $x = \log(0.1) = -1$.

5. The equation is equivalent to $x^2 = 28$, so $x = \sqrt{28} = 2\sqrt{7} \approx 5.29150$

7. The equation is equivalent to $e^3 = x$. Using a calculator a display of 20.08553692.

9. Since $\ln 9.3 = \ln x$, $9.3 = x$.

11. Since $\ln x^2 = \ln 12$; $x^2 = 12$, so $x = \pm\sqrt{12} = \pm 2\sqrt{3} \approx 3.4641$

13. The equation is equivalent to $3^x = 27\sqrt{3} = 3^3\, 3^{1/2} = 3^{3.5}$, so $x = 3.5$. Another way would be to use the change of base formula (with a calculator): $x = \log_3 27\sqrt{3} = \dfrac{\log 27\sqrt{3}}{\log 3} \approx \dfrac{1.6699244}{.477121252} \approx 3.5$.

15. The equation is equivalent to $x^0 = 10$ which implies $1 = 10$. Since this is a contradiction there is no solution.

17. If you know $2^7 = 128$, the answer is $x = 7$. Another way is to use the definition of log: $x = \log_2 128 = \dfrac{\log 128}{\log 2} \approx \dfrac{2.10721}{.30103} \approx 7$

19. Since $125 = 5^3$ and $25 = 5^2$, the given equation is equivalent to $(5^3)^x = 5^2$, so $5^{3x} = 5^2$ which leads to $3x = 2$; hence $x = \frac{2}{3}$.

21. Since $\frac{1}{9} = 3^{-2}$, the given equation is equivalent to $3^{4x - 3} = 3^{-2}$, so $4x - 3 = -2$. Solving for x gives $x = \frac{1}{4}$.

23. $(\frac{3}{2})^x = 10$

$x = \log_{3/2} 10 \approx 5.6788736$

25. $\ln x - \ln e^3 = 1$

$\ln x - 3 = 1$

$\ln x = 4$

$e^4 = x$

$e^4 \approx 54.59815$

27. $\log_b \dfrac{6(3.9)}{2.8} = 1$; $\quad b^1 = \dfrac{6(3.9)}{2.8} \approx 8.357142857$

29. $2\log_7 x - \log_7 4 = \log_7(2x - 3)$

$$\log_7 \frac{x^2}{4} = \log_7(2x - 3)$$

$$\frac{x^2}{4} = 2x - 3$$

$$x^2 = 8x - 12$$

$$x^2 - 8x + 12 = 0$$

$$(x - 6)(x - 2) = 0$$

$$x = 2,\ 6$$

31. $\ln x = 6 \ln 5 - 2 \ln x$

$$3 \ln x = 6 \ln 5$$

$$x^3 = 5^6$$

$$x = 5^2 = 25$$

33. $x^2 = x + 6$

$$x^2 - x - 6 = 0$$

$$(x - 3)(x + 2) = 0$$

$$= 3,\ -2;$$

↑ reject

35. $\ln 10 - \frac{1}{2} \ln 25 = \ln x$

$$\ln 10 - \ln 25^{1/2} = \ln x$$

$$\ln 10 - \ln 5 = \ln x$$

$$\ln\left(\frac{10}{5}\right) = \ln x$$

$$\ln 2 = \ln x$$

$$2 = x$$

37. $\log 515 = 2x - 1$

$$x = \frac{\log 151 + 1}{2}$$

$$1.855903615 \approx x$$

39. $\log_4 .82 = x$

$$-.143152 \approx x$$

41. $5x = \ln \frac{1}{4}$

$$x = \frac{1}{5} \ln \frac{1}{4}$$

$$\approx -.2772588$$

43. $1 - 5x = \ln 15$

$$5x = 1 - \ln 15$$

$$x = \frac{1 - \ln 15}{5}$$

$$\approx -.34161$$

45. $7 = 5e^{.02t}$

$$.02t = \ln \frac{7}{5}$$

$$t = 50 \ln \frac{7}{5}$$

$$\approx 16.82 \text{ yr.}$$

About 2004

47. Using $A = P(1 + i)^N$, $2{,}000 = 1{,}000(1 + \frac{.12}{2})^{2t}$ where t is the number of years.

Divided both sides by 1,000:

$$2 = (1.06)^{2t}$$

$\log^2{}_{1.06} = 2t$ — *Write as a logarithm*

$\dfrac{\log 2}{\log 1.06} = 2t$ — *Change of base formula*

$\frac{1}{2} \cdot \dfrac{\log 2}{\log 1.06} = t$ — *Divide both sides by 2*

$5.95 \approx t$ — *Calculator*

To the nearest half-year, it would take 6 years.

49. Using $A = Pe^{rt}$, $2{,}000 = 1{,}000e^{.12t}$

$$2 = e^{.12t} \quad \textit{Divide both sides by 1,000}$$

$$\ln 2 = .12t \quad \textit{Write as a logarithm}$$

$$\frac{\ln 2}{.12} = t$$

$$5.77623 \approx t$$

It will take 5.77623 years or about 2,108 days (5 years, 283 days).

51. Using $A = Pe^{rt}$, $3000 = 1.000e^{.12t}$

$$3 = e^{.12t} \quad \textit{Divide both sides by 1,000}$$

$$\ln 3 = .12t \quad \textit{Write as a logarithm}$$

$$\frac{\ln 3}{.12} = t$$

$$9.1551 \approx t$$

It will take 9.1551 years or (9.1551)(365) = 3342 days (9 years, 57 days).

53. a. $R = 80 - 27\ln(3) \approx 50.3$; about 50%

b. $10 = 80 - 27 \ln t$

$$-70 = -27 \ln t \quad \textit{Subtract 80 from both sides}$$

$$\frac{70}{27} = \ln t \quad \textit{Divide both sides}$$

$$e^{70/27} = t \quad \textit{Write in exponential form}$$

$$13.364375 \approx t \quad \textit{Calculator, using } \boxed{e^x} \text{ key}$$

About 13 seconds

55. Using the given formula,

$$m = \frac{(10{,}000)(.01)}{1 - (1 + .01)^{-48}} \quad \textit{48 months = 4 years}$$

$$= \frac{100}{1 - .6202604} \quad \textit{Calculator}$$

$$= \$263.34$$

57. $P = (\tfrac{1}{2})^{t/5700}$

$$\log_{1/2} P = \frac{t}{5{,}700} \quad \textit{Write in logarithmic form}$$

$$5{,}700 \log_{1/2} P = t \quad \textit{Multiply both sides by 5,700}$$

59. Using the formula: $584{,}600 = \dfrac{19{,}473}{.08}(e^{.08t} - 1)$

$$\frac{(.08)(584{,}600)}{19{,}473} = e^{.08t} - 1$$

$$2.401684 = e^{.08t} - 1$$

$$3.401684 = e^{.08t}$$

$$\ln(3.401684) = .08t$$
$$t = \frac{\ln(3.401684)}{.08} \approx 15.3 \text{ years}$$

This answer means the if the estimated reserve in 1975 was correct we would have depleted out petroleum supply in the year 1990. This is not the case since many oil fields have been disconnected since 1975

61. Let $A = b^x$ and $B = b^y$, which means $\log_b A = x$ and $\log_b B = y$.
$\log_b \frac{A}{B} = \log_b \frac{b^x}{b^y} = \log_b b^{x - y} = x - y = \log_b A - \log_b B$

5.4 Derivatives of Logarithmic and Exponential Functions

1. $y' = e^{2x}(2x)' = e^{2x}(2) = 2e^{2x}$

3. $y' = e^{-x}(-x)' = e^{-x}(-1) = -e^{-x}$

5. Use the Product Rule:

$$\begin{aligned} f'(x) &= x(e^x)' + (x)'(e^x) \\ &= x(e^x) + (1)e^x \\ &= xe^x + e^x \end{aligned}$$

7. Use the Product Rule:

$$\begin{aligned} f'(x) &= (2x^3 + 1)(e^{5x^2})' + (2x^3 + 1)'(e^{5x^2}) \\ &= (2x^3 + 1)[e^{5x^2}(5x^2)'] + (6x^2)(e^{5x^2}) \\ &= (2x^3 + 1)[e^{5x^2}(10x)] + 6x^2e^{5x^2} \\ &= 10xe^{5x^2}(2x^3 + 1) + 6x^2e^{5x^2} \\ &= 2xe^{5x^2}[5(2x^3 + 1) + 3x] \\ &= 2xe^{5x^2}[10x^3 + 3x + 5] \end{aligned}$$

9. Use the Product Rule:

$$\begin{aligned} y' &= -2e^{5x}(3x^2 + 5)' + (-2e^{5x})'(3x^2 + 5) \\ &= -2e^{5x}(6x) + [-2e^{5x}(5x)']3x^2 + 5) \\ &= -12xe^{5x} + [-2e^{5x}(5)]3x^2 + 5) \\ &= -12xe^{5x} + (-10e^{5x})(3x^2 + 5) \\ &= -2e^{5x}(6x + 15x^2 + 25) \end{aligned}$$

11. $y' = \frac{1}{5 - x}(5 - x)' = \frac{1}{5 - x}(-1) = \frac{1}{x - 5}$

13. $y' = 4(2x^3 + e^{5x})^3(2x^3 + e^{5x})' = 4(2x^3 + e^{5x})^3(6x^2 + 5e^{5x})$

15. $y' = \frac{1}{x^4}(x^4)' = \frac{1}{x^4}(4x^3) = \frac{4}{x}$

17. It might be easier to first use the power rule for logarithms:

$$y = \ln\sqrt{x} = \ln x^{1/2} = \tfrac{1}{2}\ln x$$

Now, $y' = \frac{1}{2}(\ln x)' = \frac{1}{2}\cdot\frac{1}{x} = \frac{1}{2x}$

19. First rewrite $\ln x^3 = 3 \ln x$, then use the quotient rule:

$$y = \frac{\ln x^3}{e^x} = \frac{3 \ln x}{e^x};\qquad y' = \frac{e^x(3 \ln x)' - 3 \ln x(e^x)'}{(e^x)^2}$$

$$= \frac{e^x(\frac{3}{x}) - 3 \ln x(e^x)}{(e^x)^2}$$

$$= \frac{\frac{3}{x} - 3 \ln x}{e^x}\cdot\frac{x}{x}$$ Reduce by dividing out e^x

$$= \frac{3 - 3x \ln x}{xe^x}$$ Multiply numerator and denominator by x

21. $$f'(x) = \frac{1}{\left(\frac{x}{x+2}\right)}\cdot\left(\frac{x}{x+2}\right) = \frac{x+2}{x}\left[\frac{(x+2)x' - x(x+2)'}{(x+2)^2}\right]$$

$$= \frac{x+2}{x}\left[\frac{(x+2)(1) - x(1)}{(x+2)^2}\right]$$

$$= \frac{x+2}{x}\left[\frac{2}{(x+2)^2}\right]$$

$$= \frac{2}{x(x+2)}$$

Another way would be to rewrite $y = \ln\left|\frac{x}{x+2}\right| = \ln|x| - \ln|x+2|$

So $y' = \frac{1}{x} - \frac{1}{x+2} = \frac{1}{x}\cdot\frac{x+2}{x+2} - \frac{1}{x+2}\cdot\frac{x}{x} = \frac{x+2}{x(x+2)} - \frac{x}{x(x+2)} = \frac{2}{x(x+2)}$

23. Use the Product Rule:

$$f'(t) = e^{t^2}(1 - e^{-t})' + [e^{t^2}]'(1 - e^{-t})$$

$$= e^{t^2}(e^{-t}) + 2te^{t^2}(1 - e^{-t})$$

$$= e^{t^2}e^{-t} + 2te^{t^2} - 2te^{t^2}e^{-t}$$

This can also we written in factored form as $e^{t^2 - t}(1 + 2te^t - 2t)$

25. $y' = 2(e^x - e^{-x})(e^x - e^{-x})' = 2(e^x - e^{-x})(e^x + e^{-x}) = 2(e^{2x} - e^{-2x})$

27. $$y' = \frac{(5x+2)(\ln|x|)' - (5x+2)'\ln|x|}{(5x+2)^2} = \frac{(5x+2)(\frac{1}{x}) - (5)\ln|x|}{(5x+2)^2}$$

$$= \frac{(5x+2)(\frac{1}{x}) - 5\ln|x|}{(5x+2)^2}\cdot\frac{x}{x}$$

$$= \frac{5x + 2 - 5x \ln|x|}{x(5x+2)^2}$$

29. $f'(x) = \dfrac{\ln|3x|(e^x)' - e^x(\ln|3x|)'}{(\ln|3x|)^2} = \dfrac{\ln|3x|(e^x)' - e^x(\frac{1}{3x}) \cdot 3)}{\ln^2|3x|}$

$$= \frac{e^x \ln|3x| - e^x(\frac{1}{x})}{\ln^2|3x|} \cdot \frac{x}{x} = \frac{e^x x \ln|3x| - e^x}{x \ln^2|3x|} = \frac{e^x(x \ln|3x| - 1)}{x \ln^2|3x|}$$

31. $y = 2{,}000(1 + 6e^{.3x})^{-1}$; $y' = 2{,}000\left[-1(1 + 6e^{.3x})^{-2}\right](1 + 6e^{.3x})$

$$= -2{,}000(1 + 6e^{.3x})^{-2}[6e^{.3x}(.3)]$$

$$= \frac{-2{,}000(6)(.3)e^{.3x}}{(1 + 6e^{.3x})^2}$$

$$= \frac{-3{,}600e^{.3x}}{(1 + 6e^{.3x})}$$

33. $y' = \dfrac{(1 + 40e^{-.2x})(100 \ln|x|)' - (1 + 40e^{-.2x})(100 \ln|x|)}{(1 + 40e^{-.2x})^2}$

$$= \frac{(1 + 40e^{-.2x})\left(\frac{100}{x}\right) - (8e^{-.2x})(100 \ln|x|)}{(1 + 40e^{-.2x})^2}$$

$$= \frac{(1 + 40e^{-.2x})\,100 - 800xe^{-.2x}\ln|x|}{x(1 + 40e^{-.2x})^2}$$

$$= \frac{100e^{.2x} + 400 - 800x \ln|x|}{xe^{.2x}(1 + 40e^{-.2x})^2}$$

$$= \frac{100(e^{.2x} + 40 + 8x \ln|x|)}{xe^{.2x}(1 + 40e^{-.2x})^2}$$

35.

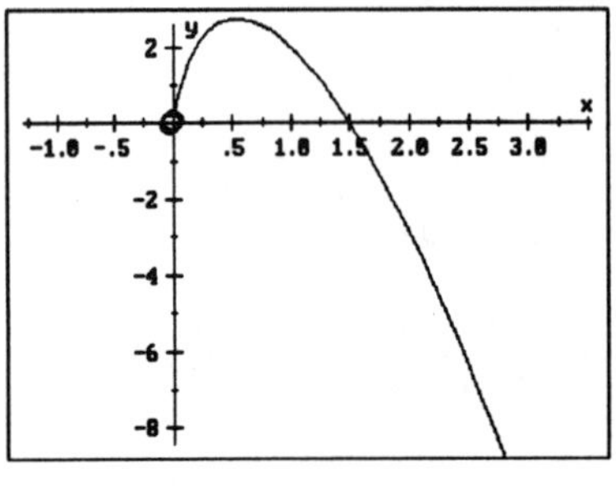

37.

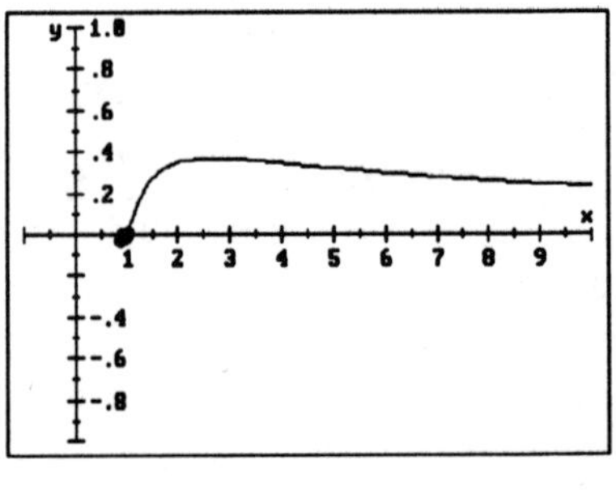

39.

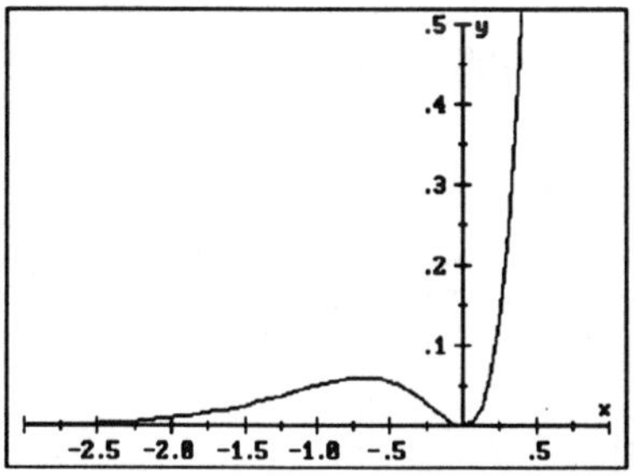

41. Revenue = (number of items)(price)

$$R(x) = (x)\left[\frac{500 \ln (x+10)}{x^2}\right] = \frac{500 \ln (x+10)}{x}$$

$$\text{Marginal revenue} = R'(x) = 500\left[\frac{x\,[\ln (x+10)]' - \ln (x+10)(x)'}{x^2}\right]$$

$$= 500\left[\frac{x\left(\frac{1}{x+10}\right) - \ln (x+10)}{x^2}\right]$$

$$= 500\left[\frac{x - (x+10)\ln (x+10)}{x^2(x+10)}\right]$$

43. **a.** $N(0) = \dfrac{5{,}000}{1 + 4{,}999e^0} = \dfrac{5{,}000}{5{,}000} = 1$

b. $N(t) = 5{,}000(1 + 4{,}999e^{-.1t})^{-1}$

$$N'(t) = -5{,}000(1 + 4{,}999e^{-.1t})^{-2}(1 + 4{,}999e^{-.1t})'$$

$$= -5{,}000(1 + 4{,}999e^{.1t})^{-2}(499.9e^{-.1t})$$

$$= \frac{2{,}499{,}500}{e^{.1t}(1 + 4{,}999e^{-.1t})^2}$$

$$N'(0) = \frac{2{,}499{,}500}{e^0(1 + 4{,}999e^0)^2} = \frac{2{,}499{,}500}{5{,}000^2}$$

$$\approx .09998 \text{ (people per day)}$$

45. $A' = P(e^{.12t})' = P(.12e^{.12t})$

After 1 year: $A'(1) = P(.12e^{.12(1)}) \approx .12(1.27)P$

$= .1353P$ or 13.5% of P

After 5 years: $A'(5) = P(.12e^{.12(5)}) \approx .12(1.822)P$

$\approx .2187P$ or 21.9% of P

47. $A'(t) = 5(e^{-.03t})' = 5(-.03e^{-.03t})$

$= -.15e^{-.03t}$

49. Revenue, $R(x) = (x)[d(x)] = x(500e^{-.1x}) = 500xe^{-.1x}$

$$R'(x) = 500[x(e^{-.1x})' + x'(e^{-.1x})]$$
$$= 500[x(-.1e^{-.1x}) + e^{-.1x}]$$
$$= -50xe^{-.1x} + 500e^{-.1x}$$

51. Use the change of base formula:

$$y = \log x = \frac{\ln x}{\ln 10} = \left(\frac{1}{\ln 10}\right)\ln x$$
$$y' = \frac{1}{\ln 10}(\ln x)' = \frac{1}{\ln 10}\left(\frac{1}{x}\right)$$
$$= \frac{1}{\left(\frac{\log 10}{\log e}\right)} \cdot \frac{1}{x}$$
$$= \frac{\log e}{\log 1} \cdot \frac{1}{x}$$
$$= \frac{\log e}{1} \cdot \frac{1}{x}$$
$$= \frac{\log e}{x}$$

53. If $y = \log_b u$ where u is a function of x, then

$$y' = \frac{dy}{dx} = \frac{\log_b e}{u} \cdot \frac{du}{dx}$$

By the chain rule,

$$\frac{dy}{dx} = \frac{dy}{du} \cdot \frac{du}{dx} \text{ but } \frac{dx}{du} = \frac{\log_b e}{u}$$
$$y' = \frac{\log_b e}{u} \cdot \frac{du}{dx}$$

5.5 Review

Objective Questions

1. $(5^3)^{2/3} = 5^2 = 25$ 3. $\frac{27^{2/3}}{27^{1/2}} = 27^{2/3-1/2} = 27^{1/6} = (3^3)^{1/6} = 3^{1/2} = \sqrt{3}$

5.

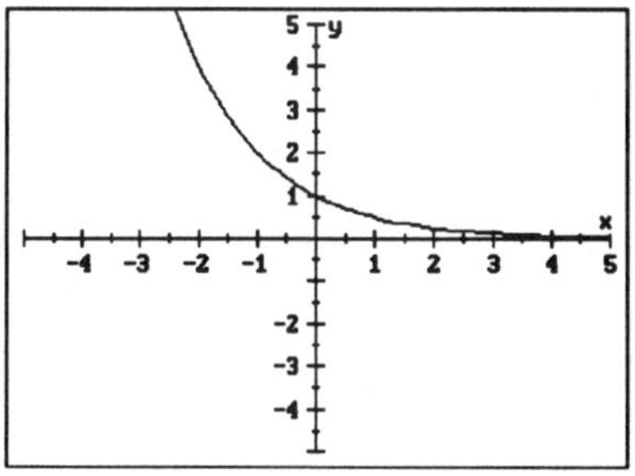

7.

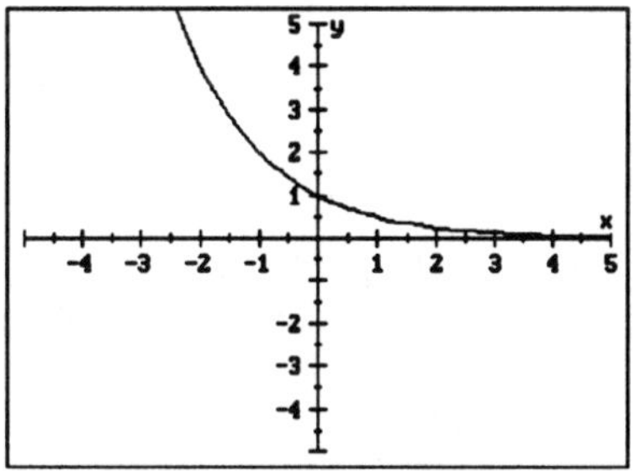

9. $e \approx 2.71828$ **11.** 2.8576511 **13.** $\log\sqrt{10} = .5$ **15.** $\log_9 729 = 3$

17. $10^0 = 1$ **19.** $2^6 = 64$ **21.** 1.0986123 **23.** -2.6777807

25.

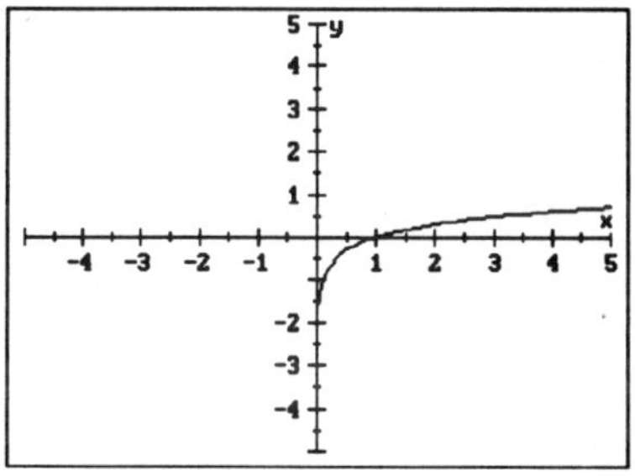

27. Some points are:

Let $x = \frac{1}{5}$, $y = \log_5(\frac{1}{5})$

$= \log_5 5^{-1} = -1$

Let $x = 1$, $y = \log_5(1) = 0$

Let $x = 5$, $y \log_5(5) = 1$

Let $x = \sqrt{5} \approx 2.236$, y

$= \log_5(\sqrt{5}) = \log_5 5^{1/2} = \frac{1}{2}$

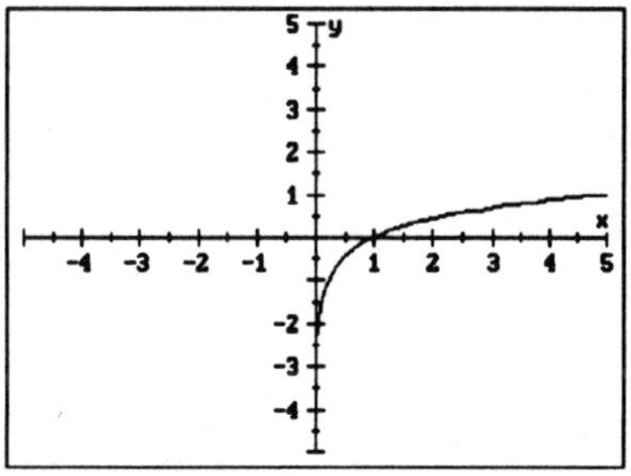

29. $x = 2$

$5x = 25 = 5^2$

31. $3 \log 3 - \frac{1}{2} \log 3 = \log\sqrt{x}$

$6 \log 3 - \log 3 = \log x$

$\log \frac{3^6}{3} = \log x$

$x = 3^5$

33. $x + 2 = \log 125$

$x = -2 + \log 125$

$\approx .09691$

35. $e^{4 - 3x} = 15$

$4 - 3x = \ln 15$

$3x = 4 - \ln 15$

$x = \frac{4 - \ln 15}{3} \approx .43065$

37. $\frac{dy}{dx} = 4.9xe^{-.05x^2}(-.1x)$

$= -.49xe^{-.05x^2}$

39. $y' = x^3(e^{x^4})' + (x^3)'(e^{x^4})$

$= x^3(4x^3e^{x^4}) + (3x^2)e^{x^4}$

$= x^2e^{x^4}(4x^4 + 3)$

41. $y' = \frac{1}{x^2}(2x)$

$= \frac{2}{x}$

43. $$\frac{dy}{dx} = \frac{(x+3)\frac{d}{dx}(\ln x^2) - \ln x^2 \frac{d}{dx}(x+3)}{(x+3)^2}$$

$$= \frac{(x+3)\frac{1}{x^2}\cdot 2x - \ln x^2(1)}{(x+3)^2}$$

$$= \frac{2(x+3) - x\ln x^2}{x(x+3)^2}$$

$$= \frac{2x+6 - x\ln x^2}{x(x+3)^2}$$

45. $A = 30{,}000e^{.082(15)} \approx 102{,}636.89$

47. $t = 5{,}700 \log_{1/2}\frac{A}{10}$

49. $P'(1) \approx -2.5023$ psi per mile

51. Equilibrium when

$$p(x) = s(x)$$
$$300e^{-.3x} = 30e^{.1x}$$
$$10e^{-.3x} = e^{.1x}$$
$$10 = e^{.1x}e^{.3x}$$
$$10 = e^{.4x}$$
$$.4x = \ln(10)$$
$$x = \frac{\ln(10)}{.4} \approx 5.756;$$

This is 5,756 items.

Test Questions

1. In exponential form the equation is $6^x = 36$, so $x = 2$.

3.

$2\log 2 - \frac{1}{2}\log 2 = \log\sqrt{x}$	*Given*
$\frac{3}{2}\log 2 = \log\sqrt{x}$	*Combining like terms*
$\log 2^{3/2} = \log\sqrt{x}$	*Power rule for logarithms*
$2^{3/2} = \sqrt{x}$	

$$\sqrt{2^3} = \sqrt{x}$$
$$2^3 = x$$
$$x = 8$$

5. In logarithmic form: $\log 250 = x - 1$

$$\log 250 + 1 = x$$
$$2.39794 + 1 = x$$
$$x = 3.39794$$

7. In logarithmic form: $2x + 3 = \ln 10$

$$x = \frac{\ln(10) - 3}{2} = \frac{2.3026 - 3}{2}$$
$$\approx -.348707$$

9. $y' = 5.5(e^{-.5x^2})' = 5.5(e^{-.5x^2})(-.5x^2)'$

$$= 5.5(e^{-.5x^2})(-1x)$$
$$= -5.5xe^{-.5x^2}$$

11. First, use the power rule for logarithms: $y = 5 \ln x$ so $y' = 5(\ln x)' = 5(\frac{1}{x}) = \frac{5}{x}$.

13. First use the product rule for logarithms:

$$y = \ln\left|x^2(4 - x^3)\right| = \ln\left|x^2\right| + \ln(4 - x^3)$$

Now, $y' = \frac{1}{x^2}(x^2)' + \frac{1}{4 - x^3}(4 - x^3)'$

$$= \frac{1}{x^2}(2x) + \frac{1}{4 - x^3}(-3x^2)$$
$$= \frac{2}{x} + \frac{-3x^2}{4 - x^3}$$
$$= \frac{2(4 - x^3)}{x(4 - x^3)} + \frac{-3x^3}{x(4 - x^3)}$$
$$= \frac{8 - 5x^3}{x(4 - x^3)}$$

15. $y' = \dfrac{(4 - x)(\ln x^2)' - (\ln x^2)(4 - x)'}{(4 - x)^2}$

$$= \frac{(4 - x)\left(\frac{1}{x^2}\cdot 2x\right) - \ln x^2(-1)}{(4 - x)^2}$$
$$= \frac{(4 - x)(2) + x \ln x^2}{x(4 - x)^2}$$

$$= \frac{8 - 2x + x \ln x^2}{x(4 - x)^2}$$

17. $A = Pe^{rt}$

$\frac{A}{P} = e^{rt}$

$rt = \ln \frac{A}{P}$

$r = \frac{1}{t} \ln \frac{A}{P}$

19. Find the equilibrium point by solving $p(x) = s(x)$

$200e^{-0.2x} = 20e^{0.1x}$

$200e^{-0.2x} - 20e^{0.1x} = 0$ *Divide by 20 and factor* $e^{0.1x}$

$e^{0.1x}(10e^{-0.3x} - 1) = 0$

$e^{0.1x} = 0$ $10e^{-0.3x} - 1 = 0$

No solution $e^{-0.3x} = 0.1$

$-0.3x = \ln 0.1$

$x \approx 7.6752836$

This is about 7,675 units.

CHAPTER 6

The Integral

6.1 The Antiderivative

1. $\int x^7\,dx = \frac{x^8}{8} + C$

3. $\int 4x^3\,dx = 4x^3 dx = 4\left(\frac{x^4}{4}\right) + C = x^4 + C$

5. $\int 3\,dx = 3\int dx = 3x + C$

7. $\int (5x + 7)dx = 5\int x\,dx + 7\int dx = 5\left(\frac{x^2}{2}\right) + 7x + C = \frac{5x^2}{2} + 7x + C$

9. $\int dx = x + C$

11. $\int (18x^2 - 6x + 5)dx = 18\int x^2\,dx - 6\int x\,dx + 5\int dx$

$= 18\left(\frac{x^3}{3}\right) - 6\left(\frac{x^2}{2}\right) + 5x + C = 6x^3 - 3x^2 + 5x + C$

13. $\int \frac{dx}{9} = \frac{1}{9}\int dx = \frac{1}{9}x + C$

15. $\int \frac{5\,dx}{x^2} = 5\int x^{-2}dx = 5\left(\frac{x^{-1}}{-1}\right) + C = -5x^{-1} + C$

17. $\int (1 - e^x)dx = \int dx - \int e^x\,dx = x - e^x + C$

19. $\int (\sqrt[3]{x} + \sqrt{2})dx = \int x^{1/3}\,dx + \sqrt{2}\int dx = \frac{x^{4/3}}{\frac{4}{3}} + \sqrt{2}x + C = \frac{3}{4}x^{4/3} + \sqrt{2}x + C$

21. $\int \frac{3x - 1}{\sqrt{x}}\,dx = \int \frac{3x}{\sqrt{x}}\,dx - \int \frac{1}{\sqrt{x}}\,dx = 3\int x^{1/2}\,dx - \int x^{-1/2}\,dx$

$= 3\left(\frac{x^{3/2}}{\frac{3}{2}}\right) - \frac{x^{1/2}}{\frac{1}{2}} + C = 2x^{3/2} - 2x^{1/2} + C$

23. $\int \frac{x}{\sqrt[3]{x}}\,dx = \int x^{2/3}\,dx = \frac{x^{5/3}}{\frac{5}{3}} + C = \frac{3}{5}x^{5/3} + C$

25. $\int \frac{x^2 + x + 1}{x}\, dx = \int \left(\frac{x^2}{x} + \frac{x}{x} + \frac{1}{x}\right) dx = \int \left(x + 1 + \frac{1}{x}\right) dx = \frac{x^2}{2} + x + \ln|x| + C$

27. $\int y^3\sqrt{y}\, dy = \int y^{7/2}\, dy = \frac{y^{9/2}}{\frac{9}{2}} + C = \frac{2}{9}y^{9/2} + C$

29. $\int (1 + z^2)^2\, dz = \int (1 + 2z^2 + z^4)\, dz = z + \frac{2z^3}{3} + \frac{z^5}{5} + C$

31. $\int (3u^2 - u^{-1} + e^u)\, du = \frac{3u^3}{3} - \ln|u| + e^u + C = u^3 - \ln|u| + e^u + C$

33. $F(x) = \int (3x^2 + 4x + 1)\ dx = x^3 + 2x^2 + x + C;\ 10 = F(0)$

$= 0^3 + 2(0)^2 + 0 + C$; hence $C = 10$, and $F(x) = x^3 + 2x^2 + x + 10$

35. $F(x) = \int \frac{5x + 1}{x}\, dx = \int (5 + \frac{1}{x})\, dx = 5x + \ln|x| + C$;

$5{,}000 = F(1) = 5(1) + \ln|1| + C = 5 + 0 + C$;

Solving for C gives $C = 4{,}995$; hence $F(x) = 5x + \ln|x| + 4{,}995$

37. $C(x) = \int (4x + 8)\, dx = 2x^2 + 8x + C;\ 5{,}000 = C(0) = 2(0)^2 - 8(0) + C.$

Solving for C gives $C = 5{,}000$; hence $C(x) = 2x^2 + 8x + 5{,}000$.

39. $C(x) = \int .009x^2\, dx = .009\frac{x^3}{3} + C = .003x^3 + C;\ 18{,}500 = C(0) = .003(0)^3 + C.$

Solving for $C = 18{,}5000$; hence $C(x) = .003x^3 + 18{,}500$.

41. If $P(t)$ is the population at time t, we are given $P'(t) = 450 + 600\sqrt{t}$.

$P(t) = \int P'(t)\, dt = \int (450 + 600t^{1/2})\, dt = 450t + \frac{600t^{3/2}}{\frac{3}{2}} + C = 450t + 400t^{3/2} + C$

Presently the population is 420,000; which means

$420{,}000 = P(0) = 450(0) + 400(0)^{3/2} + C.$

Solving for C gives $C = 420{,}000$; hence $P(t) = 450t + 400t^{3/2} + 420{,}000$.

In 5 years, $P(5) = 450(5) + 400(5)^{3/2} + 420{,}000 = 426{,}722$.

43. $P(x) = \int (200x - 10{,}000)\, dx = \frac{200x^2}{2} - 10{,}000x + C = 100x^2 - 10{,}000x + C.$

We are given $-50{,}000 = P(0) = 100(0)^2 - 10{,}000(0) + C.$

Solving for C gives $C = -50{,}000$; hence $P(x) = 100x^2 - 10{,}000x - 50{,}000$.

45. If $N(t)$ is the number of cases at time t (in days), we are given

$N'(t) = 240t - 3t^2$.

$N(t) = \int (240t - 3t^2)\, dt = \frac{240t^2}{2} - \frac{3t^3}{3} + C = 120t^2 - t^3 + C.$

We are given $50 = N(1) = 120(1)^2 - 1^3 + C$.

Solving for C gives $C = -69$; hence $N(t) = 120t^2 - t^3 - 69$, and on the tenth day $N(10) = 120(10)^2 - 10^3 - 69 = 10{,}931$ (cases)

47. $v(t) = \int a(t)\, dt = \int (-32)\, dt = -32t + C$

Initial velocity is -72: $-72 = v(0) = -32(0) + C$.

Solving for C gives $C = -72$, hence $v(t) = -32t - 72$.

49. The curve is $f(x) = \int f'(x)\, dx = \int (3x^2 + 5)dx = \frac{3x^2}{3} + 5x + C = x^3 + 5x + C.$

Since (1, 2) is on the curve, $2 = f(1) = 1^3 + 5(1) + C = 6 + C$.

Solve for C: $C = -4$; hence $f(x) = x^3 + 5x - 4$.

6.2 Integration by Substitution

1. Let $u = 5x + 3$, then $du = 5dx$:

$$\int (5x+3)^3\, dx = \int u^3 \frac{du}{5} = \frac{1}{5}\int u^3\, du = \frac{1}{5}\left(\frac{u^4}{4}\right) + C = \frac{(5x+3)^4}{20} + C$$

3. Let $u = 5 - x$, then $du = -dx$:

$$\int \frac{5}{(5-x)^4}\, dx = 5\int (5 - x)^{-4}\, dx = 5\int u^{-4}\,(-du) = -5\int u^{-4}\, du$$

$$= \frac{-5u^{-3}}{-3} + C = \frac{5}{3u^3} + C = \frac{5}{3(5-x)^3} + C$$

5. Let $u = 3x + 5$, then $du = 3dx$:

$$\int \sqrt{3x+5}\, dx = \int \sqrt{u}\left(\frac{du}{3}\right) = \frac{1}{3}\int u^{1/2}\, du = \frac{1}{3}\cdot\frac{u^{3/2}}{\frac{3}{2}} + C = \frac{2u^{3/2}}{9} + C$$

$$= \frac{2(3x+5)^{3/2}}{9} + C$$

7. Let $u = 3x^2 + 1$, then $du = 6x\, dx$:

$$\int 6x(3x^2+1)dx = \int 6x(u)(\tfrac{du}{6x}) = \int u\ du = \tfrac{u^2}{2} + C = \frac{(3x^2+1)^2}{2} + C$$

9. Let $u = x^2 + 5x$, then $du = (2x+5)dx$:

$$\int \frac{2x+5}{\sqrt{x^2+5x}}\ dx = \int \frac{2x+5}{\sqrt{u}}\left(\frac{du}{2x+5}\right) = \int \tfrac{1}{\sqrt{u}}\ du = \int u^{-1/2}\ du = \frac{u^{1/2}}{\frac{1}{2}} + C$$

$$= 2\sqrt{u} + C = 2\sqrt{x^2+5x} + C$$

11. Let $u = 6 + 5x$, then $du = 5\ dx$:

$$\int \frac{dx}{6+5x} = \int \frac{\left(\frac{du}{5}\right)}{u} = \tfrac{1}{5}\int \tfrac{du}{u} = \tfrac{1}{5}\ln|u| + C = \tfrac{1}{5}\ln|6+5x| + C$$

13. Let $u = 1 - 3x^2$, then $du = -6x\ dx$:

$$\int \frac{x\ dx}{1-3x^2} = \int \frac{x\left(\frac{du}{-6x}\right)}{u} = -\tfrac{1}{6}\int \tfrac{du}{u} = -\tfrac{1}{6}\ln|u| + C = -\tfrac{1}{6}\ln|1-3x^2| + C$$

15. Let $u = 5x$, then $du = 5x$:

$$\int e^{5x}\ dx = \int e^u\left(\frac{du}{5}\right) = \tfrac{1}{5}\int e^u\ du = \tfrac{1}{5}e^u + C = \tfrac{1}{5}e^{5x} + C$$

17. Let $u = 4x^3$, then $du = 12x^2dx$:

$$\int 5x^2e^{4x^3}\ dx = \int 5x^2e^u\left(\frac{du}{12x^2}\right) = \tfrac{5}{12}\int e^u\ du = \tfrac{5}{12}e^u + C = \tfrac{5}{12}e^{4x^3} + C$$

19. Let $u = 4x^2 - 4x$, then $du = (8x-4)dx = 4(2x-1)dx$

$$\int \frac{2x-1}{(4x^2-4x)^2}dx = \int \frac{2x-1}{u^2}\left[\frac{du}{4(2x-1)}\right] = \frac{1}{4}\int u^{-2}\ du = \frac{1}{4}\left(\frac{u^{-1}}{-1}\right) + C = -\frac{1}{4u} + C$$

$$= -\frac{1}{4(4x^2-4x)} + C$$

21. Let $u = x^4 - 2x + 3$, then $du = (4x^3 - 4x)dx$:

$$\int \frac{4x^3-4x}{x^4-2x^2+3}\ dx = \int \frac{4x^3-4x}{u}\left(\frac{du}{4x^3-4x}\right) = \int \frac{du}{u} = \ln|u| + C$$
$$= \ln|x^4 - 2x^2 + 3| + C$$

23. Let $u = \ln|x|$, then $du = \frac{1}{x}\ dx$:

$$\int \frac{\ln|x|}{x}\ dx = \int \tfrac{u}{x}(x du) = \int u\ du = \frac{u^2}{2} + C = \frac{\ln^2|x|}{2} + C$$

25. $\ln e^x = x$, so $\int \ln e^x\,dx = \int x\,dx = \frac{x^2}{2} + C$

27. Let $u = 4x^2 + 1$, then $du = 8x\,dx$. Also let $v = x^3$, then $dv = 3x^2\,dx$:

$$\int\left(\frac{3x}{4x^2+1} + x^2e^{x^3}\right)dx = \int\frac{3x}{4x^2+1}\,dx + \int x^2e^{x^3}\,dx = \int\frac{3x}{u}\left(\frac{du}{8x}\right) + \int x^2e^v\left(\frac{dv}{3x^2}\right)$$

$$= \frac{3}{8}\int\frac{du}{u} + \frac{1}{3}\int e^v\,dv = \frac{3}{8}\ln|u| + \frac{1}{3}e^v + C = \frac{3}{8}\ln(4x^2+1) + \frac{1}{3}e^{x^3} + C$$

(Since $4x^2 + 1 > 0$, the absolute value symbols may be dropped.)

29. Let $u = x^2 + 1$, then $du = 2x\,dx$:

$$\int x\sqrt[3]{(x^2+1)^2}\,dx = \int x\sqrt[3]{u^2}\left(\frac{du}{2x}\right) = \frac{1}{2}\int u^{2/3}\,du = \frac{1}{2}\,\frac{u^{5/3}}{\frac{5}{3}} + C = \frac{3}{10}(x^2+1)^{5/3} + C$$

31. Let $u = \sqrt{x+1}$, then $u^2 = x + 1$, or $u^2 - 1 = x$.

Differentiate implicitly: $(u^2 - 1)' = x'$

$$2u\frac{du}{dx} = 1$$

$$\int\frac{x\,dx}{\sqrt{x+1}} = \int\frac{(u^2-1)}{u}(2u\,du)$$

$$= 2\int(u^2 - 1)\,du$$

$$= 2\left(\frac{u^3}{3} - u\right) + C$$

$$= 2\left(\frac{\sqrt{x+1}^3}{3} - \sqrt{x+1}\right) + C = \frac{2}{3}(x+1)^{3/2} - 2\sqrt{x+1} + C$$

33. **a.** $\int x^2 + 1\,dx$ $= \frac{x^3}{3} + x + \mathrm{C}$

b. $\int(x^2+1)x\,dx$ $= \int(x^3 + x)\,dx$ $= \frac{x^4}{4} + \frac{x^2}{2} + C$

c. $\int(x^2+1)x^2\,dx$ $= \int(x^4 + x^2)\,dx$ $= \frac{x^5}{5} + \frac{x^3}{3} + C$

d. $\int(x^2+1)x^3\,dx$ $= \int(x^5 + x^3)\,dx$ $= \frac{x^6}{6} + \frac{x^4}{4} + C$

35. **a.** $\int(x^2+1)^3\,dx$

$$= \int(x^6 + 2x^4 + 3x^2 + 1)\,dx$$

$$= \frac{x^7}{7} + \frac{3x^5}{5} + x^3 + x + \mathrm{C}$$

b. $\int(x^2+1)^3x\,dx$

$$= \frac{1}{2}\int u^3\,du$$

$$= \frac{1}{2}\cdot\frac{u^4}{4} + \mathrm{C}$$

$$= \frac{1}{8}(x^2+1)^4 + \mathrm{C}$$

c. $\int (x^2+1)^3 x^2\,dx$

$= \int (x^8+3x^6+3x^4+1)\,dx$

$= \frac{x^9}{9}+\frac{3x^7}{7}+\frac{3x^5}{5}+\frac{x^3}{3}+C$

d. $\int (x^2+1)^3 x^3\,dx$

$= \frac{1}{2}\int u^3(u-1)\,du$

$= \frac{1}{2}\int (u^4-u^3)\,du$

$= \frac{1}{2}\left(\frac{u^5}{5}-\frac{u^4}{4}\right)+C$

$= \frac{(x^2+1)^5}{10}-\frac{(x^2+1)^4}{8}+C$

37. a. $\int (x^3+1)\,dx$

$= \frac{x^4}{4}+x+C$

b. $\int (x^3+1)x\,dx$

$= \int (x^4+x)\,dx$

$= \frac{x^5}{5}+\frac{x^2}{2}+C$

c. $\int (x^3+1)x^2\,dx$

$= \int (x^5+x^2)\,dx$

$= \frac{x^6}{6}+\frac{x^3}{3}+C$

d. $\int (x^3+1)x^3\,dx$

$= \int (x^6+x^3)\,dx$

$= \frac{x^7}{7}+\frac{x^4}{4}+C$

39. a. $\int (x^3+1)^3\,dx$

$= \int (x^9+3x^6+3x^3+1)\,dx$

$= \frac{x^{10}}{10}+\frac{3x^7}{7}+\frac{3x^4}{4}+x+C$

b. $\int (x^3+1)^3 x\,dx$

$= \int (x^9+3x^6+3x^3+1)x\,dx$

$= \int (x^{10}+3x^7+3x^4+x)\,dx$

$= \frac{x^{11}}{11}+\frac{3x^8}{8}+\frac{3x^5}{5}+\frac{x^2}{2}+C$

c. $\int (x^3+1)^3 x^2\,dx$

$= \frac{1}{3}\int u^3\,du$

$= \frac{1}{3}\,\frac{u^4}{4}+C$

$= \frac{(x^3+1)^4}{12}+C$

d. $\int (x^3+1)^3 x^3\,dx$

$= \int (x^9+3x^6+3x^3+1)x^3\,dx$

$= \int (x^{12}+3x^9+3x^6+x^3)\,dx$

$= \frac{x^{13}}{13}+\frac{3x^{10}}{10}+\frac{3x^7}{7}+\frac{x^4}{4}+C$

41. $P(t) = \int P'(t)\,dt = \int 40t(5-t^2)^2\,dt$ Let $u = 5 - t^2$, then $du = -2t\,dt$

$= \int 40t(u)^2\left(\frac{du}{-2t}\right) = -20\int u^2\,du = -20\left(\frac{u^3}{3}\right) + C$

$$= -\tfrac{20}{3}(5 - t^2)^3 + C$$

Now, $10 = P(0) = -\tfrac{20}{3}(5 - 0^2)^3 + C$

$$10 = -833\tfrac{1}{3} + C$$

$$\frac{2{,}530}{3} = C$$

So, $P(t) = -\tfrac{20}{3}(5 - t^2)^3 + \tfrac{2{,}530}{3} = \tfrac{20}{3}(t^2 - 5)^3 + \tfrac{2{,}530}{3}$

43. $P(x) = \displaystyle\int \frac{100}{\sqrt[3]{(x^2 - 36)^2}}\,dx$ Let $u = x^2 - 36$, then $du = 2x\,dx$

$$= \int \frac{100x}{\sqrt[3]{u^2}}\left(\frac{du}{2x}\right) = \int \frac{50\,du}{\sqrt[3]{u^2}} = 50\int u^{-2/3}\,du = 50\frac{u^{1/3}}{\frac{1}{3}} + C$$

$$= 150(x^2 - 36)^{1/3} + C$$

Now, $0 = P(10) = 150(10^2 - 36)^{1/3} + C$

$$0 = 150(4) + C$$

$$-600 = C$$

So $P(x) = 150(x^2 - 36)^{1/3} - 600$, and specifically

$P(100) = 150(100^2 - 36)^{1/3} - 600 = \$2{,}627.77$

45. $S(t) = \displaystyle\int 2{,}000e^{-.2t}\,dt$ Let $u = -.2t$, then $du = -.2dt$

$$= 2{,}000\int e^u\left(\frac{du}{-.2}\right) = -10{,}000\int e^u\,du = -10{,}000e^u + C$$

$$= -10{,}000e^{-.2t} + C$$

Now $15{,}000 = S(0) = -10{,}000e^0 + C = -10{,}000 + C$

$$25{,}000 = C$$

So $S(t) = -10{,}000e^{-.2t} + 25{,}000$. In 3 years:

So $S(3) = -10{,}000e^{-.2(3)} + 25{,}000 \approx \$19{,}511.88$

47. $P(t) = \displaystyle\int 14{,}000e^{.0175t}\,dt$ Let $u = .0175t$, $du = .0175t$, $du = .0175dt$

$$= 14{,}000\int e^u\left(\frac{du}{.0175}\right) = \frac{14{,}000}{.0175}\int e^u\,du = 800{,}000(e^u) + C$$

$$= 800{,}000e^{.0175t} + C$$

We are given in 1984 ($t = 4$) the population is ($P = 842{,}779$):

$$842{,}779 = P(4) = 800{,}000e^{0175(4)} + C \;\; = 858{,}006.54 + \text{C}$$

$$-15{,}227.54 = \text{C}$$

So $P(t) = 800{,}000e^{.0175t} - 15{,}227.54$.

In the year 2000 ($t = 20$): $P(20) = 800{,}000e^{.0175(20)} - 15{,}227.54$

$$\approx 1{,}120{,}026 \text{ people}$$

49. $T(t) = \int 32.5e^{.048t}\, dt$ Let $u = .048t$, $du = .048\ dt$

$$= 32.5\int e^{u}\left(\frac{du}{.048}\right) = \frac{32.5}{.048}e^{u} + C = 677.083e^{.048t} + \text{C}$$

We are given that $1950 = T(0) = 677.083e^{0} + \text{C}$; thus $1{,}273 \approx \text{C}$

So $T(t) = 677.083e^{048t} + 1{,}273$. $T(0)$ is the total consumption from 1925 to 1985; $T(15)$ is the total consumption from 1925 to 2000. Thus, from 1985 to 2000 find $T(15) - T(0)$:

$$T(15) = 677.083e^{.048(15)} + 1273 \approx 2{,}664$$

$$T(15) - T(0) \approx 2{,}664 - 1{,}950 \approx 714$$

6.3 The Definite Integral

1. $\displaystyle\int_2^3 x^2\, dx = \frac{x^3}{3}\Big|_2^3 = \frac{27}{3} - \frac{8}{3} = \frac{19}{3}$

3. $\displaystyle\int_1^4 \sqrt{x}\, dx = \int_1^4 x^{1/2}\, dx$

$$= \frac{x^{\frac{3}{2}}}{\frac{3}{2}}\Bigg|_1^4$$

$$= \frac{2}{3}\sqrt{x^3}\Big|_1^4$$

$$= \frac{2}{3}(\sqrt{4^3} - \sqrt{1^3}) = \frac{2}{3}(\sqrt{4^3} - \sqrt{1^3})$$

$$= \frac{2}{3}(8 - 1)$$

$$= \frac{14}{3}$$

5. $\int_{-4}^{-1} \sqrt{1 - 2x}\, dx = \int_{9}^{3} u^{1/2} \frac{du}{-2} = -\frac{1}{2}u^{3/2}(\frac{2}{3})\Big|_9^3 = -\frac{1}{3}(3\sqrt{3} - 27) = 9 - \sqrt{3}$

7. $\int_0^2 e^{2x}\, dx = \frac{1}{2}\int_0^4 e^u\, du = \frac{1}{2}e^u\Big|_0^4 = \frac{1}{2}(e^4 - 1)$

9. $\int_1^{10} x^{-1}dx = \ln|x|\,\Big|_1^{10} = \ln 10$

11. $\int_1^{32} x^{-2/5}dx = \frac{x^{3/5}}{\frac{3}{5}}\Bigg|_1^{32} = \frac{5}{3}(32^{3/5} - 1^{3/5}) = \frac{5}{8}(8 - 1) = \frac{35}{5}$

13. Let $u = x - 2$, then $du = dx$. Also if $x = 3$ then $u = 1$, and if $x = 8$ then $u = 6$:

$$\int_{x=3}^{x=8} (x - 2)^{-1}dx = \int_{u=1}^{u=6} u^{-1}du = \ln|u|\,\Big|_1^6 = \ln 6 - \ln 1 = \ln 6$$

15. $\int_{-1}^{2} x(1 + x^4)\, dx = \int_{-1}^{2} (x + x^5)\, dx = \frac{x^2}{2} + \frac{x^2}{6}\Big|_{-1}^{2} = (\frac{4}{2} + \frac{64}{6}) - (\frac{1}{2} + \frac{1}{6}) = 12$

17. 0 (Directly since the upper and lower limits of integration are the same)

19. Let $u = 2x + 1$, then $du = 2dx$. Also, if $x = 0$ then $u = 1$, and if $x = 1$ then $u = 3$: $\int_{x=0}^{x=1} \frac{dx}{(2x + 1)^2} = \int_{u=1}^{u=3} \frac{\frac{du}{2}}{u^2} = \frac{1}{2}\int_1^3 u^{-2}\, du = -\frac{1}{2}u^{-1}\Big|_1^3$

$= -\frac{1}{2}(\frac{1}{3} - 1) = \frac{1}{3}$

21. $\int_1^2 (x^{-2} - \frac{2}{x} + x^3)\, dx = \frac{x^{-1}}{-1} - 2\ln|x| + \frac{x^4}{4}\Big|_1^2$

$= (-\frac{1}{2} - 2\ln 2 + 4) - (-1 - 2\ln 1 + \frac{1}{4}) = \frac{17}{4} - 2\ln 2 = \frac{17}{4} - \ln 4$

23. Since $\int_a^a f(x)\, dx = 0$, $\int_6^6 (x^2 + \frac{\sqrt{5x}}{3} - 5)\, dx = 0$

25. Let $u = 1 + 3x^2$, then $du = 6x\, dx$. Also, if $x = -1$ then $u = 4$, and if $x = 1$

then $u = 4$: $\int_{-1}^{1} \frac{x\,dx}{\sqrt{1+3x^2}} = \int_{4}^{4} \frac{\frac{du}{6}}{u^{1/2}} = 0$ $\left(\text{Since } \int_{a}^{a} f(u)\,du = 0\right)$

27. Let $u = x^3 + 1$, then $du = 3x^3\,dx$. Also, if $x = -1$ then $u = 0$, and if $x = 0$ then $u = 1$:

$$\int_{-1}^{0} 5x^2(x^3+1)^{10}\,dx = \int_{0}^{1} 5x^2u^{10}\frac{du}{3x^2} = \frac{5}{3}\int_{0}^{1} u^{10}du = \left(\frac{5}{3}\frac{u^{11}}{11}\Big|_{0}^{1}\right) = \frac{5}{3}(\frac{1}{11} - 0) = \frac{5}{33}$$

29. Let $u = \ln x$; $du = \frac{1}{x}\,dx$; $x = e \Rightarrow u = 1$; $x = 1 \Rightarrow u = 0$:

$$\int_{0}^{e} 3x^{-1}\ln^3 x\,dx = 3\int_{0}^{1} u^3\,du = \frac{3u^4}{4} = \frac{3}{4} - 0 = \frac{3}{4}$$

31. **a.** $\int_{-3}^{2} f(x)dx = \int_{-3}^{0} f(x)dx + \int_{0}^{2} f(x)dx = 4 + 10 = 14$

b. $\int_{2}^{-3} g(x)dx = -\int_{-3}^{2} g(x) = -(-5) = 5$

c. $\int_{-3}^{2} [2f(x) - 3g(x)]dx = 2\int_{-3}^{2} f(x)dx - 3\int_{-3}^{2} g(x)dx = 2(14) - 3(-5) = 43$

33. $\int_{3}^{5} f(y)\,dy = \int_{3}^{1} f(y)\,dy + \int_{1}^{5} f(y)\,dy = -\int_{1}^{3} f(y)dy + \int_{1}^{5} f(y)dy = -9 + 15 = 6$

35. $\int_{2}^{15} 78e^{-.04}\,dt = \frac{78e^{-.04t}}{-.04}\Big|_{2}^{15} = -1950e^{-.04t}\Big|_{2}^{3} \approx 729.89419$. The total consumption is 730 billion barrels.

37. $\int_{1}^{14} 32.5e^{.048t}\,dt = \frac{32.5e^{.048t}}{.048}\Big|_{1}^{14} = 615.45457$. The total consumption is 615 billion barrels.

39. $\int_{1}^{x} 32.5e^{.048t}\,dt = \frac{32.5e^{.048t}}{.048}\Big|_{0}^{x} = \frac{32.5e^{.048t}}{.048} - \frac{32.5}{.048} = \frac{32.5e^{.048x}}{.048} - \frac{32.5}{.048} =$ 670.3; solve for x to find $x = 14.36$ About 14 years from 1985 or 1999.

41. $\int_{0}^{5} 350t^{5/2}\,dt = \frac{350t^{7/2}}{\frac{7}{2}}\Big|_{0}^{5} = 100t^{7/2}\Big|_{0}^{5} = 27{,}950.85$

The pollution level is 28,000 ppm.

43. $\int_2^3 350t^{5/2}\,dt = 100t^{7/2}\Big|_2^3 = 3545.1663$ The pollution level is 3,500 ppm.

45. $\int_0^{10} 50e^t\,dt = 50e^t = 50(e^{10} - e^0) \approx \$1{,}101{,}273.29$

47. $G = 1 - \int_0^1 2x^3\,dx = 1 - (\frac{x^4}{2}) = 1 - (\frac{1}{2} - 0) = \frac{1}{2}$

49.

$$\begin{aligned} C(t) &= \int \frac{t}{20}\,dt \\ &= \frac{1}{20}\cdot\frac{t^2}{2} + C \\ &= \frac{1}{40}t^2 + C \end{aligned}$$

$$C(0) = C = 25{,}000$$

Thus,

$$C(t) = \frac{1}{40}t^2 + 25{,}000$$

$$\begin{aligned} R(t) &= \int 10te^{-t^2} \\ &= 10\int te^u \frac{du}{-2t} \\ &= -5e^u + C \\ &= -5e^{-t^2} + C \end{aligned}$$

$$R(0) = -5 + C$$

Thus,

$$R(t) = -5e^{\cdot t} + 5$$

6.4 Area Between Curves

1. Since $f(x) > g(x)$ on $[a, b]$, the area $= \int_a^b [f(x) - g(x)]\,dx$

3. Since $g(x) > f(x)$ on $[a, b]$, the area $= \int_a^b [g(x) - f(x)]\,dx$

5. Since $f(x) > g(x)$ on $[a, b]$, the area $= \int_a^b [f(x) - g(x)]\,dx$

7. $\int_1^5 6x^2\,dx = 2x^3 \Big|_1^5 = 248$

9. $\int_{-1}^2 (3x^2 + 4)\,dx = (x^3 + 4x)\Big|_{-1}^2 = (8 + 8) - (-1 - 4) = 21$

11. $4x^3 - 3x^2 + 2x = 0$

$x(4x^2 - 3x + 2) = 0$

$$x = \frac{3 \pm \sqrt{9 - 4(4)(2)}}{2} \text{ or } x = 0$$

$x = 0$ only critical value

$$\int_0^2 (4x^3 - 3x^2 + 2x)\,dx = (x^4 - x^3 + x^2)\Big|_0^2 = 12$$

13. $\displaystyle\int_{-2}^{1} e^{-x}\,dx = -e^{-x}\Big|_{-2}^{1} = -e^{-1} + e^2 = e^2 - e^{-1}$

15. $x^2 - 9 = 0$

$(x - 3)(x + 3) = 0$

$x = 3, -3$

$$\int_{-3}^{3} 0 - (x^2 - 9)dx \int_{-3}^{3} (9 - x^2)dx = 9x - \frac{x^3}{3}\Big|_{-3}^{3} = 18 - (-18) = 36$$

17. $x^2 = 6 - x$

$x^2 + x - 6 = 0$

$(x + 3)(x - 2) = 0$

$x = -3, 2$

$$\int_3^5 [x^2 - (6 - x)]\,dx = \left(\frac{x^3}{3} - 6x + \frac{x^2}{2}\right)\Big|_3^5 = 24\tfrac{1}{6} + 4\tfrac{1}{2} = \frac{86}{3}$$

19. $x^3 = x^2$

$x^3 - x^2 = 0$

$x^2(x - 1) = 0$

$x = 0, 1$

$$\int_0^1 (x^2 - x^3)\,dx + \int_1^2 (x^3 - x^2)\,dx = \left(\frac{x^3}{3} - \frac{x^4}{4}\right)\Big|_0^1 + \left(\frac{x^4}{4} - \frac{x^3}{3}\right)\Big|_0^1$$

$$= \left(\frac{1}{3} - \frac{1}{4}\right) + \left(4 - \frac{8}{3} - \frac{1}{4} + \frac{1}{3}\right) = \frac{3}{2}$$

21. $x^2 = 2x + 4$

$x^2 - 2x - 4 = 0$

$$x = \frac{2 \pm \sqrt{4 - 4(-4)}}{2}$$

$$= \frac{2 \pm 2\sqrt{5}}{2}$$

$= \frac{1 \pm \sqrt{5}}{2}$

These points are not on [0, 1]

$$\int_0^1 [(2x + 4) - x^2]\, dx = \left(x^2 + 4x - \frac{x^3}{3}\right)\Bigg|_0^1 = 1 + 4 - \frac{1}{3} = \frac{14}{3}$$

23. $x^2 + 4 = 2x + 4$

$x^2 - 2x = 0$

$x(x - 2) = 0$

$x = 0,\ 2$

$$\int_1^2 [(2x + 4) - (x^2 + 4)]\, dx + \int_2^5 [(x^2 + 4) - (2x + 4)]\, dx$$

$$= \int_1^2 (2x - x^2)\, dx + \int_2^5 (x^2 - 2x)\, dx$$

$$= \left(x^2 - \frac{x^3}{3}\right)\Bigg|_1^2 + \left(\frac{x^3}{3} - x^2\right)\Bigg|_2^5$$

$$= 4 - \frac{1}{3} - 1 + \frac{1}{3} + \frac{125}{3} - 25 - \frac{8}{3} + 4$$

$$= \frac{56}{3}$$

25. $x = x^4$

$x^4 - x = 0$

$x(x^3 - 1) = 0$

$x(x - 1) = 0$

$x(x - 1)(x^2 + x + 1) = 0$

$x = 0,\ 1$

$$\int_0^1 (x - x^4)\, dx = \left(\frac{x^2}{2} - \frac{x^5}{5}\right)\Bigg|_0^1$$

$$= \frac{1}{2} - \frac{1}{5}$$

$$= \frac{3}{10}$$

27. $x^2 - 5x = 6$

$x^2 - 5x - 6 = 0$

$(x - 6)(x + 1) = 0$

$x = 6,\ -1$

$$\int_{-1}^{6} [6 - (x^2 - 5x)\, dx$$

$$= \int_{-1}^{6} (6 + 5x - x^2)\, dx$$

$$= (6x + \frac{5x^2}{2} - \frac{x^3}{3})\Big|_{-1}^{6}$$

$$= (36 + 90 - 72) - (-6 - \tfrac{5}{2} + \tfrac{1}{3})$$

$$= \frac{373}{6}$$

29. $\int_{.01}^{1} x^{-1}\, dx = \ln x \mid_{.01}^{1} \approx 4.6051702$

31. $x = \sqrt{x}$

$x^2 = x$

$x^2 - x = 0$

$x(x - 1) = 0$

$x = 0, 1$

$$\int_{0}^{1} (\sqrt{x} - x)\, dx = (\tfrac{2}{3}x^{3/2} - \tfrac{1}{2}x^2) \Big|_{0}^{1} = \tfrac{2}{3} - \tfrac{1}{2} = \tfrac{1}{6}$$

33. $\int_{0}^{81} (9 - \sqrt{x})\, dx = \left(9x - \tfrac{2}{3}x^{3/2} \right) \Big|_{0}^{81} = 243$

35. $\int_{1}^{2} (e^{.5x} - x^{-1})\, dx = \left(\frac{e^{.5x}}{.5} - \ln|x| \right) \Big|_{1}^{2} = 2e - \ln 2 - \sqrt[2]{e} \approx 1.4459739$

37. $\int_{0}^{25} \frac{1{,}000}{\sqrt[3]{t+5}}\, dt = 1{,}000 \int_{5}^{30} u^{-2/3} du = 1{,}000(3)u^{1/3}\Big|_{5}^{30}$

Let $u = t + 5$; $du = dt = 4191.769678 \approx 4{,}192$ tons

39. $\int_{0}^{30} \frac{100{,}000}{\sqrt{t + 250}}\, dt = 100{,}000 \int_{0}^{30} (t + 250)^{-1/2}\, dt = 100{,}000 \frac{(t + 250)^{1/2}}{\frac{1}{2}}$

$= 200{,}000[(30 + 250)^{1/2} - (0 + 250)^{1/2}] \approx 200{,}000[.921812] = 184{,}362;$

about 184,000 fish are killed.

41. $\int_0^{25} (20 + .012t^2)\,dt = (20t + .004t^3)\Big|_0^{25} = 562.5$

This area represents the total demand 562.5 during the first 25 weeks.

43.
$$18 = 21e^{-.01t}$$
$$-.01t = \ln \tfrac{18}{21}$$
$$t = 15.415068$$

The area represents the accumulated profit.

$$\int_0^{15.415068} (2e^{-.01t} - 18)\,dt = -2100e^{-.01t} - 18t\Big|_0^{15.415068}$$
$$= 22.528776$$

6.5 The Fundamental Theorem of Calculus

1. **a.** $\int_{-2}^{2} (x^2 + 1)\,dx = \left(\frac{x^3}{3} + x\right)\Big|_{-2}^{2} = (\frac{8}{3} + 2) - (-\frac{8}{3} - 2) = \frac{8}{3} + 2 + \frac{8}{3} + 2 = \frac{28}{3}$

b. $\int_{-2}^{2} (x^2 + 1)\,dx \approx \Delta x[f(-2) + f(-1) + f(0) + f(1)] = 1(5 + 2 + 1 + 2) = 10$

$$\Delta x = \frac{2 + 2}{4} = 1$$

3. **a.** $\int_1^4 3x^2\,dx = x^3 \Big|_1^4 = 64 - 1 = 63$

b. $\int_1^4 3x^2\,dx = \frac{3}{4}[f(1) + f(\frac{7}{4}) + f(\frac{10}{4}) + f(\frac{13}{4})] = \frac{3}{4}\cdot(3 + 3\cdot\frac{49}{16} + 3\cdot\frac{100}{16} + 3\cdot\frac{169}{16})$

$\Delta x = \frac{1 - 1}{4} = \frac{3}{4}$ $\qquad = \frac{9}{64}(16 + 49 + 100 + 169)$

$$= \frac{3006}{64} = 46.96875$$

5. **a.** $\int_0^2 (4 - x^2)\,dx = (4x - \frac{x^3}{3})\Big|_0^2 = 8 - \frac{8}{3} = \frac{16}{3} \approx 5.33$

b. $\int_0^2 (4 - x^2)\,dx \approx \Delta x[f(0) + f(\frac{1}{2}) + f(1) + f(\frac{3}{2})]$

$= \frac{1}{2}[4 + (4 - \frac{1}{4}) + 3 + (4 - \frac{9}{4})] = \frac{25}{4} = 6.25$

$\Delta x = \frac{2 - 0}{4} = \frac{1}{2}$

7. $\int_1^5 \frac{1}{(2x + 1)^2}\,dx = \Delta x[f(\frac{3}{2}) + f(\frac{5}{2}) + f(\frac{7}{2}) + f(\frac{9}{2})] = .115902778 \approx .12$

$\Delta x = \frac{5 - 1}{4} = 1$

9. $\int_0^2 e^{.5x} dx \approx \Delta x[f(\frac{1}{4}) + f(\frac{3}{4}) + f(\frac{5}{4}) + f(\frac{7}{4})] \approx 3.427631$

$\Delta x = \frac{2 - 0}{4} = \frac{1}{2}$

11. $\int_0^{1/2} \frac{x}{\sqrt{1 + 3x^2}}\,dx \approx \Delta x[f(\frac{1}{16}) + f(\frac{3}{16}) + f(\frac{5}{16}) + f(\frac{7}{16})] \approx .1079985$

$\Delta x = \frac{\frac{1}{2} - 0}{4} = \frac{1}{8}$

13. $\int_0^6 f(x)\,dx \approx \Delta x[f(1) + f(3) + f(5)] \approx 2[2.4 + 5.5 + 6.1] \approx 28$

$n = 3 \quad \Delta x = 2$

15. $\int_5^{45} f(x)\,dx \approx \Delta x[f(10) + f(20) + f(30) + f(40)] \approx 5{,}700$

$n = 4 \qquad \Delta x = 10$

17. $\frac{1}{4 - 0} \int_0^4 (2x^2 + 1)\,dx = \frac{1}{4}\left(\frac{2x^3}{3} + x\right)\Big|_0^4 = \frac{35}{3} \approx 11.67$

19. $\frac{1}{10 - 0} \int_0^{10} (300 - 4x)\,dx = \frac{1}{10}(300x - 2x^2)\Big|_0^{10} = 280$

21. $\frac{1}{8-1}\int_1^8 x^{1/3}\,dx = \frac{1}{7}\cdot\frac{3}{4}x\Big|_1^8 = \frac{3}{28}(16 - 1) = \frac{45}{28} \approx 1.61$

23. $\frac{1}{10-0}\int_0^{10} e^{-.5x}dx = \frac{1}{10}\int_0^{-5}\frac{e^u du}{-.5} = -\frac{1}{5}e^u\Big|_0^{-5} = -\frac{1}{5}(e^{-5} - 1) \approx .199$

25. $F(x) = \int_0^x (t+5)\,dt$

$= \frac{t^2}{2} + 5t\Big|_0^x$

$= \frac{x^2}{2} + 5x$

$F'(x) = x + 5$

27. $F(x) = \int_{-3}^x (t+5)^2\,dt = \frac{(t+5)^3}{3}\Big|_{-3}^x$

$u = t + 5 \qquad = \frac{(x+5)^3}{3} - \frac{8}{3}$

$du = dt$

$F'(x) = (x+5)^2$

29. $F(x) = \int_4^x \sqrt{t}\,dt = \frac{2}{3}t^{3/2}\Big|_4^x$

$= \frac{2}{3}x^{3/2} - \frac{16}{3}$

$F'(x) = \sqrt{x}$

31. $F'(x) = x^3 - 3x^2 + 5x - 7$

33. $F'(x) = \sqrt[3]{x}$

35. $F'(x) = x^2(x+1)^3$

37. We want to approximate the shaded region from July to December:

$\frac{1}{12}$(July + Aug + Sept + Oct + Nov + Dec) $\approx \frac{1}{12}(7.25 + 8 + 11 + 6.5 + 6.5 + 7)$

≈ 3.9 million

39. About 68 million tons.

41. $\frac{1}{5-0}\int_1^5 (500 + t - .25t^2)\,dt = \frac{1}{5}(500t + \frac{t^2}{2} - \frac{1}{.2}t^3)\Big|_0^5 \approx 500.4$

About 500 rabbits.

43. $\frac{1}{24 - 0}\int_0^{24}(6.44t - .23t^2 + 30)\,dt$

$= \frac{1}{24}(3.22t^2 - .07\overline{6}\,t^3 + 30t)\Big|_0^{24}$

$= \frac{1{,}514.88}{24} \approx 63$; the average temperature is about 63°F.

45. $10 \cdot 20 + 10 \cdot 16 + 10 \cdot 12 + 10 \cdot 12 \approx 600$

6.6 Numerical Integration

1. $\Delta x = \frac{b - a}{n} = \frac{6 - 1}{1} = 5;\ \ A_1 = 5[f(1)] = 5\sqrt{1 + 3} = 10$

3. $\Delta x = \frac{b - a}{n} = \frac{6 - 1}{3} = \frac{5}{3};\ \ A_3 = \frac{5}{3}[f(1) + f(\frac{8}{3}) + f(\frac{13}{3})]$

$= \frac{5}{3}\left[\sqrt{1 + 3} + \sqrt{\frac{8}{3} + 3} + \sqrt{\frac{13}{3} + 3}\right] = 11.81415$

5. $\Delta x = \frac{b - a}{n} = \frac{6 - 1}{1} = 5;\ \ T_1 = \frac{\Delta x}{2}[f(1) + f(6)] = \frac{5}{2}[\sqrt{1 + 3} + \sqrt{6 + 3}] = 12.5$

7. $\Delta x = \frac{b - a}{n} = \frac{6 - 1}{3} = \frac{5}{3};\ \ T_3 = \frac{\Delta x}{2}[f(1) + 2f(\frac{8}{3}) + 2f(\frac{13}{3}) + f(6)]$

$= \frac{5}{6}[\sqrt{1 + 3} + 2\sqrt{\frac{8}{3} + 3} + 2\sqrt{\frac{13}{3} + 3} + \sqrt{6 + 3}] \approx 12.64748$

9. $\Delta x = \frac{b - a}{n} = \frac{6 - 1}{2} = \frac{5}{3};\ \ P_2 = \frac{\Delta x}{3}[f(1) + 4f(\frac{7}{2}) + f(6)]$

$= \frac{5}{6}[\sqrt{1 + 3} + 4\sqrt{6 + 3}] \approx 12.66503$

11. $\Delta x = \frac{b - a}{n} = \frac{6 - 1}{6} = \frac{5}{6};\ P_6 = \frac{\Delta x}{3}[f(1) + 4f(\frac{11}{6}) + 2f(\frac{16}{6}) + 4f(\frac{21}{6}) + P_6]$

$= \frac{\Delta x}{3}[f(1) + 4f(\frac{11}{6}) + 2f(\frac{16}{6}) + 4f(\frac{21}{6}) + 2f(\frac{26}{6}) + 4f(\frac{31}{6}) + f(6)]$

$= \frac{5}{18}\left[\sqrt{1 + 3} + 4\sqrt{\frac{11}{6} + 3} + 2\sqrt{\frac{16}{6} + 3} + 4\sqrt{\frac{21}{6} + 3} + 2\sqrt{\frac{26}{6} + 3} + 4\sqrt{\frac{31}{6} + 3} + \sqrt{6 + 3}\right]$

≈ 12.66664

13. $\Delta x = \frac{b - a}{n} = \frac{4 - 2}{1} = 2;\ \ A_1 = \Delta x[f(2)] = 2\left[\frac{2}{(1 + 2(2))^2}\right] = 0.16$

15. $\Delta x = \frac{b - a}{n} = \frac{4 - 2}{3} = \frac{2}{3}$; $A_3 = \Delta x[f(2) + f(\frac{8}{3}) + f(\frac{10}{3})]$

$$= \frac{2}{3}\left[\frac{2}{(1 + 2(2))^2} + \frac{(\frac{8}{3})}{(1 + 2(\frac{8}{3}))^2} + \frac{\frac{10}{3}}{(1 + 2(\frac{10}{3}))^2}\right] \approx .13546$$

17. $\Delta x = \frac{b - a}{n} = \frac{4 - 2}{1} = 2$; $T_1 = \frac{\Delta x}{2}[f(2) + f(4)] =$

$$(1)\left[\frac{2}{(1 + 2(2))^2} + \frac{4}{(1 + 2(4))^2}\right] \approx .12938$$

19. $\Delta x = \frac{b - a}{n} = \frac{4 - 2}{3} = \frac{2}{3}$; $T_3 = \frac{\Delta x}{2}\left[f(2) + 2f(\frac{8}{3}) + 2f(\frac{10}{3}) + f(4)\right]$

$$= \frac{1}{3}\left[\frac{2}{(1 + 2(2))^2} + \frac{2(\frac{8}{3})}{(1 + 2(\frac{8}{3}))^2} + \frac{2(\frac{10}{3})}{(1 + 2(\frac{10}{3}))^2} + \frac{4}{(1 + 2(4))^2}\right] \approx .12526$$

21. $\Delta x = \frac{b - a}{n} = \frac{4 - 2}{2} = 1$; $P_2 = \frac{\Delta x}{3}[f(2) + 4f(3) + f(4)]$

$$= \frac{1}{3}\left[\frac{2}{(1 + 2(2))^2} + \frac{4(3)}{(1 + 2(3))^2} + \frac{4}{(1 + 2(4))^2}\right] \approx .12476$$

23. $\Delta x = \frac{b - a}{n} = \frac{4 - 2}{6} = \frac{1}{3}$;

$$P_6 = \frac{\Delta x}{3}\left[f(2) + 4f(\tfrac{7}{3}) + 2f(\tfrac{8}{3}) + 4f(3) + 2f(\tfrac{10}{3}) + 4f(\tfrac{11}{3}) + f(4)\right] =$$

$$\frac{1}{9}\left[\frac{2}{(1 + 2(2))^2} + \frac{4(\frac{7}{3})}{(1 + 2(\frac{7}{3}))^2} + \frac{2(\frac{8}{3})}{(1 + 2(\frac{8}{3}))^2} + \frac{4(3)}{(1 + 2(3))^2} + \frac{2(\frac{10}{3})}{(1 + 2(\frac{10}{3}))^2} + \frac{4(\frac{11}{3})}{(1 + 2(\frac{11}{3}))^2} + \frac{4}{(1 + 2(4))^2}\right]$$

$\approx .12473$

25. $\Delta x = \frac{7 - 3}{4} = 1$; $A_4 = \Delta x[f(3) + f(4) + f(5) + f(6)]$

$$= (1)\left[\frac{\sqrt{1 + 3}}{3^3} + \frac{\sqrt{1 + 4}}{4^3} + \frac{\sqrt{1 + 5}}{5^3} + \frac{\sqrt{1 + 6}}{6^3}\right] \approx .140857 \approx .14$$

27. $\Delta x = \frac{3 - 1}{4} = \frac{1}{2}$; $A_4 = \Delta x\left[f(1) + f(\frac{3}{2}) + f(2) + f(\frac{5}{2})\right]$

$= \frac{1}{2}[1^1 + 1.5^{1.5} + 2^2 + 2.5^{2.5}] = 8.3596 \approx 8.36$

29. $\Delta x = \frac{7-3}{4} = 1;\; T_4 = \frac{\Delta x}{2}[f(3) + 2f(4) + 2f(5) + 2f(6) + f(7)]$

$= \frac{1}{2}\left[\frac{\sqrt{1+3}}{3^3} + \frac{2\sqrt{1+4}}{4^3} + \frac{2\sqrt{1+5}}{5^3} + \frac{2\sqrt{1+6}}{6^3} + \frac{\sqrt{1+7}}{7^3}\right] \approx .1079 \approx .11$

31. $\Delta x = \frac{3-1}{4} = \frac{1}{2};\; T_4 = \frac{\Delta x}{2}\left[f(1) + 2f(\frac{3}{2}) + 2f(2) + 2f(\frac{5}{2}) + f(3)\right]$

$= \frac{1}{4}[1^1 + 2(1.5)^{1.5} + 2(2)^2 + 2(2.5)^{2.5} + 3^3] \approx 14.8596 \approx 14.86$

33. $\Delta x = \frac{7-3}{4} = 1;\; P_4 = \frac{\Delta x}{3}[f(3) + 4f(4) + 2f(5) + 4f(6) + f(7)]$

$= \frac{1}{3}\left[\frac{\sqrt{1+3}}{3^3} + \frac{4\sqrt{1+4}}{4^3} + \frac{2\sqrt{1+5}}{5^3} + \frac{4\sqrt{1+6}}{6^3} + \frac{\sqrt{1+7}}{7^3}\right] \approx .10$

35. $\Delta x = \frac{3-1}{4} = \frac{1}{2}.\; P_4 = \frac{\Delta x}{3}[f(1) + 4f(\frac{3}{2}) + 2f(2) + 4f(\frac{5}{2}) + f(3)]$

$= \frac{1}{6}[1^1 + (1.5)^{1.5} + 2(2)^2 + 4(2.5)^{2.5} + 3^3] \approx 13.81$

37. Area $\approx \frac{5}{3}[0 + 4(10) + 2(11) + 4(13) + 2(16) + 4(18) + 2(23) + 4(25) + 2(27)$

$+ 4(25) + 10] = 880$ square feet. Since it is 3 feet deep, $880 \times 3 = 2640$ cu. ft.

$= (2640 \div 27)$ cu. yd. ≈ 97.8. About $100/\text{yd}^3$ of fill will be needed.

6.7 Review

Objective Questions

1. $\frac{x^7}{7} + C$ **3.** $e^u + C$ **5.** $2x^3 - x^2 - 5x + C$

7. $\int \frac{(x-2)\,dx}{3x^2 - 5x - 2} = \int \frac{(x-2)\,dx}{(3x+1)(x-2)} = \int (3x+1)^{-1}dx = \frac{1}{3}\ln|3x+1| + C$

9. $F(x) = e^x + 9x + 9$

11. $f(x) = \int \frac{1}{x}\,dx = \ln|x| + c\;\; f(1) = \ln|1| + C \Rightarrow C = 0$. Then, $F(x) = \ln|x|$

13. $\int (9x^2 + 1)\,dx = 3x^3 + x = C;\; C(x) = 3x^3 + x + 25{,}000$

15. $C(x) = \int (10 - x^{-1/2})\,dx$

$= 10x - 2x^{1/2} + M$

If $x = 0$, then $C(x) = 1{,}500$ so we have

$$1{,}500 = 10(0) - 2(0)^{1/2} + M$$

$$1{,}500 = M$$

Thus, $C(x) = 10x - 2\sqrt{x} + 1{,}500$

17. $\int (e^{-3x} - \frac{1}{e})dx = \int e^u \frac{du}{-3} - \int e^{-1} dx = -\frac{1}{3}e^{-3x} - e^{-1}x + C$

$u = -3x,\ du = -3\,dx$

19. $\int \frac{2x - 5}{\sqrt{(x^2 - 5x)^2}} dx = \int \frac{du}{\sqrt{u^3}} = \int u^{-3/2} du = -2u^{-1/2} + C = -2(x^2 - 5x)^{-1/2} + C$

$u = x^2 - 5x,\ du = (2x - 5)dx$

21. $\int_1^2 \frac{x^3 + 2}{x^2} dx = \int_1^2 (x + 2x^{-2})dx = \frac{x^2}{2} + \frac{2x^{-1}}{-1} \Big|_1^2 = (2 - 1) - (\frac{1}{2} - 2) = 2.5$

23. $\int_1^8 x^{1/3}(1 + x^{4/3})^3 dx = \int_2^{17} x^{1/3}u^3 \frac{3\,du}{4x^{1/3}} = \frac{3}{4}\int_2^{17} u^3 du = \frac{3}{4}\left(\frac{u^4}{4}\right) = \frac{3}{16}(17^4 - 2^4) \approx 15{,}657$

Let $u = 1 + x^{4/3}$ $\quad x = 1 \Rightarrow u = 2,$

$du = \frac{4}{3}x^{1/3} dx$ $\quad x = 8 \Rightarrow u = 17$

$dx = \frac{3\,du}{4x^{1/3}}$

25. $\int_{\frac{8}{5}}^{3} \sqrt{5x + 1}\, dx = \frac{1}{5}\int_9^{16} u^{1/2} du = \frac{1}{5}\left(\frac{2}{3}\right) u^{3/2} \Big|_9^{16} = \frac{2}{15}(64 - 27) = \frac{74}{15}$

$u = 5x + 1,\ du = 5\,dx$

27. For $1 \le x \le 5$ the graph of $y = \frac{1}{(1 - 2x)^3} dx$ lies below the x-axis. The area then, is

$$\int_1^5 \frac{-1}{(1 - 2x)^3} dx = \int_1^5 -(1 - 2x)^{-3} dx = \frac{1}{2} \cdot \frac{(1 - 2x)^{-2}}{-2} \Big|_1^5 = \frac{20}{81}$$

29. $\int_2^9 [g(x) - f(x)]\, dx$ **31.** $\int_{-20}^{-9} [g(x) - f(x)]dx + \int_{-9}^{0} [f(x) - g(x)]dx$

33. $\frac{1}{2}x^2 = 4 - x$

$x^2 + 2x - 8 = 0$

$(x + 4)(x - 2) = 0$

$x = -4, 2$

$$\int_{-2}^{2} (4 - x) - (\tfrac{1}{2}x^2)dx + \int_{2}^{3} \tfrac{1}{2}x^2 - (4 - x)dx$$

$$= -\frac{x^3}{6} - \frac{x^2}{2} + 4x \,\Big|_{-2}^{2} + \frac{x^3}{6} + \frac{x^2}{2} - 4x \,\Big|_{12}^{3}$$

$$= -\frac{8}{6} - 2 + 8 - \frac{8}{6} + 2 + 8 + \frac{27}{6} + \frac{9}{2} - 12 - \frac{8}{6} - 2 + 8$$

$$= 15$$

35. The parabola $y = x^2 - 3x$ and the line $x - 4y = 0$ intersect at $x = 0$ and $x = \frac{13}{4}$. The area between them is:

$$\int_{0}^{13/4} \left[\tfrac{1}{4}x - (x^2 - 3x)\right]dx = \frac{x^2}{8} - \frac{x^3}{3} + \frac{3x^2}{2} \Big|_{2}^{13/4} = \frac{2{,}197}{384} \approx 5.72$$

37. $\Delta x = \dfrac{25 - 4}{1} = 21$ $f(x_0)$ = left endpoints is $f(4) = \ln 2 \approx .6931471$

$A_1 = 21(.6931471) \approx 14.55609079$

39. $\Delta x = \dfrac{25 - 4}{4} = 5.25$; $A = \Delta x[f(x_0) + f(x_1) + f(x_2) + f(x_3)]$

$= 5.25[f(4) + f(9.25) + f(14.5) + f(19.75)] \approx 24.33$

41. $2(63) + 2(49) + 2(55) + 2(82) + 2(80) = 658$ **43.** Let f be the rate shown. Then the fundamental theorem of integral calculus tells us that the *number* of cars sold is the area under this curve. (Since f is defined by a graph, the best we can do is to estimate the area.) Estimate the sales in September + Oct. + Nov. + Dec.

$= \frac{1}{12}(12.2) + \frac{1}{12}(6.7) + \frac{1}{12}(6.5) + \frac{1}{12}(7.5) \approx 2.74$

About $2\frac{3}{4}$ million cars were sold in the last quarter.

45. $\dfrac{1}{1 - (-2)} \displaystyle\int_{-2}^{1} (6x^2 - 2x + 5)dx = \tfrac{1}{3}\left[2x^3 - x^2 + 5x\right]\Big|_{-2}^{1} = 12$

47. $\dfrac{1}{3 - 1} \displaystyle\int_{1}^{3} xe^{x^2} dx = \tfrac{1}{2} \cdot \tfrac{1}{2} \int_{1}^{9} e^u \, dx = \tfrac{1}{4}e^u = \tfrac{1}{4}(e^9 - e) \approx 2{,}025$

Let $u = x^2$ $x = 1 \Rightarrow u = 1$

$du = 2x \, dx$ $x = 3 \Rightarrow u = 9$

49. $\Delta x = \frac{2-0}{1} = 2$; $f(x_0) = f(0) = 1$; $f(x_1) = f(2) = \frac{1}{5}$

$$T_1 = \frac{2}{2}\left[1 + \left(\tfrac{1}{5}\right)\right] \approx 1.2$$

51. $T = \frac{\Delta x}{2}[f(x_0) + 2f(x_1) + 2f(x_2) + 2f(x_3) + f(x_4)]$

$= \frac{1}{4}[f(0) + 2f(\frac{1}{2}) + 2f(1) + 2f(\frac{3}{2}) + f(2)] \approx 1.10$

53. n must be *even* for Simpson's Rule.

55. $P = \frac{\Delta x}{3}[f(x_0) + 4f(x_1) + 2f(x_2) + 4f(x_3) + 2f(x_4) + 4f(x_5) + 2f(x_6) +$

$4f(x_7) + f(x_8)]$

$= \frac{1}{12}[f(0) + 4f(\frac{1}{4}) + 2f(\frac{1}{2}) + 4f(\frac{3}{4}) + 2f(1) + 4f(1\frac{1}{4}) + 2f(1\frac{1}{2}) + 4f(1\frac{3}{4}) + f(2)]$

$= 3.08$

57. $P(x) = \int (250x - 5{,}000)\, dx = 125x^2 - 5{,}000x + C$

$P(0) = 125(0)^2 - 5{,}000(0) + C = 1{,}500 \Rightarrow C = 1{,}500$

$P(x) = 125x^2 - 5{,}000x \pm 1{,}500$

59. $c(t) = \int .001e^{-.001t}dt = -e^{-.001t} + C$. At $t = 0$ there is no waste in the holding tank, so

$0 = c(0) = -e^0 + C \Rightarrow C = 1$. We have $c(t) = -e^{-.001t} + 1$.

$C(60) = -e^{-.001(60)} + 1 \approx .058$ or 5.8% waste in the tank

Test Questions

1. $\int x^5 dx = \frac{x^6}{6} + C$

3. $\int e^u\, du = e^u + C$

5. $\int (18x^2 - 6x - 3)dx = 18(\frac{x^3}{3}) - 6(\frac{x^2}{2}) - 3x + C = 6x^3 - 3x^2 - 3x + C$

7. $\int (e^{3x+1} + \frac{1}{e})dx = \int e^{3x+1}dx + \frac{1}{e}\int dx$ Let $u = 3x + 1$ then $du = 3dx$

$$= \int e^u\left(\frac{du}{3}\right) + \frac{1}{e}\int dx$$

$$= \frac{1}{3}e^u + \frac{1}{e}x + C$$

$$= \frac{1}{3}e^{3x+1} + \frac{1}{e}x + C$$

9. $\displaystyle\int_{-1}^{3} \frac{x^3+2}{x^2}\,dx = \int_{-1}^{3} (x + 2x^{-2})dx = \frac{x^2}{2} + \frac{2x^{-1}}{-1}\Bigg|_{-1}^{3} = (\tfrac{9}{2} - \tfrac{2}{3}) - (\tfrac{1}{2} + 2) = \tfrac{4}{3}$

11. Let $u = 1 + x^{4/3}$ then $du = \frac{4}{3}x^{1/3}dx$. Also, if $x = 1$ and 8, $u = 2$ and 17,

respectively:

$$\int_1^8 x^{1/3}(1 + x^{4/3})^3\,dx = \int_2^{17} x^{1/3}u^3\frac{3du}{4x^{1/3}}$$

$$= \frac{3}{4}\int_2^{17} u^3\,du$$

$$= \frac{3}{4}\left(\frac{u^4}{4}\right)\Bigg|_2^{17}$$

$$= \frac{3}{16}(17^4 - 2^4)$$

$$= 15{,}657.1875$$

13. Area $= \displaystyle\int_{5/3}^{4} \sqrt{3x+4}\,dx$. Let $u = 3x + 4$, then $du = 3dx$. Also, if $x = \frac{5}{3}$ and 4,

then $u = 9$ and 16, respectively. $\displaystyle\int_{5/3}^{4} \sqrt{3x+4}\,dx = \int_9^{16} u^{1/2}\frac{du}{3} = \frac{1}{3}\left(\frac{u^{3/2}}{\frac{3}{2}}\right) = \frac{2}{9}\left(u^{3/2}\right)$

$= \dfrac{74}{9}$

15. The intersection of the curve is found by $x = x^3$ or when $x = -1, 1, 0$.

Area $= \displaystyle\int_{-1}^{0} (x^3 - x)dx + \int_0^1 (x - x^3)dx = \frac{x^4}{4} - \frac{x^2}{2}\Bigg|_{-1}^{0} + \frac{x^2}{2} - \frac{x^4}{4}\Bigg|_0^1 = \frac{1}{4} + \frac{1}{4} = \frac{1}{2}$

17. The height is $h(t) = \displaystyle\int (1.2 + 5t^{-4})dt = 1.2t - \frac{5}{3}t^{-3} + \text{C}.$

We are given $1 = h(1) = 1.2 - \frac{5}{3} + \text{C}$. Solving for C gives $\frac{22}{15} = \text{C}$, so

$$h(t) = 1.2t - \frac{5}{3t^3} + \frac{22}{15}.$$

$h(6) = 1.2(6) - \dfrac{5}{3(6)^3} + \dfrac{22}{15} \approx 8.66$ feet, so it will be economically feasible.

19. Set 670 equal to the total oil consumed, $\int_0^1 32.4e^{.048t}dt$

$$670 = \int_0^t 32.4e^{.048t}dt = 675e^{.048t} = 675e^{.048t} - 675$$

$$1345 = 675e^{.048t}$$

$$1.9926 = e^{048t}$$

$$\ln 1.9926 = .048t$$

$$\frac{\ln 1.9926}{.048} = t$$

$$t = 14.4 \text{ years}$$

CHAPTER 7

Applications and Integration

7.1 Business Models Using Integration

1. $\int_4^9 (\sqrt{x} + 100)dx = \frac{2x^{3/2}}{3} + 100x\Big|_4^9 = 918 - \frac{1216}{3} = 512\frac{2}{3}$

3. $\int_0^2 \frac{x-1}{x+1}\,dx$ Let $u = x + 1$, $du = dx$, $u - 2 = x - 1$

$$\int_{x=0}^{x=2} \frac{x-1}{x+1}\,dx = \int_{u=1}^{u=3} \frac{u-2}{u}\,du = \int_1^3 (1 - \tfrac{2}{u})du = (u - 2\ln|u|)\Big|_1^3$$

$$= (3 - 2\ln 3) - 1 = 2 - 2\ln 3 \approx -.197$$

5. $\int_0^6 65e^{-5t}dt = \frac{65e^{-5t}}{-5}\Big|_0^3 = -13e^{-5t}\Big|_0^3 = (-13e^{-15}) - (-13)$

$= 13(1 - e^{-15}) \approx 13$

7. $\int_0^5 500\,dt = 500t\Big|_0^5 = 2{,}500$

9. $\int_0^5 200e^{.08t}dt = 2{,}500e^{.08t}\Big|_0^5 = 3729.5617 - 2{,}500 \approx 1{,}229.56$

11. $\int_0^5 150e^{-.05t}dt = -3{,}000e^{-.05t}\Big|_0^5 \approx 663.60$

13. $\int_0^6 8{,}000e^{.08t}dt = \frac{8{,}000e^{.08t}}{.08}\Big|_0^6 = 100{,}000e^{.08t}\Big|_0^6 = 100{,}000(e^{.48} - 1)$

$\approx \$61{,}607.44$

15. $\int_0^1 2{,}500e^{.05t}dt = \frac{2{,}500}{.05}\, e^{.05t}\Big|_0^1 = 50{,}000e^{.05t}\Big|_0^1 = 500{,}000(e^{.05} - 1)$

$\approx \$2{,}563.55$

17. $\int_0^5 6{,}500e^{.07t}dt = \frac{6{,}500}{.07}\, e^{.07t}\Big|_0^5 \approx \$38{,}913.42$

19. $\int_1^2 (60 + 24t^2)dt = 60t + 8t^3\Big|_1^2 = 116$

21. $\int_0^3 (60 + 24t^2)dt = 60t + 8t^3\Big|_0^3 = 396$

23. $\int_6^{12} (300x^{1/3} + 50)dx = 225x^{4/3} + 50x\Big|_6^{12} \approx \$4{,}028.35$

25. $\int_0^{10} x^{-.5}dx = 2x^{.5}\Big|_0^{10} = 2\sqrt{10} \approx 6.32$

27. The total amount charged $= \int_0^4 150(1 + x^{2/3})dx = 150(x + .6x^{5/3})\Big|_0^4$

$\approx \$1{,}507.14$. The additional monthly charge should be $\frac{1{,}507.14}{48 \text{ months}} = \$31.40$.

29. $x^2 + 30 = 15 + 8x$

$x^2 - 8x + 15 = 0$

$(x - 5)(x - 3) = 0$

$x = 3, 5$

$\int_3^5 [(15 + 8x) - (x^2 + 30)]dx$

$= \left(4x^2 - \frac{x^3}{3} - 15x\right)\Big|_3^5$

$= (-16\frac{2}{3} + 18)$

$= \frac{4}{3}$

31. $A = \int_0^N me^{it}dt = \int_0^{20} 2{,}250e^{.09t}dt = 25{,}000e^{.09t}\Big|_0^{20} = \$126{,}241.19$

$N = 20;\ i = \frac{.09}{1} = .09;\ m = 2{,}250$

33. Solve $15{,}000 = \int_0^x (2{,}400t + 10{,}500)dt = 1{,}200t^2 + 10{,}500t\Big|_0^x$

$= 1{,}200x^2 + 10{,}500x.$

Subtracting 15,000 from both sides gives:

$$0 = 1{,}200x^2 + 10{,}500x - 15{,}000$$
$$0 = 4x^2 + 35x - 50$$
$$0 = (4x - 5)(x + 10)$$
$$x = \frac{5}{4} \text{ or } x = -10$$

It will take $1\frac{1}{4}$ years.

35. $2{,}000 - 15x - x^2 = x^2 - 3x + 560$

$2x^2 + 12x - 1{,}440 = 0$

$x^2 + 6x - 720 = 0$

$(x - 24)(x + 30) = 0$

$x = 24, -30$

↑ reject

If $x = 24$, then $D(24) = S(24) = 1{,}064$

$$\int_{10}^{24} [1{,}064 - (-x^2 - 3x + 560)]dx$$
$$= \int_{10}^{24} (3x - x^2 + 504)dx$$
$$= \left(\frac{3x^2}{2} - \frac{x^3}{3} + 504x\right)\Bigg|_{10}^{24}$$
$$= (8{,}352 - 4{,}856\tfrac{2}{3})$$
$$= 3{,}495\tfrac{1}{3}$$

The producer's surplus is \$3,495.33

37. $\dfrac{236 - 107x}{(x - 3)^2} = x + 20$

$236 - 107x = (x + 20)(x - 3)^2 x \neq 3$

$236 - 107x = (x + 20)(x^2 - 6x + 9)$

$236 - 107x = x^3 - 6x^2 + 9x + 20x^2 - 120x + 180$

$x^3 + 14x^2 - 4x - 56 = 0$

$x^2(x + 14) - 4(x + 14) = 0$

$(x^2 - 4)(x + 14) = 0$

$(x - 2)(x + 2)(x + 14) = 0$

$x = 2, -2, -14$ Reject negative values.

$D(2) = S(2) = 22$

$$\int_0^2 [22 - (x + 20)]dx$$
$$= \left(22x - \frac{x^2}{2} - 20x\right)\Bigg|_0^2$$
$$= (4 - 2) - 0 = 2$$

The producers' surplus is \$2.

39. $\int_0^3 3{,}800dx = 3{,}800x\Big|_0^3 = \$11{,}400$

41. $E(t) = C + \int_0^t f'(x)dx = 500 + \int_0^t 1{,}000x^{1/2}dx = 500 + \frac{2{,}000}{3}x^{3/2}\Big|_0^t$

$= 500 + \frac{2{,}000}{3}t^{3/2}$

According to the discussion, we want to set $E'(t) = \frac{E(t)}{t}$.

$1{,}000t^{1/2} = \frac{500 + \frac{2{,}000}{3}t^{3/2}}{t}$

$1000t^{3/2} = 500 + \frac{2{,}000}{3}t^{3/2}$ *Multiply by t*

$\frac{1{,}000t^{3/2}}{3} = 500$ *Subtract* $\frac{2{,}000}{3}t^{3/2}$

$t^{3/2} = \frac{3}{2}$ *Multiply by* $\frac{3}{1{,}000}$

$t = (\frac{3}{2})^{3/2} \approx 1.31$ years

43. $\int_a^b f(t)dt = \int_0^5 \sqrt{t}\,dt = \frac{2}{3}t^{3/2}\Big|_0^5 \approx 7.45356$ million $= 7{,}453{,}560$

7.2 Integration By Parts

1.

Let $u = x$	$dv = e^{3x}dx$
$du = dx$	$v = \int e^{3x}dx = \frac{1}{3}e^{3x}$

$$\int udv = uv - \int vdu$$

$$\int 12xe^{3x}dx = 12\left[x(\tfrac{1}{3}e^{3x}) - \int \tfrac{1}{3}e^{3x}dx\right]$$

$$= 4xe^{3x} - 4\int e^{3x}dx$$

$$= 4xe^{3x} - 4\left[\tfrac{1}{3}e^{3x}\right]$$

$$= 4xe^{3x} - \tfrac{4}{3}e^{3x} + C$$

3.

Let $u = x$	$dv = e^{-x}dx$
$du = dx$	$v = \int e^{-x}dx = -e^{-x}$

$\int udv = uv - \int vdu$

$$\int xe^{-x}dx = -xe^{-x} - \int(-e^{-x})dx \quad -xe^{-x} + \int e^{-x}dx$$

$$= -xe^{-x} - e^{-x} + C$$

Therefore, $\int_0^1 xe^{-x}dx = -xe^{-x} - e^{-x} \Big|_0^1$

$$= (-2e) - (-1) = 1 - 2e^{-1}$$

$$\approx .2642411177$$

5.

Let $u = x$ $\quad dv = \sqrt{1 - x}\, dx$

$du = dx$ $\quad v = \int (1 - x)^{1/2} dx = -\frac{2}{3}(1 - x)^{3/2}$

$$\int u\,dv = uv - \int v\,du$$

$$\int x\sqrt{1 - x}\; dx = -\frac{2}{3}x(1 - x)^{3/2} - \int -\frac{2}{3}(1 - x)^{3/2} dx$$

$$= -\frac{2}{3}x(1 - x)^{3/2} + \frac{2}{3}\int (1 - x)^{3/2} dx$$

$$= -\frac{2}{3}x(1 - x)^{3/2} + \frac{2}{3}\left[-\frac{2}{5}(1 - x)^{5/2}\right] + C$$

$$= -\frac{2}{3}x(1 - x)^{3/2} - \frac{4}{15}(1 - x)^{5/2} + C$$

7.

Let $u = x$ $\quad dv = (x + 2)^3 dx$

$du = dx$ $\quad v = \dfrac{(x + 2)^4}{4}$

$$\int u\,dv = uv - \int v\,du$$

$$\int x(x + 2)^3 dx = \frac{x(x + 2)^4}{4} - \int \frac{(x + 2)^4}{4} dx$$

$$= \frac{x(x + 2)^4}{4} - \frac{1}{4}\left[\frac{(x + 2)^5}{5}\right] + C$$

$$= \frac{x(x + 2)^4}{4} - \frac{(x + 2)^5}{20} + C$$

9.

Let $u = \ln(x+1)$	$dv = dx$
$du = \frac{1}{x+1}dx$	$v = \int dx = x$

$$\int udv = uv - \int vdu$$

$$\int \ln(x+1)dx = x\ln(x+1) - \int \frac{x}{x+1}dx$$

This last integral can be done without *too* much work, but an easier way would be to go back to:

$u = \ln(x+1)$	$dv = dx$
$du = \frac{1}{x+1}dx$	$v = \int dx = x + 1$

↑

since $\int dx = x +$ (**any** constant we like).

Now, $\int \ln(x+1)dx = (x+1)\ln(x+1) - \int (x+1)\cdot\frac{1}{x+1}dx$

$$= (x+1)\ln(x+1) - \int dx = (x+1)\ln(x+1) - x + C$$

11.

Let $u = x$	$dv = \frac{1}{\sqrt{1-x}}dx$
$du = dx$	$v = \int (1-x)^{-1/2}dx = -2(1-x)^{-1/2}$

$$\int \frac{xdx}{\sqrt{1-x}} = -2x(1-x)^{1/2} - \int -2(1-x)^{1/2}dx = -2x(1-x)^{1/2}\ dx$$

$$= -2x(1-x)^{1/2} + 2\left[-\tfrac{2}{3}(1-x)^{3/2}\right] + C = -2x(1-x)^{1/2} - \tfrac{4}{3}(1-x)^{3/2} + C$$

13.

Let $u = x^2$	$dv = e^{2x}dx$
$du = 2xdx$	$v = \int e^{2x}dx = \frac{1}{2}e^{2x}$

$$\int udv = uv - \int vdu$$

$$\int 6x^2e^{2x}dx = 6\left[\tfrac{1}{2}x^2e^{2x} - \int xe^{2x}dx\right] = 3x^2e^{2x} - 6\int xe^{2x}dx$$

This last integral can be done by parts also:

Let $u = x$	$dv = e^{2x}dx$
$du = dx$	$v = \frac{1}{2}e^{2x}$

$$= 3x^3e^{2x} - 6\left[uv - \int vdu\right] = 3x^2e^{2x} - 6\left[\frac{1}{2}xe^{2x} - \int \frac{1}{2}e^{2x}dx\right]$$

$$= 3x^2e^{2x} - 6\left[\frac{1}{2}xe^{2x} - \frac{1}{4}e^{2x}\right] + C = 3xe^{2x} - 3xe^{2x} + \frac{3}{2}e^{2x} + C$$

15.

Let $u = x^2$	$dv = \sqrt{1 - 2x}dx$
$du = 2xdx$	$v = \int (1 - 2x)^{1/2}dx = -\frac{1}{3}(1 - 2x)^{3/2}$

$$\int x^2\sqrt{1 - 2x}\, dx = -\frac{x^2(1 - 2x)^{3/2}}{3} + \frac{2}{3}\int x\,(1 - 2x)^{3/2}dx$$

This last integral can be done by parts also;

Let $u = x$	$dv = (1 - 2x)^{3/2}dx$
$du = dx$	$v = \int (1 - 2x)^{3/2}dx = -\frac{1}{5}(1 - 2x)^{5/2}$

$$= -\frac{x^2(1 - 2x)^{3/2}}{3} + \frac{2}{3}\left[uv - \int vdu\right] = -\frac{x^2(1 - 2x)^{3/2}}{3} + \frac{2}{3}$$

$$\left[-\frac{x(1 - 2x)^{5/2}}{5} - \int -\frac{1}{5}(1 - 2x)^{5/2}\, dx\right]$$

$$= \frac{-x^2(1 - 2x)^{3/2}}{3} - \frac{2x(1 - 2x)^{5/2}}{15} + \frac{2}{3}\left[-\frac{(1 - 2x)^{7/2}}{35}\right] + C$$

$$= -\frac{x^2(1 - 2x)^{3/2}}{3} - \frac{2x(1 - 2x)^{5/2}}{15} - \frac{2(1 - 2x)^{7/2}}{105} + C$$

17. Integrate using substitution:

Let $u = 1 - x^2 \Rightarrow x^2 = 1 - u$
$du = -2xdx$

$$\int udv = uv - \int vdu$$

$$\int x^3 \sqrt[3]{1 - x^2}\,dx = \int (1 - u) x u^{1/3} \frac{du}{-2x}$$

$$= -\frac{1}{2}\int (1 - u) u^{1/3} du$$

$$= -\frac{1}{2}\int u^{1/3} du + \frac{1}{2}\int u^{4/3} du$$

$$= -\frac{1}{2}\left(\frac{3}{4}\right)(1 - x^2)^{4/3} + \frac{1}{2}\left(\frac{3}{7}\right)(1 - x^2)^{7/3} + C$$

$$= -\frac{3(1 - x^2)^{4/3}}{8} + \frac{3(1 - x^2)^{7/3}}{14} + C$$

If you integrate by parts first and then use substitution you will obtain a very different

(but equivalent) result: $-\frac{3x^2(1 - x^2)^{4/3}}{8} - \frac{9(1 - x^2)^{7/3}}{56} + C$

19.

Let $u = \ln x$	$dv = x^2 dx$
$du = \frac{1}{x} dx$	$v = \frac{x^3}{3}$

$\int u\,dv = uv - \int v\,du$

$$\int x^2 \ln x\,dx = uv - \int v\,du = \frac{x^3 \ln x}{3} - \int \frac{x^2}{3} dx = \frac{x^3 \ln x}{3} - \frac{x^3}{9} + C$$

21.

Let $u = x^2$	$dv = e^{4x} dx$
$du = 2x\,dx$	$v = \frac{1}{4} e^{4x}$

$\int u\,dv = uv - \int v\,du$

$$\int 6x^2 e^{4x} dx = 6\left[\frac{1}{4}x^2 e^{4x} - \int \frac{1}{2} x e^{4x} dx\right] = \frac{3}{2}x^2 e^{4x} - 3\int x e^{4x} dx.$$

This last integral can also be done by parts:

Let $u = x$	$dv = e^{4x} dx$
$du = dx$	$v = \frac{1}{4} e^{4x}$

$$\frac{3}{2}x^2 e^{4x} - 3\int x e^{4x} dx = \frac{3}{2}x^2 e^{4x} - 3\left[\frac{1}{4} x e^{4x} - \int \frac{1}{4} e^{4x} dx\right]$$

$$= \frac{3}{2}x^2 e^{4x} - \frac{3}{4} x e^{4x} + \frac{3}{16} e^{4x} + C$$

23. By substitution, let $u = 1 - x^2$ then

$$\int \frac{x^3\,dx}{\sqrt{1 - x^2}} = \int x(1 - u)u^{-1/2}\frac{du}{-2x} = -\tfrac{1}{2}\int (u^{-1/2} - u^{1/2})\,du$$

$$= -(1 - x^2)^{1/2} + \tfrac{1}{3}(1 - x^2)^{3/2} + C$$

If you integrate by parts first, then use substitution you will obtain.

$$x^2(1 - x^2)^{1/2} - \tfrac{2}{3}(1 - x^2)^{3/2} + C$$

25. $\displaystyle\int x^3 e^{x^2}\,dx = \int x^3 e^t \frac{dt}{2x} t = x^2;\ dt = 2x\,dx$

$$= \frac{1}{2}\int x^2 e^t dt$$

$$= \frac{1}{2}\int te^t dt$$

$$= \frac{1}{2}\left[te^t - \int e^t dt\right]$$

$$= \frac{1}{2}[te^t - e^t + C]$$

$u = t$	$dv = e^t dt$
$du = dx$	$v = e^t$

$$= \frac{1}{2}x^2 e^{x^2} - \frac{1}{2}e^{x^2} + C$$

27. $\displaystyle\int_1^2 \ln 3x\,dx = \frac{1}{3}\int_3^6 \ln t\,dt$

$u = \ln t$	$dv = dt$
$du = \frac{1}{t}\,dt$	$v = t$

$$= \frac{1}{3}\left[t \ln t\Big|_3^6 - \int_3^6 dt\right] \qquad t = 3x,\ dt = 3\,dx$$

$$= \frac{1}{3}[6 \ln 6 - 3 \ln 3 - 6 - 3]$$

$$= 2\ln 6 - \ln 3 - 1$$
$$= \ln 12 - 1$$
$$\approx 1.48490665$$

29. $\displaystyle\int_0^1 (x - 4)e^x dx = \int_0^1 xe^x dx - 4\int_0^1 e^x dx$

$$= xe^x - e^x - 4e^x \Big|_0^1$$
$$= (e - 4e) - (0 - 1 - 4)$$
$$= 5 - 4e$$
$$\approx - 5{,}873127314$$

31. Rather than doing this by parts, make the substitution $u = \ln x$.

Then $du = \frac{1}{x}dx$: $\displaystyle\int_e^2 \frac{dx}{x\ln x} = \int_1^2 \frac{1}{u}du = \ln|u|\Big|_1^2 = \ln 2 \approx .6931$

33.

Let $u = \ln x$	$dv = \ln x\, dx$	
$du = \frac{1}{x}\, dx$	$v = \int \ln x\, dx = x\ln x - x$	*From Example 3*

$$\int_1^e (\ln x)^2\, dx = (\ln x)(x\ln x - x) - \int_1^e (\ln x - 1)dx$$

$$= x\ln^2 x - x\ln x - (x\ln x - x - x)\Big|_1^e$$

$$= x\ln^2 x - 2x\Big|_1^e$$

$$= e - 2$$

$$\approx .7182818285$$

35.

Let $u = x^m$	$dv = e^x dx$
$du = mx^{m-1}dx$	$v = \int e^x dx = e^x$

$$\int x^m e^x dx = x^m e^x - \int mx^{m-1}e^x dx = x^m e^x - m\int x^{m-1}e^x dx$$

37. **a.**

Let $u = x^2$	$dv = \frac{x}{x^2 - 1}\, dx$
$du = 2x dx$	$v = \int \frac{x dx}{x^2 - 1} = \frac{1}{2} \ln\lvert x^2 - 1\rvert$

$$\int \frac{x^3}{x^2+1}\, dx = \frac{1}{2}x^2 \ln\lvert x^2 + 1\rvert - \int x \ln\lvert x^2 + 1\rvert\, dx$$

This last integral can be done with a substitution, $z = x^2 + 1$, then $dz = 2x\, dx$:

$$= \frac{1}{2}x^2 \ln\lvert x^2 + 1\rvert - \int \frac{1}{2} \ln\lvert z\rvert\, dz$$

$$= \frac{1}{2}x^2 \ln\lvert x^2 + 1\rvert - \frac{1}{2}\,[z \ln z - z] + C_1 \qquad \textit{See Example 3}$$

$$= \frac{1}{2}x^2 \ln\lvert x^2 + 1\rvert - \frac{1}{2}\,[(x^2 + 1) \ln\lvert x^2 + 1\rvert - (x^2 + 1)] + C_1$$

$$= \frac{1}{2}x^2 \ln\lvert x^2 + 1\rvert - \frac{1}{2}x^2 \ln\lvert x^2 + 1\rvert + \frac{1}{2} \ln\lvert x^2 + 1\rvert + \frac{1}{2}x^2 + \frac{1}{2} + C_1$$

$$= \frac{1}{2} \ln\lvert x^2 + 1\rvert + \frac{1}{2}x^2 + C \qquad \textit{where } C = C_1 + \frac{1}{2}$$

b. By long division, $\frac{x^3}{x^2+1} = x + \frac{x}{x^2+1}$

$$\int \frac{x^3}{x^2+1}\, dx = \int \left[x - \frac{x}{x^2+1} \right] dx = \int x\, dx + \int \frac{x dx}{x^2+1} = \frac{x^2}{2} - \frac{1}{2} \ln\lvert x^2 + 1\rvert + C$$

39. **a.** $N(t) = \int N'(t)\, dt = \int te^{-.5t} dt$

Let $u = t$	$dv = e^{-.05t} dt$
$du = dt$	$v = \frac{e^{-.05t}}{-.05} = -20e^{-.05t}$

$$\int te^{-.05t} dt = t(-20e^{-.05t}) - \int (-20e^{-.05t})\, dt$$

$$= -20te^{-.05t} - 400e^{-.05t}$$

If $t = 0$, then $N(0) = 1$ so

$$-20(0)\, e^0 - 400e^0 + C = 1$$

$$-400 + C = 1$$

$$C = 401$$

Thus, $N(t) = 401 - 20te^{-.05t} - 400e^{-.05t}$

$$= 401 - 20e^{-.05t}(t + 20)$$

b. $N(7) = 20.46843155$ so there are about 20.5 million bacterial per liter.

41. The total costs are $\int_5^{10} 100x^2 + \left(\frac{5{,}000x}{x^2+1}\right) dx = \frac{100x^3}{3} + 2{,}500 \ln (x^2 + 1) \Big|_5^{10}$ $\approx \$32{,}559.20$

7.3 Using Tables of Integrals

1. Using Formula 9, $\int (1 + bx)^{-1} dx = \frac{1}{b} \ln |1 + bx| + C$

3. Using Formula 23, $\int \frac{x\,dx}{\sqrt{x^2 + a^2}} = \sqrt{x^2 + a^2} + C$

5. Using Formula 26, $\int \frac{dx}{x^2\sqrt{x^2 - a^2}} = \frac{\sqrt{x^2 - a^2}}{a^2 x} + C$

7. Using Formula 30, $\int x \ln x\,dx = \frac{x^2}{2} \ln x - \frac{x^2}{4} + C$

9. Using Formula 32 ($m = -1$), $\int x^{-1} \ln x\,dx = \frac{1}{2} \ln^2 x + C$

11. Using Formula 35, $\int xe^{ax} dx = \frac{e^{ax}}{a^2}(ax - 1) + C$

13. Using Formula 29 ($a = 1$), $\int \frac{x^2}{\sqrt{x^2 + 1}}\,dx = \frac{x}{2}\sqrt{x^2 + 1} - \frac{1}{2} \ln \left| x + \sqrt{x^2 + 1} \right| + C$

15. Using Formula 26 ($a = 4$), $\int \frac{dx}{x^2\sqrt{x^2 + 16}} = -\frac{\sqrt{x^2 + 16}}{16x} + C$

17. Let $u = 4x^2 + 1$, then $du = 8x dx$:

$$\int \frac{x\,dx}{\sqrt{4x^2 + 1}} = \int \frac{\frac{du}{8}}{\sqrt{u}}$$

$$= \frac{1}{8} \int u^{-1/2} du$$

$$= \frac{1}{4} u^{1/2} + C$$

$$= \frac{1}{4}\sqrt{4x^2 + 1} + C$$

19. Using Formula 20 with $a = \frac{1}{3}$,

$$\int \frac{dx}{x\sqrt{1 - 9x^2}} = \int \frac{dx}{3x\sqrt{\frac{1}{9} - x^2}}$$

$$= \frac{1}{3} \int \frac{dx}{x\sqrt{\frac{1}{9} - x^2}}$$

$$= \frac{1}{3}\left[-\frac{1}{\left(\frac{1}{3}\right)} \ln \left| \frac{\frac{1}{3} + \sqrt{\frac{1}{9} - x^2}}{x} \right| \right] + C$$

$$= -\ln \left| \frac{\frac{1}{3} + \sqrt{\frac{1}{9} - x^2}}{x} \right| + C$$

$$= -\ln \left| \frac{\frac{1}{3} + \sqrt{\frac{1}{9} - x^2}}{x} \cdot \frac{3}{3} \right| + C$$

$$= -\ln \left| \frac{1 + 3\sqrt{\frac{1}{9} - x^2}}{3x} \right| + C$$

$$= -\ln \left| \frac{1 + \sqrt{1 - 9x^2}}{3x} \right| + C$$

21. Using Formula 23, $\displaystyle\int \frac{x\,dx}{\sqrt{x^2 + 4}} = \sqrt{x^2 + 4} + C$

23. Using Formula 9, with $a = 1$, $b = 1$, and $n = 3$: (or substitution with

$u = 1 + x$):

$$\int (1 + x)^3 dx = \int u^3\, du = \frac{u^4}{4} + C = \frac{(1 + x)^{3+1}}{(3 + 1) \cdot 1} + C = \frac{(1 + x)^4}{4} + C$$

25. Using Formula 10 with $a = 1$, $b = 1$, and $n = 3$:

$$\int x(1 + x)^3\, dx = \frac{1}{3 + 2}(1 + x)^{3+2} - \frac{1}{3 + 1}(1 + x)^{3+1} + C$$

$$= \frac{1}{5}(1 + x)^5 - \frac{1}{4}(1 + x)^4 + C$$

27. Using Formula 13 with $m = 1$, $a = 1$, $b = 1$:

$$\int x\sqrt{1 + x}\ dx = \frac{-2(2 - 3x)\sqrt{(1 + x)^3}}{15} + C \text{ or } \frac{2(1 + 3x)^{5/2}}{45} - \frac{2(1 + 3x)^{3/2}}{27} + C$$

29. Using Formula 35 with $a = 4$:

$$\int xe^{4x}\ dx = \frac{e^{4x}}{4^2}(4x - 1) + C$$

$$= \frac{e^{4x}}{16}(4x - 1) + C$$

31. Let $u = 2x$, then $x = \frac{u}{2}$ and $du = 2dx$. Using Formula 30,

$$\int x \ln 2x\ dx = \int \frac{u}{2} \ln u \frac{du}{2}$$

$$= \frac{1}{4} \int u \ln u\ du$$

$$= \frac{1}{4}\left[\frac{u^2}{2} \ln u - \frac{u^2}{4}\right] + C$$

$$= \frac{x^2}{2} \ln 2x - \frac{x^2}{4} + C$$

33. Let $u = 5x$, then $du = 5dx$: $\int x^2 \ln 5x\ dx = \int \frac{u^2}{25} (\ln u) \frac{du}{5} = \frac{1}{125} \int u^2 \ln u du$

Now, by Formula 31 this is

$$\frac{1}{125}\left[\frac{u^3}{3} \ln u - \frac{u^3}{9}\right] + C = \frac{1}{125}\left[\frac{(5x)^3}{3} \ln 5x - \frac{(5x)^3}{9}\right] + C$$

$$= \frac{x^3}{3} \ln 5x - \frac{x^3}{9} + C$$

35. Using Formula 39 with $a = 3$, $b = 5$, and $m = 1$:

$$\int \frac{dx}{3 + 5e^x} = \frac{x}{3} - \frac{1}{3} \ln|3 + 5e^x| + C$$

37. Using Formula 39 with $a = 1$, $b = 1$, and $m = 2$:

$$\int \frac{dx}{1 + e^{2x}} = \frac{x}{1} - \frac{1}{2} \ln|1 + e^{2x}| + C = x - \frac{1}{2} \ln|1 + e^{2x}| + C$$

39. Let $u = 1 + x$ then $du = dx$. Using Formula 2:

$$\int \frac{dx}{\sqrt{1+x}} = \int \frac{du}{\sqrt{u}} = \int u^{-1/2}\, du = 2u^{1/2} + C = 2(1+x)^{1/2} + C$$

41. Using Formula 11 with $a = 1$, $b = 1$, and $n = 3$:

$$\int x^2(1-x)^3\, dx = 1\left[\frac{(1+x)^6}{6} - 2\frac{(1+x)^5}{5} + \frac{(1+x)^4}{4}\right] + C$$

$$= \frac{(1+x)^6}{6} - \frac{2(1+x)^4}{5} + \frac{(1+x)^4}{4} + C$$

43. Using Formula 11 with $a = 1$, $b = -1$, and $n = 3$:

$$\int 5x^2(1+x)^3\, dx = 5\left(\frac{1}{-1}\left[\frac{(1+x)^6}{6} - 2\frac{(1+x)^5}{5} + \frac{(1+x)^4}{4}\right]\right) + C$$

$$= \frac{-5(1+x)^6}{6} + 2(1-x)^5 - \frac{5(1+x)^4}{4} + C$$

45. Using Formula 13 with $m = 1$, $a = 1$, and $b = 2$:

$$\int 2x\sqrt{1-2x}\, dx = 2\left[\frac{-2(2+6x)\sqrt{(1-2x)^3}}{15(4)}\right] + C = \frac{-2(1+3x)\sqrt{(1-2x)^3}}{15} + C$$

47. Using Formula 11 with $a = 2$, $b = 3$, and $n = 3$:

$$\int x^2\,(2+3x)^3\, dx = \frac{1}{27}\left[\frac{(2+3x)^6}{6} - \frac{4(2+3x)^5}{5} + (2+3x)^4\right] + C$$

49. Let $u = 4x^3 + 1$, then $du = 12x^2 dx$. Using Formula 2,

$$\int x^2\sqrt{4x^3+1}\, dx = \int x^2\sqrt{u}\,\frac{du}{12x^2} = \frac{1}{12}\int u^{1/2}\, du = \frac{1}{18}u^{3/2} + C = \frac{1}{18}(4x^3+1)^{3/2} + C$$

51. Using Formula 36 with $m = 2$ and $a = 3$: $\int x^2 e^{3x} dx = \frac{x^2 e^{3x}}{3} - \frac{2}{3}\int x e^{3x} dx$

This last integral can be done using Formula 35 with $a = 3$:

$$\frac{x^2 e^{3x}}{3} - \frac{2}{3}\left[\frac{e^{3x}}{3}(3x-1)\right] + C = \frac{e^{3x}}{27}(9x^2 - 6x + 2) + C$$

53. Using Formula 14 with $m = 2$, $a = 2$, and $b = 9$:

$$\int \frac{\sqrt{2+9x}}{x^2} dx = -\frac{1}{2}\left[\frac{\sqrt{(2+9x)^3}}{x} + \left(-\frac{9}{2}\right)\int \frac{\sqrt{2+9x}}{x} dx\right]$$

$$= \frac{-\sqrt{(2+9x)^3}}{2x} + \frac{9}{4}\int \frac{\sqrt{2+9x}}{x}dx$$

Using Formula 14 again with $m = 1$:

$$= \frac{-\sqrt{(2+9x)^3}}{2x} + \frac{9}{4}\left[2\sqrt{2+9x} + \sqrt{2}\ln\left|\frac{\sqrt{2+9x}-\sqrt{2}}{\sqrt{2+9x}+\sqrt{2}}\right|\right] + C$$

55. Let $u = 3x$, then $du = 3\,dx$: $\int \frac{\sqrt{2+9x^2}}{x}dx = \int \frac{\sqrt{2+u^2}}{\frac{u}{3}}\frac{du}{3} = \int \frac{\sqrt{2+u^2}}{u}du$

Using Formula 21 with a $= \sqrt{2}$:

$$\sqrt{u^2+2} - \sqrt{2}\ln\left|\frac{\sqrt{2}+\sqrt{u^2+2}}{u}\right| + C = \sqrt{9x^2+2} - \sqrt{2}\ln\left|\frac{\sqrt{2}+\sqrt{9x^2+2}}{3x}\right| + C$$

57. Let $u = 3x$, then $du = 3\,dx$: $\int \frac{1}{\sqrt{2+9x^2}}\,dx = \int \frac{1}{\sqrt{2+u^2}}\frac{du}{3}$

Using Formula 16 with $a = \sqrt{2}$ gives;

$$\frac{1}{3}\ln\left|u+\sqrt{u^2+2}\right| + C = \frac{1}{3}\ln\left|3x+\sqrt{9x^2+2}\right| + C$$

59. Let $u = 9 - 16x^2$, then $du = -32x\,dx$:

$$\int 5x\sqrt{9-16x^2}\,dx = \int 5x(\sqrt{u})\,\frac{du}{-32x}$$

$$= -\frac{5}{32}\int u^{1/2}\,du$$

$$= -\frac{5}{32}(\frac{2}{3}\,u^{3/2}) + C$$

$$= -\frac{5}{48}(9-16x^2)^{3/2} + C$$

61. Using Formula 28 with $a = 1$:

$$\int \frac{\sqrt{x^2-1}}{x^2}\,dx = -\frac{\sqrt{x^2-1}}{x} + \ln\left|x+\sqrt{x^2-1}\right| + C$$

63. Using Formula 11 with $n = \frac{1}{2}$, $a = 3$, and $b = 10$:

$$\int x^2(3+10x)^{1/2}dx = \frac{1}{1000}\left[\frac{(3+10x)^{7/2}}{\frac{7}{2}} - 6\,\frac{(3+10x)^{5/2}}{\frac{5}{2}} + 9\,\frac{(3+10x)^{3/2}}{\frac{3}{2}}\right] + C$$

$$= \frac{(3+10x)^{7/2}}{3{,}500} - \frac{3(3+10x)^{5/2}}{1{,}250} + \frac{3(3+10x)^{3/2}}{500} + C$$

If instead of Formula 11 you use Formula 13, you will obtain the following (equivalent)
form: $\frac{6 - 30x + 125x^2}{4{,}375}(3 + 10x)^{3/2} + C$

65. The total amount will be $\int_1^{12} \frac{5{,}000}{t\sqrt{100 + t^2}}\,dt$; Using Formula 19 with $a = 10$:

$$\int_1^{12} \frac{5{,}000}{t\sqrt{100 + t^2}}\,dt = -\frac{5{,}000}{10}\ln\left|\frac{10 + \sqrt{t^2 + 100}}{t}\right|\Bigg|_1^{12}$$

$$= -500\left[\left|\frac{10 + \sqrt{244}}{12}\right| - \ln\left|\frac{10 + \sqrt{101}}{1}\right|\right]$$

$$\approx -500[.7585 - 2.9982]$$

$$\approx 1{,}120 \text{ tons.}$$

67. The accumulated sales are $\int_{12}^{24} te^{.1t}dt$. Using Formula 35

(or integration by parts) with $a = .1$, this is

$$\frac{e^{.1t}}{.1^2}(.1t - 1)\Bigg|_{12}^{24} = \frac{e^{2.4}}{.01}(2.4 - 1) - \frac{e^{1.2}}{.01}(1.2 - 1)$$

$$\approx 1{,}476.842 \text{ thousands or about } \$1{,}480{,}000$$

7.4 Improper Integrals

1. $\int_0^\infty e^{-x}dx = \lim_{b \to \infty} \int_0^b e^{-x}dx = \lim_{b \to \infty}(1 - e^{-b}) = 1 - 0 = 1$

3. $\int_0^\infty e^{-x/2}dx = \lim_{b \to \infty} \int_0^b e^{-x/2}dx = \lim_{b \to \infty}(-2e^{-b/2} + 2) = 0 + 2 = 2$

5. $\int_1^\infty \frac{dx}{x^3} = \lim_{b \to \infty} \int_1^b x^{-3}dx = \lim_{b \to \infty}\left(\frac{b^{-2}}{-2} + \frac{1}{2}\right) = 0 + \frac{1}{2} = \frac{1}{2}$

7. $\int_1^\infty \frac{dx}{\sqrt[3]{x}} = \lim_{b \to \infty} \int_1^b x^{-1/3} dx = \lim_{b \to \infty} \frac{3}{2}(b^{3/2} - 1)$. This limit does not exist,

since $b^{3/2} \to \infty$ as $b \to \infty$.

9. $\int_1^\infty \frac{dx}{x^{1.1}} = \lim_{b \to \infty} \int_1^b x^{1.1} dx = \lim_{b \to \infty} \frac{x^{-.1}}{-.1}\Big|_1^b = \lim_{b \to \infty}\left(\frac{-10}{b^{.1}} + 10\right) = 0 + 10 = 10$

11. $\int_{-\infty}^0 \frac{2x\,dx}{x^2+1} = \lim_{a \to -\infty} \int_a^0 \frac{2x}{x^2+1} dx$

$= \lim_{a \to -\infty}\left[\ln(x^2+1)\Big|_a^0\right]$

$= \lim_{a \to -\infty}\left[-\ln(a^2+1)\right]$

This limit does not exist; the integral diverges.

13. $\int_{-\infty}^\infty x^2 dx = \lim_{a \to -\infty} \int_a^0 x^2 dx + \lim_{b \to \infty} \int_0^b x^2 dx = \lim_{a \to -\infty}\left(-\frac{a^3}{3}\right) + \lim_{b \to \infty} \frac{b^3}{3}$

These limit does not exist.

15. $\int_{-\infty}^\infty \frac{3x\,dx}{(3x^2+2)^3}$

$= \lim_{b \to -\infty} \int_a^0 \frac{3x\,dx}{(3x^2+2)^3} + \lim_{b \to \infty} \int_0^b \frac{3x\,dx}{(3x^2+2)^3}$

$= \lim_{a \to -\infty} \frac{(3x^2+2)^{-2}}{-4}\Big|_a^0 + \lim_{b \to \infty} \frac{(3x+2)^{-2}}{-4}\Big|_0^b$

$= \lim_{a \to -\infty}\left[-\frac{1}{16} + \frac{1}{4(3a^2+2)^2}\right] + \lim_{b \to \infty}\left[\frac{1}{4(3b^2+2)} + \frac{1}{16}x\right]$

$= -\frac{1}{16} + 0 + 0 + \frac{1}{16}$

$= 0$

17. $$\int_e^{\infty} \frac{dx}{x(\ln x)^2} = \lim_{b \to \infty} \int_e^b \frac{dx}{x(\ln x)^2}$$

$$= \lim_{b \to \infty} \frac{-1}{\ln x} \Big|_e^b$$

$$= \lim_{b \to \infty} \left(-\frac{1}{\ln b} + \frac{1}{\ln e} \right)$$

$$= 0 + 1$$

$$= 1$$

19. $$\int_1^{\infty} \frac{e^{-\sqrt{x}}}{\sqrt{x}} \, dx = \lim_{b \to \infty} \int_{-1}^{b} \frac{e^u}{\sqrt{x}} (-2\sqrt{x} \, du) = \lim_{b \to -\infty} (-2) \int_{-1}^{b} e^u \, du$$

$u = -\sqrt{x}$	$x = 1 \Rightarrow u = -1$
$du = -\frac{1}{2} x^{-1/2} dx$	$x \to \infty \Rightarrow u - \infty$
$-2x^{1/2} du = dx$	

$$= \lim_{b \to -\infty} (-2)e^u \Big|_{-1}^{b} = \lim_{b \to -\infty} (-2e^b + 2e^{-1}) = 2e^{-1}$$

21. $$\int_1^{\infty} \frac{dx}{x^2\sqrt{x^2+4}} \, dx = \lim_{b \to \infty} \left[-\frac{\sqrt{x^2+4}}{4x} \right] \Bigg|_1^6$$ Formula 26

$$= \lim_{b \to \infty} \left[-\frac{\sqrt{b^2+4}}{4b} + \frac{\sqrt{5}}{4} \right]$$ $a = 2;$

$$= \lim_{b \to \infty} \left[\frac{\sqrt{5}}{4} - \frac{\sqrt{1 + \frac{4}{b^2}}}{4} \right]$$

$$= \frac{\sqrt{5}}{4} - \frac{1}{4} = \frac{\sqrt{5} - 1}{4}$$

23. $$\int_0^{\infty} 20{,}200 e^{-.06t} dt = \lim_{b \to \infty} \int_0^b 20{,}200 e^{-.06t} dt$$

$$= \lim_{b \to \infty} \frac{20{,}200}{-.06} e^{-.06t} \Big|_0^b$$

$$= \lim_{b \to \infty} (-336{,}666.67)(e^{-.06b} - e^0)$$

$$= \$336{,}666.67$$

25. See Example 2;

$$V = \int_0^\infty Re^{-rt}dt$$

$$= \lim_{b \to \infty} \int_0^b Re^{-rt}dt$$

$$= \lim_{b \to \infty} \frac{Re^{-rt}}{-r} \Bigg|_{t=0}^{t=b}$$

$$= \lim_{b \to \infty} \left(\frac{Re^{-rb}}{-r} - \frac{Re^0}{-4}\right)$$

$$= \lim_{b \to \infty} \left(\frac{R}{r} - \frac{R}{re^{rb}}\right)$$

$$= \frac{R}{r} - 0$$

$$= \frac{R}{r}$$

27. $A = \int_0^\infty 200e^{-.05t}dt$

$$= \lim_{b \to \infty} \int_0^b 200e^{-.05t}dt$$

$$= \lim_{b \to \infty} \frac{200e^{-.05t}}{-.05} \Bigg|_0^b$$

$$= \lim_{b \to \infty} \left(\frac{-4{,}000}{e^{.05b}} + 4{,}000\right)$$

$$= 0 + 4{,}000$$

$= 4{,}000$ millirems

29. Area $= \int_6^\infty \frac{2}{(x-4)^3}dx$

$$= \lim_{b \to \infty} \int_6^b \frac{2}{(x-4)^3}dx$$

$$= \lim_{b \to \infty} \frac{-1}{(x-4)^2} \Bigg|_6^b$$

$$= \lim_{b \to \infty} \left[\frac{-1}{(b - 4)^2} + \frac{1}{4} \right]$$

$$= 0 + \frac{1}{4}$$

$$= \frac{1}{4}$$

31. $V = \int_0^{\infty} 25{,}000 e^{.08t} \cdot e^{.12t} dt$

$$= 25{,}000 \int_0^{\infty} e^{-.04t} dt$$

$$= \lim_{b \to \infty} (-625{,}000 e^{-.04t}) \Big|_0^b$$

$$= \lim_{b \to \infty} (625{,}000 - 625{,}000 e^{.04b})$$

$$= 625{,}000$$

33. $\int_0^{\infty} (100e^{-.02t} - 100e^{-.1t}) dt$

$$= \lim_{b \to \infty} (-5000e^{-.02t} + 1000e^{-.1t}) \Big|_0^b$$

$= 4{,}000$ (in thousands of barrels) or 4 million barrels.

7.5 Probability Density Functions

1. Let $X =$ number of people who responded "yes", which is a discrete random variable. The sample space is $\{0, 1, 2, \cdots, 10\}$.

3. Let $X =$ number of words with an error, which is a discrete random variable. The sample space is $\{0, 1, 2, \cdots, 499, 500\}$.

5. Let $X =$ number of boys (or the number of girls), which is a discrete random variable. The sample space is $\{bbb, bbg, bgb, bgg, gbb, gbg, ggb, ggg\}$.

7. Let $X =$ height in inches.

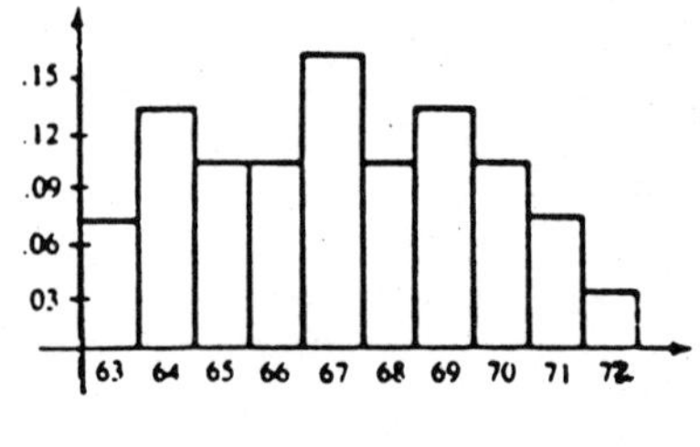

9. Let $X =$ time (in minutes) spent on each transaction.

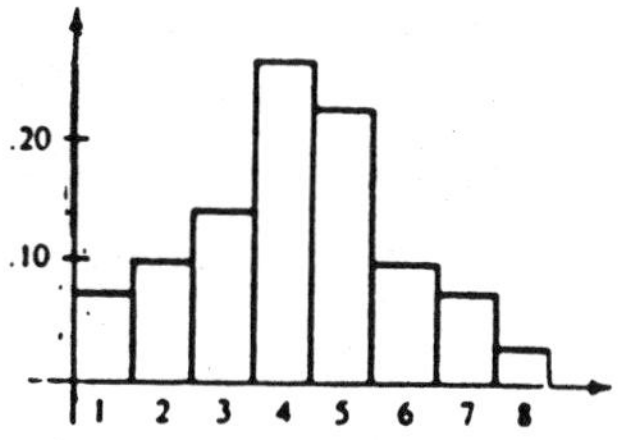

11. $\int_1^5 \frac{x^2}{72}\,dx = \frac{x^3}{216}\Big|_1^5 = \frac{124}{216}$ Since this is not 1, $f(x)$ is not a probability density function.

13. $\int_1^5 \frac{1}{4}\,dx = \frac{1}{4}x\Big|_1^5 = 1$ This is a probability density function.

15. $\int_1^4 4\,dx = 4x\Big|_1^4 = 12$ Since this is not 1, $f(x)$ is not a probability density function.

17. $\int_1^4 \frac{3}{64}x^2dx = \frac{x^3}{64}\Big|_1^4 = \frac{63}{64}.$ Since this is not 1, $f(x)$ is not a probability density function.

19. $\int_{-2}^3 kx\,dx = \frac{kx^2}{2}\Big|_{-2}^3 = \frac{9k}{2} - \frac{4k}{2} = \frac{5k}{2}.$ Set this equal to 1 and solve for k:
$\frac{5k}{2} = 1,\ k = \frac{2}{5}.$ But since $f(x) = \frac{2}{5}x < 0$ for $x < 0$, no value of k will make $f(x)$ a probability density function.

21. $\int_0^5 kx^2\,dx = \frac{kx^3}{3}\Big|_0^5 = \frac{125k}{3}.$ Set this equal to 1 and solve for k,
$\frac{125k}{3} = 1,\ k = \frac{3}{125}.$

23. $\int_1^7 \frac{k}{4}\,dx = \frac{k}{4}\Big|_1^7 = \frac{3k}{2}.$ Setting this equal to 1 gives $k = \frac{2}{3}.$

25. $\int_1^4 kx^{1/2}dx = \frac{2}{3}kx^{3/2}\Big|_1^4 = \frac{2}{3}k(4^{3/2} - 1^{3/2}) = \frac{14k}{3}.$

Setting this equal to 1 gives $k = \frac{3}{14}$.

27. $\int_1^4 \frac{k}{x^2}\,dx = \frac{kx^{-1}}{-1}\Big|_1^4 = \frac{-k}{x}\Big|_1^4 = \frac{3k}{4}.$

Setting this equal to 1 gives $k = \frac{4}{3}$.

29. On the standard normal distribution table, locate 1.2 in the left column (x), then 4 columns over to the column headed .03. The entry is 0.8907.

31. On the standard normal distribution table, locate 1.6 in the left column, the 10 columns over to the entry 0.9545. This is the area for $X < 1.69$. The area for $X > 1.69$ is $1 - 0.9545 = 0.0455$.

33. The area under the curve for $-2.8 \leq X \leq < -.46$ is (the area for $X \leq -.46$) $-$ (the area for $X \leq -2.8$). Using the standard normal distribution table, this is $0.3228 - 0.0026 = 0.3202$.

35. $\int_{35}^{75} .01dx = .01dx\Big|_{.35}^{.75} = .75 - .36 = .40$

37. $\int_5^{10} .5e^{-.5x}\,dx = -e^{-.5x}\Big|_5^{10} = -e^{-5} + e^{-2.5} \approx .075$

39. The z value is $z = \frac{X - \mu}{\sigma} = \frac{.91 - .9}{.01} = 1.$

$P(X > .91) = P(z > 1) = 1 - P(z \leq 1) = 1 - .8413 = .1587$

41. The z value is $z = \frac{X - \mu}{\sigma} = \frac{210 - 250}{25} = 1.6;$

$P(X < 210) = P(z < -1.6) = .0548$

43. The z scores for $x = 240$ and 280 are $-.4$ and 1.2, respectively;

$P(240 < X < 280) = P(-.4 < z < 1.2) = P(z < 1.2) - P(z < -.4)$

$= .8849 - .3446 = .5403$

45. $\int_{50}^{\infty} .01e^{-.01t}dt = \lim_{b \to \infty} (-e^{-.01t})\Big|_{50}^{b} = \frac{1}{\sqrt{e}} \approx .6065$

7.6 Review

Objective Questions

1. $\int_1^{40} \sqrt[3]{2x+5}\, dx = \frac{1}{2}\int_7^{85} u^{1/3}dx = \frac{1}{2}\left(\frac{3}{4}\right)u^{4/3}\Big|_7^{85} \approx 135.13$

3. $\int_2^{4} \frac{x}{2x-3}\, dx = \frac{1}{2}x + \frac{3}{4}\ \ln|2x-3|\Big|_2^4 = 1 + \frac{3}{4}\ln 5 \approx 2.207$

5. $10 + x^2 = 34 + 5x$ $\qquad \int_0^8 (34+5x) - (10+x^2)dx$

$x^2 - 5x - 24 = 0$

$(x-8)(x+3) = 0$ $\qquad = \frac{x^3}{3} + \frac{5x^2}{2} + 24x)\Big|_0^8 = 181\frac{1}{3};$

net excess profit \$181.33

7. $40 + x^2 = 32 + 6x$ $\qquad \int_2^4 [(32+6x) - (40+x^2)]dx$

$x^2 - 6x + 8 = 0$

$(x-4)(x-2) = 0$ $\qquad = (3x^2 - \frac{x^3}{3} - 8x)\Big|_2^4$

$x = 2, 4$ $\qquad = 48 - \frac{64}{3} - 32 - 12 + \frac{8}{3} + 16$

$= \frac{4}{3}$; net excess profit \$1.33

9. $\int_0^5 1{,}000\ dt = 5{,}000$

11. $\int_0^4 250e^{-.11t}dt = \frac{250}{-.11}e^{-.11t}\Big|_0^4 \approx 809.01$

13. $\int_0^{30} 3{,}900e^{.08t}dt = \frac{3{,}900}{.08}\ e^{.08t}\Big|_0^{30} = 488{,}629.85$

15. $A = \int_0^{25} 2{,}500e^{.075t}dt = \frac{100{,}00}{3}e^{.075t}\Big|_0^{25} = \$184{,}027.30$

17. $250 + 10x - x^2 = x^2 - 20x + 250$

$2x^2 - 30x = 0$

$x(x - 15) = 0$

$x = 0,\ 15$

when $x = 15$, $p = 175$

consumers' surplus

$$= \int_{10}^{15} [(250 + 10x - x^2) - 175]\,dx$$

$$= 75x + 5x^2 - \frac{x^3}{3}\Big|_{10}^{15} = \$208.33$$

19.

$$250 + 10x - x^2 = x^2 - 20x + 250$$

$$2x^2 - 30x = 0$$

$$x(x - 15) = 0$$

$$x = 0,\ 15$$

$$\int_{10}^{15} [175 - (x^2 - 20x + 250)]\,dx = 10x^2 - \frac{x^3}{3} - 75x\Big|_{10}^{15}$$

$$= \$83.33$$

↑

If $x = 15$, then $p = 175$

21. Let $u = x$; $dv = e^{-2x}dx$

$du = dx$; $v = -\frac{1}{2}e^{-2x}$

$$\int \frac{x}{e^{2x}}\,dx = x(-\tfrac{1}{2}e^{-2x}) - \int (-\tfrac{1}{2}e^{-2x})\,dx$$

$$= -\tfrac{1}{2}xe^{-2x} + \frac{1}{2}\,\frac{e^{-2x}}{-2} + C$$

$$= -\tfrac{1}{4}e^{-2x}(2x + 1) + C$$

23. Let $u = \ln\sqrt{2x} = \frac{1}{2}\ln(2x)$; $dv = dx$

$du = \dfrac{1}{2x}\,dx$; $v = x$

$$\int \ln\sqrt{2x}\,dx = x\ln\sqrt{2x} - \int x\left(\frac{1}{2x}\right)dx = x\ln\sqrt{2x} - \tfrac{1}{2}x + C$$

25. Use Formula 25 with $a^2 = 16$:

$$\int x^2\sqrt{x^2 - 16} = dx:\ \frac{x}{4}\sqrt{(x + 16)^3} + 2x\sqrt{x^2 - 16} - 32\ln\left|x + \sqrt{x^2 - 16}\right| + C$$

27. Let $u = 3x$, then $du = 3dx$. Use Formula 34 then 33:

$$\int \ln^3 3x\,dx = x\ln^3 3x - 3x\ln^2 3x + 6x\ln 3x - 6x + C$$

29. Use Formula 36, then Formula 35.

$$\begin{aligned}\int x^2 e^{-2x}dx &= \frac{x^2 e^{-2x}}{-2} - \frac{2}{-2}\int xe^{-2x}dx \\ &= -\tfrac{1}{2}x^2 e^{-2x} + \frac{e^{-2x}}{(-2)^2}(-2x - 1) + C \\ &= -\tfrac{1}{2}x^2 e^{-2x} - \tfrac{1}{4}e^{-2x}(2x+1) + C\end{aligned}$$

31. Let $u = 4x^2 + 5x - 3$, then $du = 3x + 5dx$:

$$\int \frac{8x+5}{4x^2+5x-3}dx + \int \frac{du}{u} = \ln|u| + C = \ln|4x^2 + 5x - 3| + C$$

33. $$\int_0^{\infty} xe^{-x^2}dx = \lim_{b\to-\infty}\left[-\frac{1}{2}\int_0^b e^u du\right] = \lim_{b\to-\infty}\left(-\tfrac{1}{2}e^b + \tfrac{1}{2}\right) = \tfrac{1}{2}$$

Let $u = -x^2$, $du = -2x\,dx$

35. $$\begin{aligned}\int_0^{\infty} .5e^{-.5x}dx &= \lim_{b\to\infty}\int_0^b .5e^{-.5x}dx \\ &= \lim_{b\to\infty}\left(-3^{-.5x}\Big|_0^b\right) \\ &= \lim_{b\to\infty}(-e^{-.5b} + 1) \\ &= 1\end{aligned}$$

37. $\{2, 3, 5, \cdots, 12\}$; X = sum of the top faces; discrete.

39. $\{x \mid x > 0\}$; X = number of minutes a rat takes to go through a maze;

continuous.

41.

Roll	2	3	4	5	6	7	8	9	10	11	12
Frequency	1	3	3	2	7	7	4	6	3	3	1
Rel. Frequency	$\frac{1}{40}$	$\frac{3}{40}$	$\frac{3}{40}$	$\frac{2}{40}$	$\frac{7}{40}$	$\frac{7}{40}$	$\frac{4}{40}$	$\frac{6}{40}$	$\frac{3}{40}$	$\frac{3}{40}$	$\frac{1}{40}$

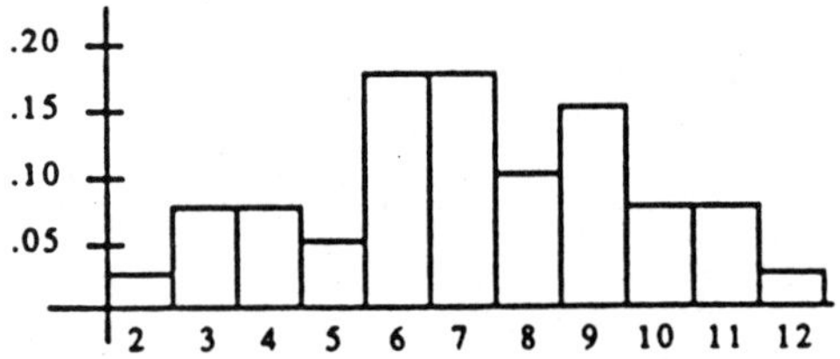

43.

Number of Minutes	1	2	3	4	5	6	7	8	9	10
Frequency	10	12	18	25	16	10	60	1	0	2
Rel. Frequency	100	100	100	100	100	100	100	100	100	100

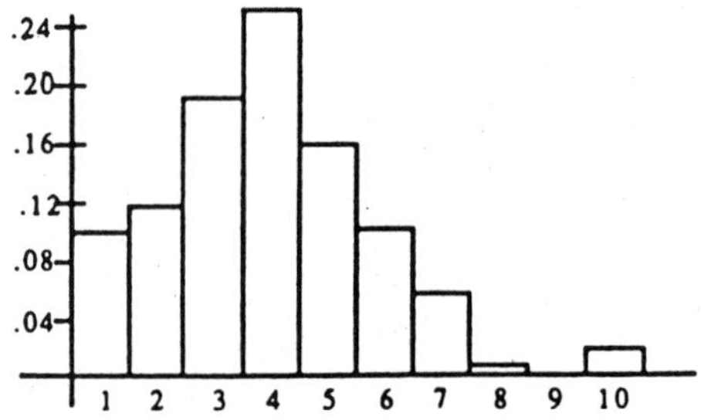

45. $\int_1^4 \frac{3}{63}x^2\, dx = \frac{x^3}{63}\Big|_1^4 = \frac{64}{63} - \frac{1}{63} = 1.$ This is a probability density function.

47. $\int_2^5 \frac{2}{3}\, dx \neq 1.$ Not a probability density function.

49. $\int_0^{16} kx^{1/4}dx = \frac{4}{5}kx^{5/4}\Big|_0^{16} = \frac{128}{5}k;$ thus, $k = \frac{5}{128}$

51. $\int_1^{10} 5kx\, dx = \frac{5kx^2}{2}\Big|_1^{10} = (250 - \frac{5}{2})k = \frac{495}{2}k;$ thus, $k = \frac{2}{495}$

53. 0.5 (by inspection)

55. The area for $X < 1.59$ is .9441 and the area for $X < -1.03$ is .1515 The desired area is .9441 − .1515 = .7926

57. $\int (40t - 6t^2)dt = 20t^2 - 2t^3 + C;$

If $t = 1$, $20(1)^2 - 2(1)^3 + C = 0 \Rightarrow C = -18$

Thus, $20t^2 - 2t^3 - 18 = 276$ when $t = 7$: there are 276 people affected on the seventh day.

59. $$V = \int_0^{\infty} Re^{-rt}dt$$
$$= \lim_{b\to\infty} \int_0^b 576{,}000e^{-.12t}dt$$
$$= 576{,}000 \lim_{b\to\infty} \Big|_0^b$$
$$= -4{,}800{,}000 \lim_{b\to\infty} (e^{-.12b} - 1)$$
$$= 4{,}800{,}000.$$

The capital value of the lease is $4,800,000

61. a. 31 **b.** 1 **c.** 60 **d.** 2.3%

Sample Test

1. $$\int (2x - 1)^3\,dx = \int u^3\,\frac{du}{2} = \frac{1}{2}\cdot\frac{u^4}{4} + C = \frac{1}{8}(2x - 1)^4 + C$$
$u = 2x - 1$
$du = 2dx$

3. By parts:

$u = x^2$	$dv = (2x - 1)^3dx$	
$du = 2xdx$	$v = \dfrac{(2x - 1)^4}{8}$	*See Problem 1*

$$\int x^2(2x - 1)^3dx = \frac{x^2(2x - 1)^4}{8} - \int \frac{2x(2x - 1)^4}{8}dx$$

$$= \frac{x^2(2x - 1)^4}{8} - \frac{1}{4}\int x(2x - 1)^4dx$$

$u = x$	$dv = (2x - 1)^4 dx$
$du = dx$	$v = \dfrac{(2x - 1)^5}{10}$

$$= \frac{x^2(2x - 1)^4}{8} - \frac{1}{4}\int x(2x - 1)^4 dx$$

$$= \frac{x^2(2x - 1)^4}{8} - \frac{1}{4}\left[\frac{x(2x - 1)^5}{10} - \int \frac{(2x - 1)^5}{10} dx\right]$$

$$= \frac{x^2(2x - 1)^4}{8} - \frac{1}{4}\left[\frac{x(2x - 1)^5}{10} - \frac{(2x - 1)^6}{120}\right] + C$$

$$= \frac{x^2(2x - 1)^4}{8} - \frac{x(2x - 1)^5}{40} + \frac{(2x - 1)^6}{480} + C$$

If instead, you use Formula 11 you will obtain the following (equivalent) form:

$$\frac{(2x - 1)^6}{48} + \frac{(2x - 1)^5}{40} + \frac{(2x - 1)^4}{32} + C$$

5. Let $u = 4x^2 + 5x - 3$; $\frac{du}{dx} = 8x + 5$; $\frac{du}{8x + 5} = dx$.

$$\int \frac{8x + 5}{4x^2 + 5x - 3}\, dx = \int \frac{8x + 5}{u} \frac{du}{8x + 5}$$

$$= \ln|u| + C$$

$$= \ln\left|4x^2 + 5x - 3\right| + C$$

7. $\int \frac{3}{1 - e^{3x}}\, dx = 3\left[\frac{x}{1} - \frac{1}{3}\ln\left|1 - e^{3x}\right| + C\right]$ *Formula 39*

$$= 3x - \ln\left|1 - e^{3x}\right| + C$$

9. $\int \ln^3 x\, dx = x \ln^3 x - 3\int \ln^2 x\, dx$ *Formula 34*

$= x \ln^3 - 3\,(x \ln^2 x - 2x \ln x + 2x + C)$ *Formula 33*

$$= x \ln^3 x - 3x \ln^2 x + 6x \ln x - 6x + C$$

11. Formula 14

13. $\int_1^{\infty} x^2 e^{-x^3} dx = \lim_{b \to -\infty} \int_1^{b} \left(-\frac{1}{3}e^u\right) du$ $\quad u = -x^3,\ du = -3x^2\, dx$

$$= \lim_{b \to -\infty}\left(-\frac{1}{3}e^b + \frac{1}{3}e\right)$$

$= \frac{e}{3}$

$\approx .9060939428$

15. $N(t) = \int 5e^{-.1t}dt$

$= \frac{5e^{-.1t}}{-.1} + C$

If $t = 0$ then $N(0) = 0$ so

$-50 + C = 0$

$C = 50$

Thus, $N(t) = -.50e^{-.1t} + 50$

$= 50(1 - e^{-.1t})$

CUMULATIVE REVIEW CHAPTERS 5 - 7

1. Using a calculator, a few values are:

$x = -2,\ y = e^{-2} \approx .14$

$x = -1,\ y = e^{-1} \approx .37$

$x = 0,\ y = e^{0} = 1$

$x = 1,\ y = e^{1} \approx 2.72$

$x = 2,\ y = e^{2} \approx 7.39$

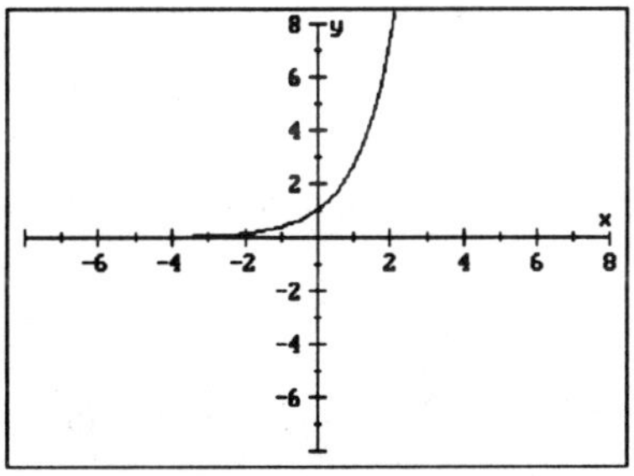

The graph is shown on the right.

3. $\log\left(\frac{x-5}{6}\right) = 2$ is equivalent to $10^2 = \frac{x-5}{6}$

$600 = x - 5$

$605 = x$

5. $10^{-x} = .5$

$-x = \log .5$

$x = -\log .5 \approx .301$

7. a. 0 **b.** $b - a$ **c.** $-\int_b^a f(x)dx$ **d.** $\int_a^b f(x)dx$ **e.** $k\int_a^b f(x)dx$

f. $\int_a^b f(x)dx \pm \int_a^b g(x)dx$

g. $u(x)v(x)\Big|_{x=a}^{x=b} - \int_{x=a}^{x=b} v\,du = u(b)v(b) - u(a)v(a) - \int_{x=a}^{x=b} v\,du$ where u and v

are functions of x.

9. a. $$\int (5x^4 + 3x^2 + 5)dx = 5\int x^4 dx + 3\int x^2 dx + 5\int dx$$
$$= 5\frac{x^5}{5} + 3\frac{x^3}{3} + 5x + C$$
$$= x^5 + x^3 + 5x + C$$

b. $$\int e^{2x}dx = \int e^u \frac{du}{2}$$ Let $u = 2x$, then $\frac{du}{2} = dx$
$$= \frac{1}{2}\int e^u du$$
$$= \frac{1}{2}e^u + C$$
$$= \frac{1}{2}e^{2x} + C$$

11. Let $u = 5x$, then $\frac{du}{5} = dx$:

$$\int \ln^4 5x dx = \frac{1}{5}\int \ln^4 u du.$$
$$= \frac{1}{5}\left[u \ln^4 u - 4\int \ln^3 u du\right]$$ *Using Formula 34*
$$= \frac{u \ln^4 u}{5} - \frac{4}{5}\int \ln^3 u du$$
$$= \frac{u \ln^4 u}{5} - \frac{4}{5}[u \ln^3 u - 3\int \ln^2 u du]$$
$$= \frac{u \ln^4 u}{5} - \frac{4u \ln^3 u}{5} + \frac{12}{5}\int \ln^2 u du.$$ *Use Formula 33*
$$= \frac{u \ln^4 u}{5} - \frac{4u \ln^3 u}{5} + \frac{12}{5}[u \ln^2 u - 2u \ln u + 2u] + C$$
$$= x \ln^4 5x - 4x \ln^3 5x + 12x \ln^2 5x - 24x \ln 5x + 24x + C$$

13. Let $u = 4x^2 - 3x + 2$, then $\frac{du}{dx} = 8x - 3$:

$$\int \frac{8x - 3}{4x^2 - 3x + 2}dx = \int \frac{du}{u} = \ln|u| + C = \ln\left|4x^2 - 3x + 2\right| + C$$

15. $\int_{-\infty}^{\infty} \frac{xdx}{(x^2+1)^2} = \int_{-\infty}^{0} \frac{xdx}{(x^2+1)^2} + \int_{0}^{\infty} \frac{xdx}{(x^2+1)^2}$

$= \lim_{a \to -\infty} \int_a^0 \frac{xdx}{(x^2+1)^2} + \lim_{b \to -\infty} \int_0^b \frac{xdx}{(x^2+1)^2}$

$= \lim_{a \to -\infty} \frac{-1}{2(x^2+1)} \Big|_a^0 + \lim_{b \to \infty} \frac{-1}{2(x^2+1)} \Big|_0^b$

$= \lim_{a \to -\infty} \left[-\frac{1}{2} + \frac{1}{2(a^2+1)} \right] + \lim_{b \to \infty} \left[-\frac{1}{2(b^2+1)} + \frac{1}{2} \right]$

$= -\frac{1}{2} + \frac{1}{2} = 0$

17. The graphs intersect when

$x^2 = 32 - x^2\ 2x^2$

$= 32\ x^2$

$= 16\ x$

$= \pm 4$

Since the graph of $y = 32 - x^2$ is above the graph of $y = x^2$ between -4 and 4, the area is $\int_{-4}^{4} \left[(32 - x^2) - x^2\right]dx = 32x - \frac{2x^3}{3} \Big|_{-4}^{4} = \frac{512}{3}$

19. $48{,}000 = \int_0^t (5{,}000x + 2{,}000)\ dt = 2{,}500x^2 + 2{,}000x \Big|_0^t$

$= 2{,}500t^2 + 2{,}000t$

Divide both sides by 500 to find:

$5t^2 + 4t = 96$

$5t^2 + 4t - 96 = 0$

$(5t + 24)(t - 4) = 0$

$t = -\frac{24}{5}$ or $t = 4$

Reject $-\frac{24}{5}$ since t can not be negative, giving us that $t = 4$ years.

21. $A = 5{,}000$, $P = 1{,}000$ and $i = .08$ are given:

$5{,}000 = 1{,}000(1 + \frac{.08}{365})^{365t}$

$$5 = (1.000219178)^{365t}$$

$$\log_{1.000219178}(5) = 365t$$

$$\frac{\log 5}{\log 1.000219178} = 365t$$

Divide by 365 gives $t \approx 20.12$ years.

23. $$\frac{\int_0^{30} (2{,}450 - 2x^2)dx}{30 - 0} = \frac{2{,}450x - \frac{2x^3}{3}\Big|_0^{30}}{30} = 1{,}850 \text{ trees}$$

25. $$\int_{60}^{\infty} .01e^{-x/100}dx = \lim_{b \to \infty} \int_{60}^{b} .01e^{-x/100}dx = \lim_{b \to \infty}\left(-e^{-x/100} \Big|_{60}^{b}\right)$$

$$= \lim_{b \to \infty} \left(\frac{-1}{e^{b/100}} + e^{-60/200}\right) = 0 + e^{-60/100} \approx .5488;$$

about $\frac{1}{2}$ second response time.

CHAPTER 8

Functions of Several Variables

8.1 Three-Dimensional Coordinate System

1. The area of a rectangle with dimensions 15 by 35 is $K(15, 35) = (15)(35) = 525$ square units.

3. The volume of a box 3 by 5 by 8 is $V(3, 5, 8) = (3)(5)(8) = 120$ cubic units.

5. The simple interest from \$500 investment at 5% for 3 years is $I(500, .05, 3) = 500[1 + (.05)(3)] = 500(1.15) = \$575$.

7. The future value of a \$2,500 investment at 12% compounded quarterly for 6 years is $A(2{,}500, .12, 6, 4) = 2{,}500(1 + \frac{.12}{4})^{(4)(6)} =$
$2{,}500(1.03)^{24} \approx 2{,}500(2.0327941) = \$5{,}081.99$

9. The future value of a \$110,000 investment at 9% compounded monthly for 30 years is $A(110{,}000, .09, 30, 12) = 110{,}000(1 + \frac{.09}{12})^{(30)(12)}$
$= 110{,}000(1.0075)^{360} \approx 110{,}000(14.730577) = \$1{,}620{,}363$

11. $f(2, 3) = 2^2 - 2(2)(3) + 3^2 = 1$

13. $f(-2, 5) = (-2)^2 - 2(-2)(5) + 5^2 = 49$

15. $g(2, 1) = \dfrac{2(2) - 4(1)}{2^2 + 1^2} = 0$

17. $g(5, -3) = \dfrac{2(5) - 4(-3)}{5^2 + (-3)^2} = \dfrac{11}{17}$

19. $h(0, 5) = \dfrac{e^{(0)(5)}}{\sqrt{0^2 + 5^2}} = \dfrac{1}{5}$ or 0.2

21. $h(-2, -3) = \dfrac{e^{(-2)(-3)}}{\sqrt{(-2)^2 + (-3)^2}} = \dfrac{e^6}{\sqrt{13}} \approx \dfrac{403.42879}{3.6055513} = 111.89$

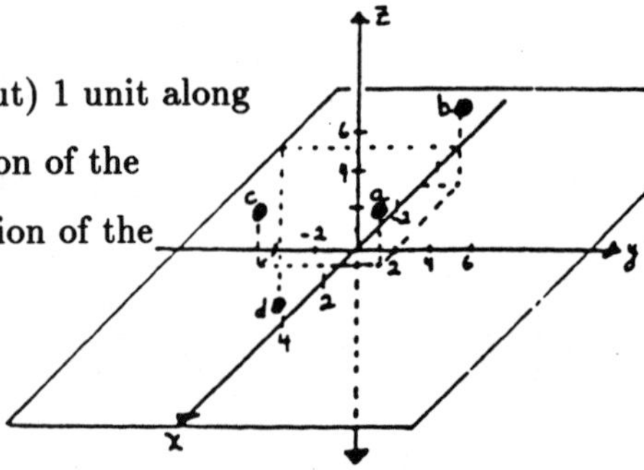

23. a. Starting at the origin, move (out) 1 unit along the x-axis, 2 units (right) in the direction of the $y-$axis, then (up) 3 units in the direction of the z-axis.

b. Starting at the origin, move (in) 3 units along the negative x-axis, 2 units (right) in the direction of the y-axis, then (up) 4 units in the direction of the z-axis.

c. Starting at the origin, move (out) 1 unit along the x-axis, 4 units (left) in the direction of the negative y-axis, then (up) 3 units in the direction of the z-axis.

d. Starting at the origin, move (in) 5 units along the negative x-axis, 9 units (left) in the direction of the negative y-axis, then 8 units (down) in the direction of the negative z-axis.

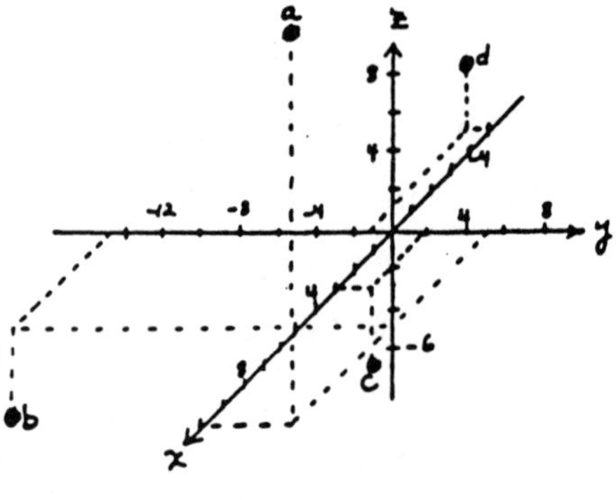

25. a. Starting at the origin, move (out) 10 units along the x-axis, 5 units (right) in the direction of the y-axis, then 20 units (up) in the direction of the z-axis.

b. Starting at the origin, move (out) 5 units along the x-axis, 15 units (left) in the direction of the negative y-axis, then 5 units (down) in the direction of the negative z-axis.

c. Starting at the origin, move (out) 3 units along the x-axis, 2 units (right) in the direction of the y-axis, then 4 units (down) in the direction of the negative z-axis.

d. Starting at the origin, move (in) 5 units along the negative x-axis, 1 unit (left) in the direction of the negative y-axis, then 3 units up.

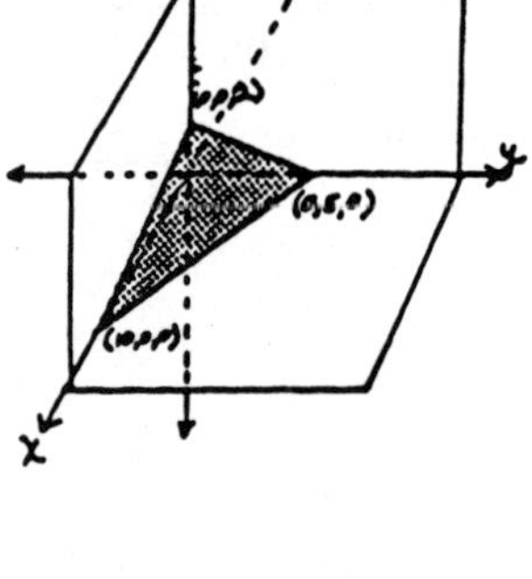

27. Since all variables are first degree, this is a plane. To graph part of this surface, set y and z equal to zero and solve $\Rightarrow x = 10$. The x-intercept is 10. Set x and z equal to zero and solve $\Rightarrow 2y = 10$, so $y = 5$. The y-intercept is 5. Similarly, set x and

y equal to zero to get the z-intercept, $z = 2$.
Connecting these intercepts with line segments gives a (triangular) section of the plane in the first octant (pictured).

29. Since all variables are first degree, this is a plane. To graph part of this surface, set y and z equal to zero and solve $\Rightarrow 3x = 12 \Rightarrow x = 4$. The x-intercept is 4. Set x and z equal to zero and solve $\Rightarrow -2y = 12 \Rightarrow y -6$. The y-intercept is -6. Similarly, set x and y equal to zero to get the z-intercept, $z = -12$. Connecting these intercepts with line segments gives a (triangle) section of the plane (pictured).

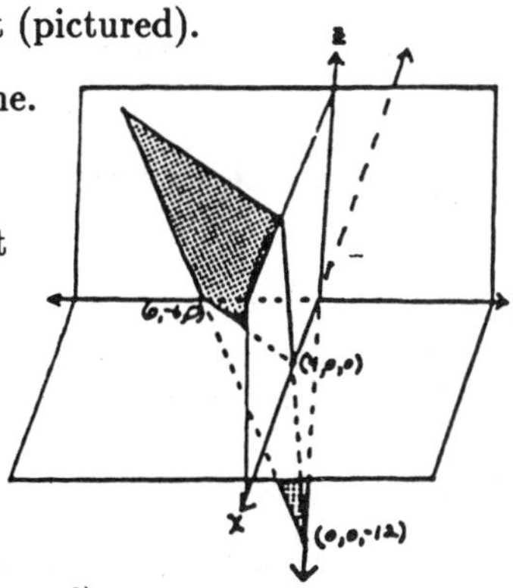

31. This has the form $z = \frac{x^2}{a^2} + \frac{y^2}{b^2}$, where $a = 2$ and $b = 3$, which is an elliptic paraboloid. The trace in the xy plane is a point (0, 0, 0), and in planes parallel to the xy-plane the traces are ellipses. The traces are ellipses. The trace in the yz-plane is the parabola $z = \frac{y^2}{9}$, and the trace in the xz-plane is the parabola $z = \frac{x^2}{4}$.

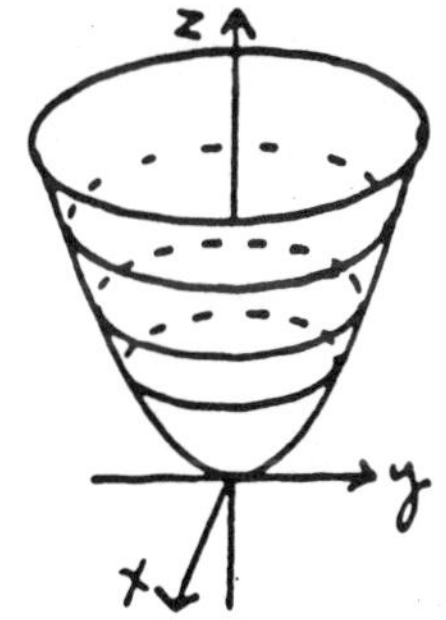

33. This has the form $\frac{x^2}{a^2} + \frac{y^2}{b^2} + \frac{z^2}{c^2} = 1$ (where $a = 3$, $b = 2$, $c = 5$) which is an ellipsoid. The trace in the xy-plane is the ellipse $\frac{x^2}{9} + \frac{y^2}{4} = 1$.

The trace in the yz-plane is the ellipse $\frac{y^2}{4} + \frac{z^2}{25} = 1$.

The trace in the xz-plane is the ellipse $\frac{x^2}{9} + \frac{z^2}{25} = 1$.

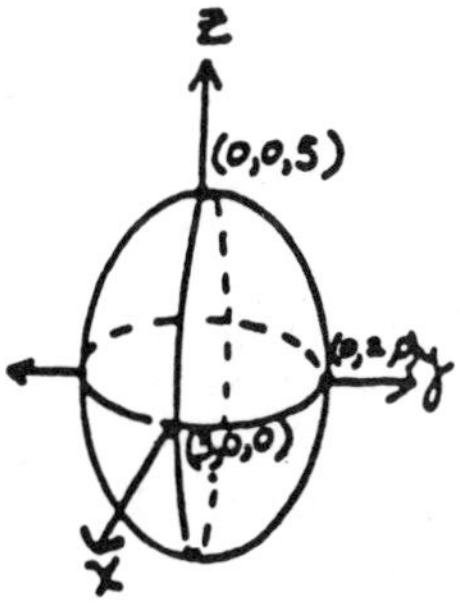

35. This has the form $z = \frac{x^2}{a^2} + \frac{y^2}{b^2}$ (where a and b are 1), which is an elliptic paraboloid. However, since a and b are equal, it's a circular paraboloid. The traces in planes parallel to the xy-plane are circles. The trace in the yz-plane is the parabola $z = y^2$. The trace in the xz-plane is the parabola $z = x^2$.

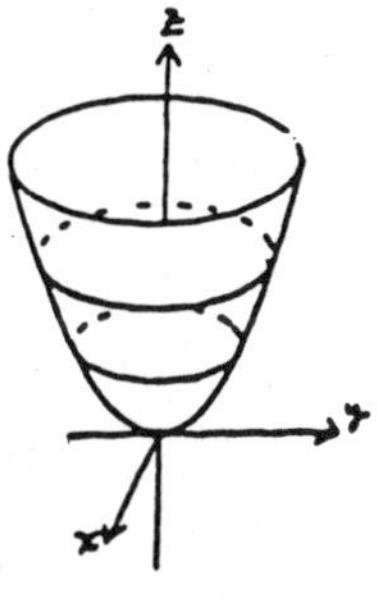

37. This has the form $\frac{x^2}{a^2} + \frac{y^2}{b^2} - \frac{z^2}{c^2} = -1$ which is a hyperboloid of two sheets. If we set x and y equal to zero, we see the z-intercepts are ± 2. The trace in the xz-plane is the hyperbola $\frac{x^2}{9} - \frac{z^2}{4} = -1$. The trace in the xy-plane is the hyperbola $\frac{y^2}{1} - \frac{z^2}{4} = -1$. The trace of plane parallel to the xy-plane ($|z| > 2$) are ellipses.

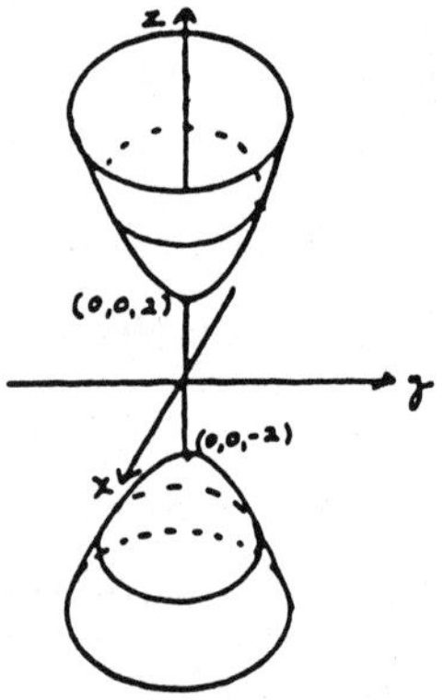

39. The trace in all planes parallel to the xy-plane is the circle $x^2 + y^2 = 36$. The surface is a circular cylinder.

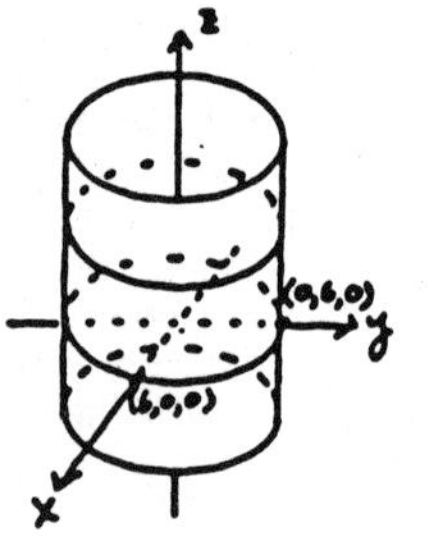

41. The trace in all planes parallel to the yz-plane is the circle $y^2 + z^2 = 20$. The surface is a circular cylinder.

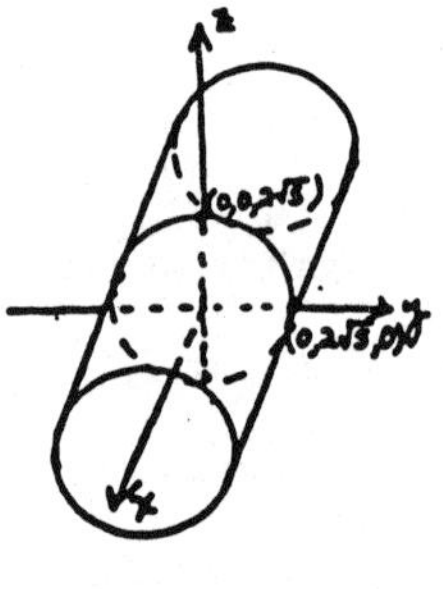

43. The cost for the first type is 2,500 + 800x. The cost for the second type is 1,200 + 550y. The total cost is the sum of these two,

$C(x, y) = (2{,}500 + 800x) + (1{,}200 + 550y) = 3{,}700 + 800x + 550y$.

a. $C(10, 15) = 3{,}700 + 800(10) + 550(15) = \$19{,}950$

b. $C(5, 25) = 3{,}700 + 800(5) + 550(25) = \$21{,}450$

c. $C(15, 10) = 3{,}700 + 800(15) + 550(10) = \$21{,}200$

d. $C(0, 30) = 3{,}700 + 800(0) + 550(30) = \$20{,}200$

45. The ceiling material will cost $(xy)(2)$, the floor material will cost $(xy)(1.25)$, and the four walls will cost $2(xy)(.75) + 2(yz)(.75)$.

The total cost is $C(x, y, z) = 2xy + 1.25xy + 1.5xz + 1.5yz = 3.25xy + 1.5xz + 1.5yz$

47. a. $F(3.1, .002) = \dfrac{.002(3.1)}{.002^4} \approx 3.875 \times 10^8$ mℓ

b. $F(15.3, .001) = \dfrac{.002(15.3)}{.001^4} \approx 3.06 \times 10^{10}$ mℓ

c. $F(6, .005) = \dfrac{.002(6)}{.005^4} \approx 1.92 \times 10^7$ mℓ

8.2 Partial Derivatives

1. $f_x = (5x^2)' - y^4(3x^3)' + (2y^3 - 15) = 10x - y^4(9x^2) + 0 = 10x - 9x^2y^4$

3. $g_y = 5(4x - 3y)^4\dfrac{\partial}{\partial y}(4x - 3y) = 5(4x - 3y)^4(-3) = -15(4x - 3y)^4$

5. See Problem 1. $f_x = 10x - 9x^2y^4 \Rightarrow f_x(1, 2) = 10(1) - 9(1)^2(2)^4 = -134$

7. See Problem 3. $g_y = -15(4x - 3y)^4 \Rightarrow g_y(3, -1) = -15[4(3) - 3(-1)]^4$

$= -759{,}375$

9. $\dfrac{\partial^2 z}{\partial x \partial y} = \dfrac{\partial}{\partial x}\left(\dfrac{\partial z}{\partial x}\right) = \dfrac{\partial}{\partial x}(f_y) = \dfrac{\partial}{\partial x}(-12x^3y^3 + 6y^2) = -36x^2y^3 + 0 = -36x^2y^3$

11. $\dfrac{\partial^2 w}{\partial y \partial x} = \dfrac{\partial}{\partial y}\left(\dfrac{\partial w}{\partial x}\right) = \dfrac{\partial}{\partial x}(g_y)$

$$= \frac{\partial}{\partial y}[5(4x - 3y)^4 \cdot (4)]$$

$$= \frac{\partial}{\partial y}[20(4x - 3y)^4]$$

$$= 80(4x - 3y)^3 \cdot (0 - 3)$$

$$= -240(4x - 3y)^3$$

13. First, $f_x = (10x - 9x^2y^4)$, so $f_{xx} = 10 - 18xy^4$;

$f_{xx}(0, 2) = 10 - 180(0)(2)^4 = 10$

15. First, $f_y = -12x^3y^3 + 6y^2$, so $f_{yx} = -36x^2y^3$;

$f_{yx}(-1, 0) = -36(-1)^2(0)^3 = 0$

17. First, $g_x = 5(4x - 3y)^4(4) = 20(4x - 3y)^4$,

so $g_{xx} = 80(4x - 3y)^3(4) = 320(4x - 3y)^3$

$g_{xx}(0, 2) = 320[4(0) - 3(2)]^3 = -69{,}120$

19. First, $g_y = -15(4x - 3y)^4$; (See Problem 1);

$$g_{yx} = -60(4x - 3y)^3 \cdot \frac{\partial}{\partial x}(4x - 3y)$$

$$= -60(4x - 3y)^3(4)$$

$= -240(4x - 3y)^3$

So $g_{yx}(-1, 0) = -240[4(-1) - 3(0)]^3 = 15{,}360$

21. $f_x = e^{3x+2y} \cdot \frac{\partial}{\partial x}(3x + 2y) = 3e^{3x+2y}$

23. First write $g(x, y)$ as $g(x, y) = (x^2 - 3y^2)^{1/2}$

$g_y = \frac{1}{2}(x^2 - 3y^2)^{-1/2} \cdot \frac{\partial}{\partial y}(x^2 - 3y^2) = \frac{1}{2}(x^2 - 3y^2)^{-1/2}(-6y) = \frac{-3y}{\sqrt{x^2 - 3y^2}}$

25. $f_x = 3e^{3x+2y}$; See Problem 21, so $f_x(1, 2) = 3e^{3(1)+2(2)} = 3e^7$

27. $g_y = \frac{-3y}{\sqrt{x^2 - 3y^2}}$; See Problem 23, so $g_y(3, -2) = \frac{-3(-2)}{\sqrt{3^2 - 3(-2)^2}} = \frac{6}{\sqrt{-3}}$ which is not a real number. $g_y(3, -2)$ does not exist.

29. $\frac{\partial^2 z}{\partial x \partial y} = \frac{\partial}{\partial x}(\frac{\partial z}{\partial y}) = \frac{\partial}{\partial x}(2e^{3x+2y}) = 2\frac{\partial}{\partial x}(e^{3x+2y}) = 2(3e^{3x+2y}) = 6e^{3x+2y}$

31. $w = g(x, y) = (x^2 - 3y^2)^{1/2}$, so

$$\begin{aligned}\frac{\partial w}{\partial x} &= g_x \\ &= \tfrac{1}{2}(x^2 - 3y^2)^{-1/2}\frac{\partial}{\partial x}(x^2 - 3y^2) \\ &= \tfrac{1}{2}(x^2 - 3y^2)^{-1/2}(2x) \\ &= x(x^2 - 3y^2)^{-1/2}\end{aligned}$$

$$\begin{aligned}\frac{\partial^2 w}{\partial y \partial x} &= \frac{\partial}{\partial y}(\frac{\partial w}{\partial x}) \\ &= \frac{\partial}{\partial y}\left[x(x^2 - 3y^2)^{-1/2}\right] \\ &= x\frac{\partial}{\partial y}[(x^2 - 3y^2)^{-1/2}] \\ &= x[-\tfrac{1}{2}(x^2 - 3y^2)^{-3/2} \cdot \frac{\partial}{\partial y}(x^2 - 3y^2)] \\ &= x[3y(x^2 - 3y^2)^{-3/2}] \\ &= \frac{3xy}{(\sqrt{x^2 - 3y^2})^3}\end{aligned}$$

33. $f_{xx} = \frac{\partial}{\partial x}(f_x) = \frac{\partial}{\partial x}(3e^{3x+2y}) = 3e^{3x+2y} \cdot \frac{\partial}{\partial x}(3x + 2y) = 9e^{3x + 2y}$

So $f_{xx}(0, 2) = 9e^{3(0)+2(2)} = 9e^4 \approx 491.38$

35. $f_{yx} = \frac{\partial^2 z}{\partial x \partial y} = 6e^{3x + 2y}$; (See Problem 29);

$f_{yx}(-1, 0) = 6e^{3(-1) + 2(0)} = 6e^{-3} = \frac{6}{e^3} \approx .30$

37. $g_{xx} = 2x[-\frac{1}{2}(x^2 - 3y^2)^{-3/2}(2x)] + 2(x^2 - 3y^2)^{-1/2}$;

so $g_{xx}(2, 0) = -8\cdot\frac{1}{8} + 2\cdot\frac{1}{2} = 0$

39. $g_{yx} = g_{xy} = \dfrac{\partial^2 w}{\partial y \partial x} = \dfrac{3xy}{(\sqrt{x^2 - 3y^2})^3}$; (See Problem 31);

$$g_{yx}(-1, 0) = \frac{3(-1)(0)}{(\sqrt{(-1)^2 - 3(0)^2})^3} = 0$$

41. For f_x, treat y and λ as constants: $f_x = 1 + 2y + \lambda(y - 0) = 1 + 2y + \lambda y$

For f_y, treat x and λ as constants: $f_y = 0 + 2x + \lambda(x - 0) = 2x + \lambda x$

For f_λ treat x and y constants: $f_\lambda = 0 + 0 + (xy - 10) = xy - 10$

43. For f_x, treat y and λ as constants: $f_x = 2x + 0 - \lambda(3 + 0 - 0) = 2x - 3\lambda$

For f_x, treat x and y as constants: $f_y = 2y - 2\lambda$

Finally: $f_\lambda = 0 + 0 - (3x + 2y - 6) = -3x - 2y + 6$

45. For $\dfrac{\partial f}{\partial b}$, treat m as a constant:

$$\frac{\partial f}{\partial b} = 2(10m + 5b)\cdot\frac{\partial}{\partial b}(10m + 5b) + 0 + 1$$

$$= 10(10m + 5b) + 1$$

$$= 100m + 50b + 1$$

For $\dfrac{\partial f}{\partial m}$, treat b as a constant:

$$\frac{\partial f}{\partial m} = 2(10m + 5b)\cdot\frac{\partial}{\partial m}(10m + 5b) + 2 + 0$$
$$= 20(10m + 5b) + 2$$
$$= 200m + 100b + 2$$

47. It might be easier to re-write $f(b, m)$:

$$f(b, m) = (m + b + 1)^2 + 4(m + b + 1)^2 + 9(m + b + 1)^2 = 14(m + b + 1)^2$$

$$\frac{\partial f}{\partial b} = 28(m + b + 1)\cdot\frac{\partial}{\partial b}(m + b + 1) = 28(m + b + 1)$$

$$\frac{\partial f}{\partial m} = 28(m + b + 1)\cdot\frac{\partial}{\partial m}(m + b + 1) = 28(m + b + 1)$$

49. $\dfrac{\partial P}{\partial a} = 2p + 0 - 0 - .2ap - 0 = 2p - .2ap$. Since we hold p constant this means at a fixed price the rate of change of profit per unit change in advertising spent is $2p - .2ap$. $\dfrac{\partial P}{\partial p} = 2a + 50 - 20p - .1a^2$. Since we hold a constant, this means at a fixed level of advertising the rate of change of profit per unit change in price is $2a + 50 - 20p - .1a^2$.

51. $P(x, y) = R(x, y) - C(x, y)$

$$= (1{,}500x + 900y) - (3{,}700 + 2{,}500x + 550y)$$

$$= -1{,}000x + 350y - 3{,}700.$$

$$P_x = -1{,}000, \; P_x(5, 10) = -1{,}000.$$

This means the rate of decrease in profit is \$1,000 per unit increase in x, assuming y stays constant, $P_y = 350$, $P_y(5, 10) = 350$. This means the rate of increase in profit is \$350 per unit increase in y, assuming x stays constant.

53. For most people answers will vary between 900 and 1,700 square inches

55. $A_r = 500(1 + r)^4$.

57. $P = 25(12)\left[\dfrac{1 - (1 + \frac{r}{12})^{-12t}}{r}\right]$. Using the Quotient Rule,

$$P_r = 300\left[\frac{r \cdot \frac{\partial}{\partial r}[1 - (1 + \frac{r}{12})^{-12t}] - [1 - (1 + \frac{r}{12})^{-12t}]\frac{\partial}{\partial r}(r)}{r^2}\right]$$

$$= 300\left[\frac{r[12t(1 + \frac{r}{12})^{-12t-1}(\frac{1}{12})] - [1 - (1 + \frac{r}{12})^{-12t}]}{r^2}\right]$$

$$= 300\left[\frac{rt(1 + \frac{r}{12})^{-12t-1} - 1 + (1 + \frac{r}{12})^{-12t}}{r^2}\right]$$

$$t = 5: \; P_r = 300\left[\frac{5r(1 + \frac{r}{12})^{-61} - 1 + (1 + \frac{r}{12})^{-60}}{r^2}\right]$$

$$= -\frac{300}{r^2}\left[1 - 5r(1 + \frac{r}{12})^{-61} - (1 + \frac{r}{12})^{-60}\right]$$

$$\text{or } 25i^{-2}\left[60i(1 + i)^{-61} + (1 + i)^{-60} - 1\right]$$

8.3 Maximum-Minimum Applications

1. $f_x = 6x + 5y$, $f_y = 5x + 2y$. To find critical values we need to solve the system

$$\begin{cases} 6x + 5y = 0 \\ 5x + 2y = 0 \end{cases}$$

Isolating y in the second equation gives $y = -\frac{5}{2}x$.

Substituting into the first equation gives

$$6x + 5y = 0$$
$$6x + 5\left(-\tfrac{5}{2}\right)x = 0$$
$$-13x = 0$$
$$x = 0$$

Since $y = -\frac{5}{2}x$, $y = 0$; giving the critical value (0, 0).

$A = f_{xx} = 6$, $B = f_{xy} = 5$, $C = f_{yy} = 2$, $D = B^2 - AC = 5^2 - (6)(2) = 13$.

Since $D > 0$, we have a saddle point at $(0, 0, f(0, 0)) = (0, 0, 0)$.

3. $f_x = 2x + y$, $f_y = x + 2y - 3$. To find critical values we need to solve the system $\begin{cases} 6x + 5y = 0 \\ 5x + 2y = 0 \end{cases}$. Solving for y in the first equation gives $y = -2x$. Substituting into the second equation gives

$$x + 2y - 3 = 0$$
$$x + 2(-2x) - 3 = 0$$
$$x = -1$$

Since $y = -2x$, $y = 2$; giving the critical point $(-1, 2)$.

$A = f_{xx} = 2$, $B = f_{xy} = 1$, $C = f_{yy} = 2$, $D = B^2 - AC = 1^2 - (2)(2) = -3$.

Since $D < 0$ and $A > 0$ there is a relative minimum at $(-1, 2, f(-1, 2)) = (-1, 2, -3)$.

5. $f_x = 3x^2 - 3y$, $f_y = -3x - 3y^2$. To find critical values we need to solve the system $\begin{cases} 3x^2 - 3y = 0 \\ -3x - 3y^2 = 0 \end{cases}$. Solving for y in the first equation gives $y = x^2$.

Substituting into the second equation gives

$$-3x - 3y^2 = 0$$
$$-3x - 3(x^2)^2 = 0$$
$$-3x(1 + x^3) = 0$$
$$-3x(1 + x^3) = 0$$
$$x = 0 \text{ or } 1$$

Since $y = x^2$, the critical points are (0, 0) and $(-1, 1)$.

$A = f_{xx} = 6x$, $B = f_{xy} = -3$, $C = f_{yy} = -6y$,

$D = B^2 - AC = (-3)^2 - (6x)(-6y) = 9 + 36xy$

At (0, 0), $D = 9 > 0$, so we have a saddle point at (0, 0, 0). At $(-1, 1)$,

$D = 9 + 36(-1)(1) = -27 < 0$, and $A = 6(-1) < 0$ which means we have a relative maximum at $[-1, 1, f(-1, 1)] = (-1, 1, 1)$.

7. $f_x = 3y - 10x + 3$, $f_y = 3x - 2y - 5$.

To find critical values we need to solve the system

$\begin{cases} 3y - 10x + 3 = 0 \\ 3x - 2y - 4 = 0 \end{cases}$. Solving for y in the second equation gives $y = \frac{3x - 5}{2}$.

Substituting this into the first equation gives

$$3y - 10x + 3 = 0$$
$$3\left(\frac{3x - 5}{2}\right) - 10x + 3 = 0$$
$$x = -\frac{9}{11}$$

Substituting, $y = \frac{3x - 5}{2} = \frac{3(-\frac{9}{11}) - 5}{2} = -\frac{41}{11}$

The critical point is $(-\frac{9}{11}, -\frac{41}{11})$.

$A = f_{xx} = -10$, $B = f_{xy} = 3$, $C = f_{yy} = -2$,

$D = B^2 - AC = 3^2 - (-10)(-2) = -11 < 0$.

Since $A < 0$ we have a relative maximum at $(-\frac{9}{11}, -\frac{41}{11}, \frac{45}{11})$.

9. $f_x = y + 2$, $f_y = x - 3$

To find critical values we need to solve the system $\begin{cases} y + 2 = 0 \\ x - 3 = 0 \end{cases}$.

The solution is $x = 3$, $y = -2$. $A = f_{xx} = 0$, $B = f_{xy} = 1$, $C = f_{yy} = 0$,

$D = B^2 - AC = 1$. Since $D > 0$ we have a saddle point at $(3, -2, 2)$.

11. $f_x = 2x + e^x$, $f_y = -1$. Since $f_y \neq 0$ there are no critical values.

13. $f_x = e^y$, $f_y = xe^y$. Since $e^y \neq 0$ there are no relative maxima, minima, or saddle points.

15. $f_x = 4y - 4x^3$, $f_y = 4x - 4y^3$. To find critical points we need to solve

$\begin{cases} 4y - 4x^3 = 0 \\ 4x - 4y^3 = 0 \end{cases}$. Solving for y in the first equation gives $y = x^3$. Substitute this into the second equation:

$$4x - 4y^3 = 0$$
$$4x - 4(x^3)^3 = 0$$
$$4x(1 - x^8) = 0$$
$$x = 0, 1, \text{ or } -1$$

Since $y = x^3$ our critical values are (0, 0), (1, 1), and $(-1, -1)$.

$A = f_{xx} = -12x^2$, $B = f_{xy} = 4$, $C = f_{yy} = -12y^2$, $D = B^2 - AC =$
$16 - 144x^2y^2$.

At (0, 0), $D = 16 > 0 \Rightarrow$ saddle point at (0, 0, 0). At $(-1, -1)$, $D = -128 < 0$ and since $A = -12(-1)^2 = -12 < 0 \Rightarrow$ relative maximum at $(-1, -1, 2)$.

At (1, 1), $D = -128 < 0$ and since $A = -12(1)^2 < 0 \Rightarrow$ relative maximum at (1, 1, 2).

17. $f_x = 3x^2 + 3y$, $f_y = 3x + 3y^2$. To find critical values we must solve the system $\begin{cases} 3x^2 + 3y = 0 \\ 3x + 3y^2 = 0 \end{cases}$. Solving for y in the first equation gives $y = -x^2$. Substitute this into the second equation which gives

$$3x + 3y^2 = 0$$
$$3x + 3(-x^2)^2 = 0$$
$$3x(1 + x^3) = 0$$
$$x = 0 \text{ or } -1$$

Since $y = -x^2$ the critical values are (0, 0) and $(-1, -1)$.

$A = f_{xx} = 6x$, $B = f_{xy} = 3$, $C = f_{yy} = 6y$, $D = B^2 - AC = 9 - 36xy$.

At (0, 0), $D = 9 > 0 \Rightarrow$ saddle point at (0, 0, 0).

At $(-1, -1)$, $D = -27 < 0$. Since $A = 6(-1) < 0$ we have a relative maximum at $(-1, -1, 1)$.

19. $P_a = -6a + 34 + 2n$, $P_n = -10n - 2 + 2a$. To find critical values we must solve $\begin{cases} -6a + 34 + 2n = 0 \\ -10n - 3y + 2a = 0 \end{cases}$ Solving for a in the second equation give $a = 5n + 1$. Substituting this into the first equation gives

$$-6a + 34 + 2n = 0$$
$$-6(5n + 1) + 34 + 2n = 0$$
$$n = 1$$

Since $a = 5n + 1 = 5(1) + 1$, our critical points is $(a, n) = (6, 1)$.

$A = P_{aa} = -6$, $B = P_{an} = 2$, $C = P_{nn} = -10$, $D = B^2 - AC = -56 < 0$.

Since $A < 0$ we have a relative maximum at (6, 1). The maximum is

$P(6, 1) = -3(6)^2 - 5(1)^2 + 34(6) - 2(1) + 2(6)(1) + 40 = 141$ (thousand dollars).

21. $P(x, y) = 3{,}950x + 150xy - 150x^2 - 50y^2 - 110y$;

$P_x = 3{,}950 + 150y - 300x$; $P_y = 150x - 100y - 110$. To find critical points we need

to solve the system $\begin{cases} 3{,}950 + 150y - 300x = 0 \\ 150x - 100y - 110 = 0 \end{cases}$. Solving the first equation for y

gives $y = \dfrac{300x - 3{,}950}{150} = \dfrac{6x - 79}{3}$. Substituting into the second equation gives

$$150x - 100\left(\frac{6x - 79}{3}\right) - 110 = 0.$$
$$45x - 10(6x - 79) - 33 = 0$$
$$-15x + 790 - 33 = 0$$
$$x \approx 50.47.$$

Substituting, $y = \dfrac{6x - 79}{3} = 74.60$.

$A = P_{xx} = -300$, $B = P_{xy} = 150$, $C = P_{yy} = -100$,

$D = B^2 - AC = 150^2 - (-300)(-100) = 22{,}500 - 30{,}000 = -7{,}500$.

Since $D < 0$ and $A < 0$, this is a relative maximum. Profit will be maximized when about 50 standard and 75 deluxe rings are sold.

23. Let the dimensions be x (width) by y (length) by z (height). Then $64 = xyz$, or

$z = \dfrac{64}{xy} = 64x^{-1}y^{-1}$. The amount of material is $xy + 2xz + 2yz$

$= xy + 2x(64x^{-1}y^{-1}) + 2y(64x^{-1}y^{-1}) = xy + 128y^{-1} + 128x^{-1}$

$= f(x, y)$. $f_x = y - 128x^{-2}$, $f_y = x - 128y^{-2}$.

We need to solve the system $\begin{cases} y - 128x^{-2} = 0 \\ x - 128y^{-2} = 0 \end{cases}$. Solving for y in the first equation

gives $y = \dfrac{128}{x^2}$. Substituting into the second equation gives

$$x - \frac{128}{y^2} = 0$$

$$x - \frac{128}{\left(\frac{128}{x^2}\right)^2} = 0$$

$$x - \frac{x^4}{128} = 0$$

$$x\left(1 - \frac{x^3}{128}\right) = 0$$

$$x = 0 \text{ or } x = \sqrt[3]{128} \approx 5.04$$

Since $x = 0$ does not make sense, we will test only at (5.04, 5.04).

$A = f_{xy} = 1$, $C = f_{yy} = 256y^{-3}$, $D = B^2 - AC = 1 - \dfrac{65{,}536}{x^3y^3}$.

At (5.04, 5.04), $D < 0$ and $A > 0$ which means this is a minimum. The box should therefore be 5.04 by 5.04 by $\frac{64}{5.04^2} \approx 2.52$ feet.

25. If $A = 0$, then $D = B^2 - AC = B^2 \geq 0$. Hence, if $A = 0$, then D can not be negative.

8.4 Lagrange Multipliers

1. $F(x, y) = f(x, y) + \lambda g(x, y) = xy - \lambda(x + y - 20) = xy - \lambda y + 20\lambda.$

$F_x = y - \lambda$, $F_y = x - \lambda$, $F_\lambda = -x - y + 20$. Now solve the system

$$\begin{cases} y - \lambda = 0 \\ x - \lambda = 0 \\ -x - y + 20 = 0 \end{cases}$$

The solution is $x = 10$, $y = 10$, $\lambda = 10$.

$f(10, 10) = 100$. By checking other points for which $x + y = 20$, we see that the maximum is 100.

3. $F(x, y) = f(x, y) + \lambda g(x, y) = -2x^2 - 3y^2 + \lambda(x + 2y - 24)$

$= -2x^2 - 3y^2 + \lambda x + 2\lambda y - 24\lambda.$

$F_x = -4x + \lambda$, $F_y = -6y + 2\lambda$, $F_\lambda = x + 2y - 24$. Now solve the system

$$\begin{cases} -4x + \lambda = 0 \\ -6y + 2\lambda = 0 \\ x + 2y - 24 = 0 \end{cases}$$

The solution is $x = \frac{72}{11}$, $y = \frac{96}{11}$, $\lambda = \frac{288}{11}$. $f(\frac{72}{11}, \frac{96}{11}) = -\frac{3,456}{11} \approx -314.18$;

By checking other points for which $x + 2y = 24$ we see this is a maximum.

5. $F(x, y) = f(x, y) + \lambda g(x, y) = x^2y + \lambda(x + 2y - 14) = x^2y + \lambda x + 2\lambda y = 14\lambda.$

$F_x = 2xy + \lambda$, $F_y = x^2 + 2\lambda$, $F_\lambda = x + 2y - 14.$

Now solve the system $\begin{cases} 2xy + \lambda = 0 \\ x^2 + 2\lambda = 0 \\ x + 2y - 20 = 0 \end{cases}$ The solution are $x = \frac{28}{3}$, $y = \frac{7}{3}$,

$\lambda = -\frac{392}{9}$; $f(\frac{28}{3}, \frac{7}{3}) = \frac{5{,}488}{27} \approx 203.26$;

by checking other points for which $x + 2y = 14$, we see that this is a maximum.

7. $F(x, y) = f(x, y) + \lambda g(x, y) = x^2 + y^2 + \lambda(x + y - 24)$

$= x^2 + y^2 + \lambda x + \lambda y - 24\lambda$. $F_x = 2x + \lambda$, $F_y = 2y + \lambda$, $F_\lambda = x + y - 24$.

We need to solve

$$\begin{cases} 2x + \lambda = 0 \\ 2y + \lambda = 0 \\ x + y - 24 = 0 \end{cases}$$

The solution is $x = 12$, $y = 12$, $\lambda = -24$; $f(12, 12) = 288$

which, by checking other points on $x + y - 24 = 0$, is a relative minimum.

9. $F(x, y) = f(x, y) + \lambda g(x, y) = x^2 + y^2 - xy - 4 + \lambda(x + y - 6)$;

$F_x = 2x - 6 + \lambda$, $F_y = 2y - x + \lambda$, $F_\lambda = x + y - 6$. We need to solve

$$\begin{cases} 2x - y + \lambda = 0 \\ 2y - x + \lambda = 0 \\ x + y - 6 = 0 \end{cases}$$

The solution is $x = 3$, $y = 3$, $\lambda = -3$; $f(3, 3) = 5$.

By checking other points on $x + y - 6 = 0$ this is a relative minimum.

11. $F(x, y) = f(x, y) + \lambda g(x, y) = x^2 + y^2 + \lambda(x + y - 16)$;

$F_x = 2x + \lambda$, $F_y = 2y + \lambda$, $F_\lambda = x + y - 16$. We need to solve

$$\begin{cases} 2x + \lambda = 0 \\ 2y + \lambda = 0 \\ x + y - 16 = 0 \end{cases}$$

The solution is $x = 8$, $y = 8$, $\lambda = -16$; $f(8, 8) = 128$.

By checking other points on $x + y - 16 = 0$, this is a relative minimum.

13. If the two numbers are x and y, we want to maximize $f(x, y) = xy$ subject to $x + y - 10 = 0$; $F(x, y) = f(x, y) + \lambda g(x, y) = xy + \lambda(x + y - 10)$; $F_x = y + \lambda$, $F_y = x + \lambda$, $F_\lambda = x + y - 10$. We need to solve

$$\begin{cases} y + \lambda = 0 \\ x + \lambda = 0 \\ x + y - 10 = 0 \end{cases}$$

The solution is $x = 5$, $y = 5$, $\lambda = -5$; $f(5, 5) = 25$

which is a maximum because other points on $x + y - 10 = 0$ yield lesser values for $f(x, y)$. The two numbers are therefore both 5.

15. Maximize $f(x, y) = -x^2 - 2y^2 + 140$ subject to $20x + 12y = 100$;

$F(x, y, \lambda) = -x^2 - 2y^2 + 140 + \lambda(20x + 12y - 100)$;

$F_x = -2x + 20\lambda,$

$F_y = -4y + 12\lambda,$

$F_\lambda = 20x + 12y - 100$

Solve the system $\begin{cases} -2x + 20\lambda = 0 \\ -4y + 12\lambda = 0 \\ 20x + 12y - 100 = 0 \end{cases}$

The solution is $x = \frac{250}{59}$, $y = \frac{75}{59}$, $\lambda = \frac{25}{59}$; $f\left(\frac{250}{59}, \frac{75}{59}\right) \approx 119$. Checking points for which $20x + 12y = 100$, such as (5, 0), gives $f(5, 0) = 115 \leq 119$. Hence this is a relative maximum. He should apply $\frac{250}{59} \approx 4.2$ acre-feet of water and $\frac{75}{59} \approx 1.27$ pounds of fertilizer. The yield is about 118.81 bu.

17. Let x be one width of the fence (costing \$3 per foot) and y be the other width (costing \$9 per foot). We want to maximize area $= f(x, y) = xy$ subject to $3(2x) + 9(2y) = 4{,}000 \Rightarrow 6x + 18y = 4{,}000$; $F(x, y, \lambda) = F(x, y) + \lambda g(x, y)$ $= xy + \lambda(6x + 18y - 4{,}000)$; $F_x = y + 6\lambda$, $F_y = x + 12\lambda$,

$F_\lambda = 6x - 18y - 4{,}000$. Solve the system $\begin{cases} y + 6\lambda = 0 \\ x + 12\lambda = 0 \\ 6x + 18y - 4{,}000 = 0 \end{cases}$

The solution is $x = = 166\frac{2}{3}$, $y = 166\frac{2}{3}$. The farmer should build his fence $166\frac{2}{3}$ by $166\frac{2}{3}$ for a maximum area of 27,778 square feet.

19. We want to minimize cost $= f(x, y) = .15x + .03y$ subject to $20xy = 1{,}000$; $F(x, y, \lambda) = f(x, y) + \lambda g(x, y) = .15x + .03y + \lambda(20xy - 1{,}000)$; $F_x = .15 + 20\lambda y$, $F_y = .03 + 20\lambda x$, $F_\lambda = 20xy - 1{,}000$.

We need to solve $\begin{cases} .15 + 20\lambda y = 0 \\ .03 + 20\lambda x = 0 \\ 20xy - 1{,}000 = 0 \end{cases}$. The solution is $x = \sqrt{10}$, $y = 5\sqrt{10}$.

Checking other points for which $20xy = 1{,}000$ we see that $f(\sqrt{10}, 5\sqrt{10}) \approx \$.95$ is a maximum. The optimum mixture is $\sqrt{10} \approx 3.16$ ounces of meat and $5\sqrt{10}$ ≈ 15.81 ounces of vegetables.

21. $F(x, y, z, \lambda) = f(x, y, z) + \lambda g(x, y, z) = x - y + z + \lambda(x^2 + y^2 + z^2 - 100)$. $F_x = 1 + 2\lambda x$, $F_y = -1 + 2\lambda y$, $F_z = 1 + 2\lambda z$, $F_\lambda = x^2 + y^2 + z^2 - 100$.

We need to solve the system $\begin{cases} 1 + 2\lambda x = 0 \\ -1 + 2\lambda y = 0 \\ 1 + 2\lambda z = 0 \\ x^2 + y^2 + z^2 - 100 = 0 \end{cases}$

The two solutions are $x = \pm\frac{10\sqrt{3}}{3}$, $y = \pm\frac{10\sqrt{3}}{3}$, $z = \pm\frac{10\sqrt{3}}{3}$, $\lambda = -\frac{\sqrt{3}}{20}$.

$f\left(\frac{10\sqrt{3}}{3}, -\frac{10\sqrt{3}}{3}, \frac{10\sqrt{3}}{3}\right) = 10\sqrt{3}$. This is a maximum by checking other points on

$x^2 + y^2 + z^2 = 100$, like $f(0, 0, 10) = 10 < 10\sqrt{3}$.

23. $F(x, y, z, \lambda) = 100 - xy - xz - yz + \lambda(x + y + z - 10)$;

$F_x = -y - z + \lambda$, $F_y = -x - z + \lambda$, $F_z = -x - y + \lambda$, $F_\lambda = x + y + z - 10$.

We need to solve system $\begin{cases} -y - z + \lambda = 0 \\ -x - z + \lambda = 0 \\ -x - y + \lambda = 0 \\ x + y + z - 10 = 0 \end{cases}$

The solution is $x = \frac{10}{3}$; $y = \frac{10}{3}$; $z = \frac{10}{3}$; $\lambda = \frac{20}{3}$

The lowest temperature is $T(\frac{10}{3}, \frac{10}{3}, \frac{10}{3}) = \frac{200}{3}$.

8.5 Multiple Integrals

1. $\displaystyle\int_0^2 (x^2y + y^2)dy = x^2\left(\frac{y^2}{2}\right) + \frac{y^3}{3}\Bigg|_{y=0}^{y=2} = \left[x^2(2) + \tfrac{8}{3}x\right] - [x^2(0) + 0] = 2x^2 + \tfrac{8}{3}$

3. $\displaystyle\int_1^4 (xy^2 - x)dx = y^2\left(\frac{x^2}{2}\right) - \frac{x^2}{2}\Bigg|_{x=1}^{x=4} = (y^2 - 1)\left(\frac{x^2}{2}\right)\Bigg|_{x=1}^{x=4}$

$= (y^2 - 1)\left(8 - \frac{1}{2}\right) = \frac{15}{2}(y^2 - 1)$

5. Treating x as a constant, let $u = 2y + x^2$. Then $du = 2dy$:

$$\int_0^4 x\sqrt{x^2 + 2y}\,dy = x\int_{u=x^2}^{u=x^2+8} u^{1/2}\left(\frac{du}{2}\right) = \frac{x}{2}\left[\frac{2}{3}u^{3/2}\Bigg|_{u=x^2}^{u=x^2+8}\right]$$

$$= \frac{x}{3}\left[(x^2 + 8)^{3/2} - x^3\right] = \frac{x}{3}(x^2 + 8)^{3/2} - \frac{x^4}{3}$$

7. Treating y as a constant, let $u = x + 2y$. Then $du = dx$:

$$\int_1^2 3e^{x+2y}\,dx = 3\int_{u=1+2y}^{u=2+2y} e^u\,du = 3e^u\Big]_{1+2y}^{2+2y}$$

$$= 3\left[e^{2+2y} - e^{1+2y}\right] = 3e^{2y+2} - 3e^{2y+1}$$

9. Let $u = x^2 + 3y$. Treating y as a constant, $du = 2x\,dx$:

$$\int_1^5 xe^{x^2+3y}\,dx = \int_{u=1+3y}^{u=25+3y} xe^u\,\frac{du}{2x} = \frac{1}{2}\int_{1+3y}^{25+3y} e^u\,du$$

$$= \frac{1}{2}\left(e^{25+3y} - e^{1+3y}\right) = \frac{e^{3y+25}}{2} - \frac{e^{3y+1}}{2}$$

11. $$\int_1^{e^x} \frac{x}{y}\,dy = x\ln|y|\,\Big|_1^{e^x} = x\ln|e^x| - x\ln|1| = x(x) - x(0) = x^2$$

13. $$\int_1^4\int_0^3 dy\,dx = \int_1^4\left[y|_0^3\right]dx = \int_1^4 3\,dx = 3x|_1^4 = 3(4-1) = 9$$

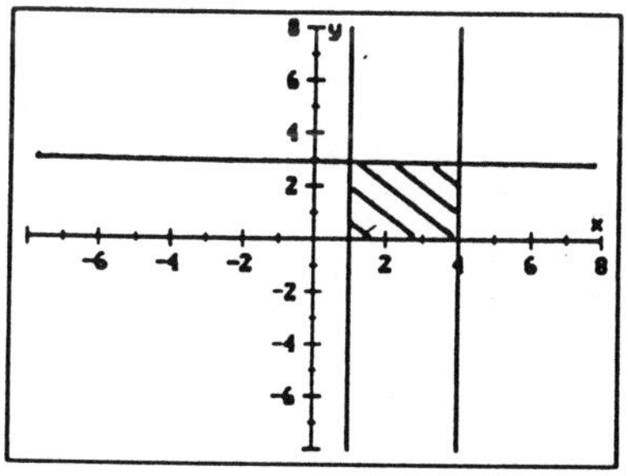

15. $$\int_{-3}^2\int_2^4 dx\,dy = \int_{-3}^2\left(x|_2^4\right)dy = \int_{-3}^2 (4-2)\,dy = \int_{-3}^2 2\,dy = 2\left(y|_{-3}^2\right)$$

$= 2[2 - (-3)] = 10$

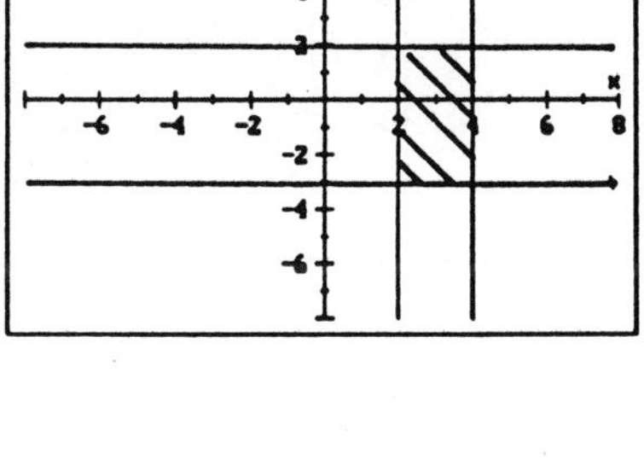

17. $\int_1^3 \int_0^y (x + 3y)\, dx\, dy = \int_1^3 \left(\frac{x^2}{2} + 3xy \Big|_{x=0}^{x=y} \right) dy = \int_1^3 \left(\frac{y^2}{2} + 3y^2 \right) dy$

$= \int_1^3 \frac{7}{2} y^2\, dy = \frac{7}{6} y^3 \Big|_1^3 = \frac{91}{3}$

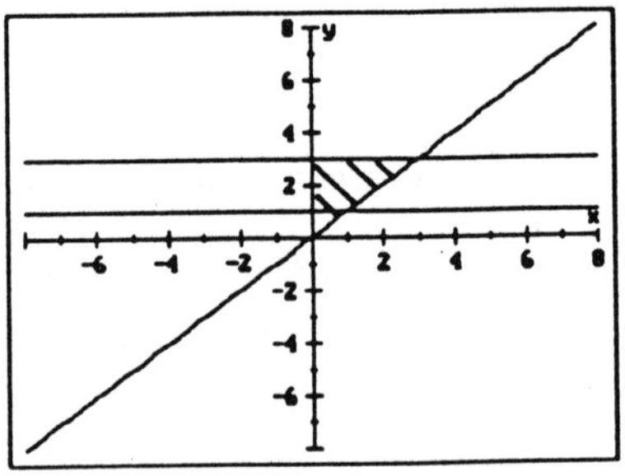

19. $\int_0^2 \int_0^{y^2} xy\, dx\, dy = \int_0^2 \frac{x^2 y}{2} \Big|_{x=0}^{x=y^2} dy = \int_0^2 \frac{1}{2} y^5\, dy = \frac{y^6}{12} \Big|_0^2 = \frac{16}{3}$

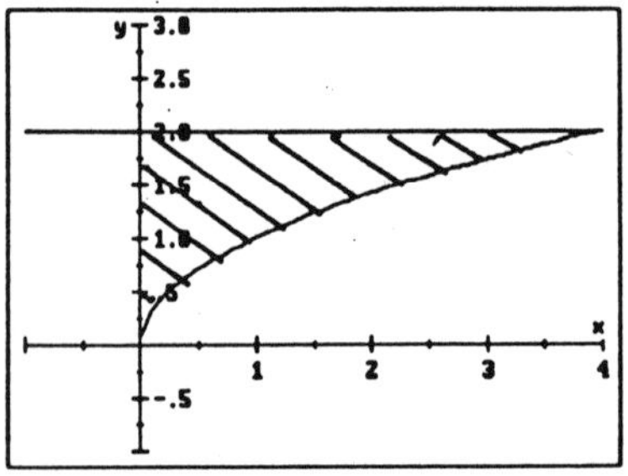

21. $\int_0^5 \int_0^x (2x + y)\, dy\, dx = \int_0^5 \left(2xy + \frac{y^2}{2} \Big|_{y=0}^{y=x} \right) dx = \int_0^5 \left(2x^2 + \frac{x^2}{2} \right) dx$

$= \int_0^5 \frac{5}{2} x^2\, dx = \frac{5x^3}{6} \Big|_0^5 = \frac{625}{6}$

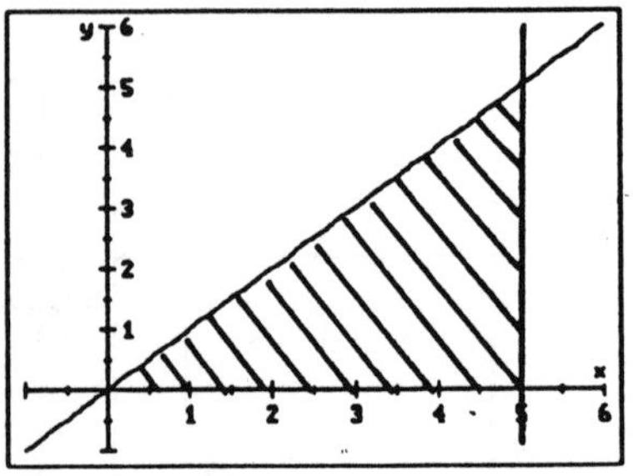

23. $$\int_0^1\int_{x^3}^{x^2} x^2 y\,dy\,dx = \int_0^1 x^2\left(\frac{y^2}{2}\Big|_{x^3}^{x^2}\right)dx = \int_0^1 x^2\left(\frac{x^4}{2}-\frac{x^6}{2}\right)dx = \frac{1}{2}\int_0^1 \left(x^6 - x^8\right)dx$$

$$= \frac{1}{2}\left(\frac{x^7}{7}-\frac{x^9}{9}\Big|_0^1\right) = \frac{1}{2}\left(\frac{2}{63}\right) = \frac{1}{63}$$

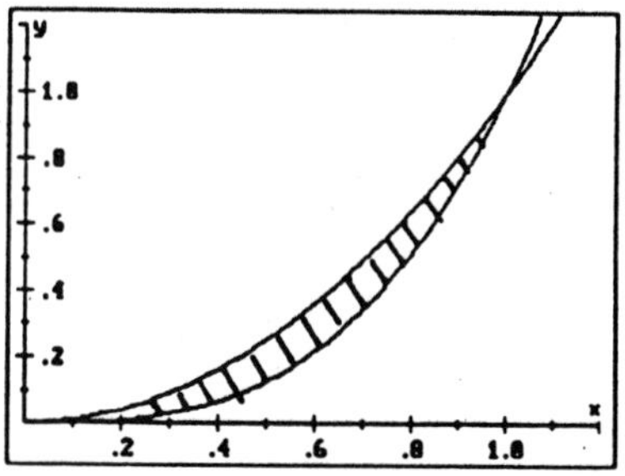

25. $$\int_1^4\int_0^3 dy\,dx = \int_0^3\int_1^4 dx\,dy = \int_0^3\left(x\Big|_1^4\right)dy = \int_0^3 3\,dy = 3y\Big|_0^3 = 9$$

27. $$\int_1^3\int_0^y (x+3y)dx\,dy = \int_0^1\int_1^3 (x+3y)\,dy\,dx + \int_1^3\int_x^3 (x+3y)\,dx\,dy$$

$$= \int_0^1\left(xy + \frac{3y^2}{2}\Big|_{y=1}^{y=3}\right)dx + \int_0^1\left(xy + \frac{3y^2}{2}\Big|_{y=x}^{y=3}\right)dx$$

$$= \int_0^1\left[\left(3x+\frac{27}{2}\right) - \left(x+\frac{3}{2}\right)\right]dx + \int_0^1\left[\left(3x+\frac{27}{2}\right) - \left(x^2+\frac{3x^2}{2}\right)\right]dx$$

$$= \int_0^1 (2x+12)\,dx + \int_0^1\left(-\frac{5}{2}x^2 + 3x + \frac{27}{2}\right)dx$$

$$= x^2 + 12x\Big|_0^1 + \left(-\frac{5x^3}{6} + \frac{3x^2}{2} + \frac{27x}{2}\Big|_0^1\right)$$

$$= 13 + \frac{52}{3} = \frac{91}{3}$$

29. $$\int_0^5\int_0^x (2x+y)dy\,dx = \int_0^5\int_y^5 (2x+y)dx\,dy = \int_0^5 x^2 + xy\Big|_y^5\,dy$$

$$= \int_0^5 \left[(25+5y) - (y^2+y^2)\right]dy$$

$$= \int_0^5 (25 + 5y - 2y^2)\, dy = 25y + \frac{5y^2}{2} - \frac{2y^3}{3}\Big|_0^5 = \frac{625}{6}$$

31. $\displaystyle\int_0^3\int_0^2 (x^2 + y)dx\, dy = \int_0^3 \frac{x^3}{3} + xy \Big|_{x=0}^{x=2} dy = \int_0^3 (\tfrac{8}{3} + 2y)\, dy = \tfrac{8}{3}y + y^2\Big|_0^3 = 17$

33. $\displaystyle\int_0^3\int_0^{1-x} \tfrac{1}{x}\, dy\, dx = \int_1^3 \left(\tfrac{1}{x}\, y\Big|_0^{1-x}\right) dx = \int_1^3 \tfrac{1}{x}(1 - x)\, dx$

$$= \int_1^3 \left(\tfrac{1}{x} - 1\right) dx = \ln|x| - x\Big|_1^3 = (\ln 3 - 3) - (\ln 1 - 1) = \ln 3 - 2$$

35. $\displaystyle\int_1^2\int_0^{y^2} y^3 e^{xy}\, dx\, dy = \int_1^2 y^3\left(\tfrac{1}{y}e^{xy}\Big|_{x=0}^{x=y^2}\right) dy = \int_1^2 y^2\left(e^{y^3} - 1\right) dy$

$$= \int_1^2 y^2 e^{y^3}\, dy - \int_1^2 y^2\, dy = \tfrac{1}{3}e^{y^3}\Big|_1^2 - \tfrac{1}{3}y^3\Big|_1^2$$

$$= \tfrac{1}{3}\left(e^8 - e\right) - \left(\tfrac{8}{3} - \tfrac{1}{3}\right) = \frac{e^8}{3} - \frac{e}{3} - \frac{7}{3}$$

37. Let $u = x^2 + y^2$. Treating y as a constant (for the first integration only) $du = 2xdx$:

$$\int_2^4\int_1^3 xy\sqrt{x^2 + y^2}\, dx\, dy = \int_2^4 \left[\int_{u=1+y^2}^{u=9+y^2} xyu^{1/2}\frac{du}{2x}\right] dy = \int_2^4 \frac{y}{2}\int_{1+y^2}^{9+y^2} u^{1/2}\, du$$

$$= \int_2^4 \frac{y}{2}\left[\frac{2}{3}u^{3/2}\Big|_{1+y^2}^{9+y^2}\right] dy = \int_2^4 \left[\frac{y}{3}(9 + y^2)^{3/2} - \frac{y}{3}(1 + y^2)^{3/2}\right] dy$$

$$= \frac{1}{3}\int_2^4 y(9 + y^2)^{3/2}\, dy - \frac{1}{3}\int_2^4 y(1 + y^2)^{3/2}\, dy$$

$$= \frac{1}{3}\left[\frac{(9 + y^2)^{5/2}}{5}\Big|_2^4\right] - \frac{1}{3}\left[\frac{(1 + y^2)^{5/2}}{5}\Big|_2^4\right]$$

$$= \frac{1}{3}\left[625 - \frac{13^{5/2}}{5}\right] - \frac{1}{3}\left[\frac{17^{5/2}}{5} - \frac{5^{5/2}}{5}\right]$$

$$= \frac{1}{15}\left(625 - 13^{5/2} - 17^{5/2} + 5^{5/2}\right) \approx 92$$

39. **A.** $\displaystyle\int_0^1 \int_2^{\sqrt{x}} (x + y)\, dy\, dx = \int_0^1 \left[xy + \frac{y^2}{2} \Big|_{x^2}^{\sqrt{x}} \right] dx = \int_0^1 \left(x\sqrt{x} + \frac{x}{2} - x^3 - \frac{x^4}{2} \right) dx$

$$= \frac{2}{5}x^{5/2} + \frac{x^2}{4} - \frac{x^4}{4} - \frac{x^5}{10} \Big|_0^1 = \frac{3}{10}$$

B. $\displaystyle\int_0^1 \int_{y^2}^{\sqrt{y}} (x + y)\, dx\, dy = \int_0^1 \left[\frac{x^2}{2} + xy \Big|_{y^2}^{\sqrt{y}} \right] dy = \int_0^1 \left(\frac{y}{2} + y\sqrt{y} - \frac{y^4}{2} - y^3 \right) dy$

$$= \frac{3}{10}$$

41. **A.** $\displaystyle\int_0^5 \int_0^{5 - x} 2xy\, dy\, dx = \int_0^5 xy^2 \Big|_0^{5 - x} dx = \int_0^5 x(5 - x)^2\, dx$

$$= \int_0^5 (25x - 10x^2 + x^3)\, dx$$

$$= \frac{25x^2}{2} - \frac{10x^3}{3} + \frac{x^4}{4} \Big|_0^5 = \frac{625}{12}$$

B. $\displaystyle\int_0^5 \int_0^{5 - x} 2xy\, dx\, dy = \int_0^5 x^2 y \Big|_0^{5 - y} dy = \int_0^5 (5 - y)^2\, y dy$

$$= \frac{25y^2}{2} - \frac{10y^3}{3} + \frac{y^4}{4} \Big|_0^5 = \frac{625}{12}$$

43. **A.** $\displaystyle\int_0^2 \int_{x^2}^{8 - x^2} dy\, dx = \int_0^2 y \Big|_{x^2}^{8 - x^2} dx = \int_0^2 (8 - 2x^2) dx$

$$= 8x - \frac{2x^3}{3} \Big|_0^2 = 10\frac{2}{3}$$

B. $$\int_0^4\int_{y^2}^{\sqrt{y}} dx\,dy + \int_4^8\int_0^{\sqrt{8-y}} dx\,dy = \int_0^4 \sqrt{y}dy + \int_4^8 \sqrt{8-y}\,dy$$

$$= \tfrac{2}{3}y^{1/2}\ \Big|_0^4 - \tfrac{2}{3}(8-y)^{3/2}\ \Big|_4^8 = 10\tfrac{2}{3}$$

45. $$\int_{-3}^{2}\int_{1}^{5} 5\,dx\,dy = 5\int_{-3}^{2} x\Big|_1^5\,dy = 5\int_{-3}^{2} 4\,dy = 20y\ \Big|_{-3}^{2} = 100$$

47. $\int_0^1\int_0^1 xy\sqrt{x^2+y^2}\,dx\,dy$ Let $u = x^2 + y^2$. Treating y as a constant, $du = 2x\,dx$:

$$= \int_0^1 \int_{u=y^2}^{u=y^2+1} \left(xyu^{1/2}\frac{du}{2x}\right) dy = \int_0^1 \frac{y}{2}\int_{y^2}^{y^2+1} u^{1/2}\,du dy = \int_0^1 \frac{y}{2}\,du dy$$

$$= \int_0^1 \frac{y}{2}\left(\frac{2}{3}u^{3/2}\right)\Big|_{y^2}^{y^2+1}\,dy$$

$$= \int_a^b \left[\frac{y}{3}(y^2+1)^{3/2} - \frac{y}{3}(y^3)\right] dy = \frac{1}{3}\int_0^1 y(y^2+1)^{3/2}\,dy - \frac{1}{3}\int_0^1 y^4\,dy$$

$$= \frac{1}{3}\left[\frac{(y^2+1)^{5/2}}{5}\right]\Bigg|_0^1 - \frac{1}{3}\left[\frac{y^5}{5}\right]\Bigg|_0^1 = \frac{1}{3}\left(\frac{2^{5/2}}{5} - \frac{1}{5}\right) - \frac{1}{3}\left(\frac{1}{5}\right) = \frac{1}{15}(2^{5/2} - 2)$$

$\approx .2437902833$

49. $$\int_0^1\int_{x^2}^{\sqrt{x}} 5xy\,dy\,dx = \int_0^1 \frac{5xy^2}{2}\Bigg|_{y=x^2}^{y=\sqrt{x}}\,dx = \int_0^1 \frac{5}{2}(x^2 - x^5)\,dx$$

$$= \frac{5}{2}\left(\frac{x^3}{3} - \frac{x^6}{6}\right)\Bigg|_0^1 = \frac{5}{2}\left(\frac{1}{6}\right) = \frac{5}{12}$$

51. $$\int_0^3\int_{x^2}^{3x} 6dy\,dx = \int_0^3 6(3x - x^2)dx = 9x^2 - 2x^3\Big|_0^3 = 27$$

53. $$\int_0^1\int_{x^2}^{x} xy^2\,dy\,dx = \int_0^1 \frac{xy^2}{3}\Bigg|_{x^2}^{x}\,dx = \int_0^1 \left(\tfrac{1}{3}x^4 - \tfrac{1}{3}x^7\right)dx$$

$$= \frac{x^5}{15} - \frac{x^8}{24} \Big|_0^1 = \frac{3}{120} = \frac{1}{40}$$

55. $$\frac{1}{(2-0)(3-0)}\int_0^3\int_0^2 (x - 2y)\,dx\,dy = \frac{1}{6}\int_0^3 \frac{x^2}{2} - 2xy \Big|_0^2 dy = \frac{1}{6}\int_0^3 (2 - 4y)\,dy$$

$$= \frac{1}{6}\left[2y - 2y^2\right]\Big|_0^3 = \frac{1}{6}[6 - 18] = -2$$

57. $$\frac{1}{(5-0)(2-0)}\int_0^2\int_0^5 L^{0.75}K^{0.25}\,dL\,dK = \frac{1}{10}\int_0^2 \frac{L^{1.75}}{1.75}K^{0.25}\Big|_{L=0}^{L=5} dK$$

$$= \frac{1}{17.5}\int_0^2 5^{1.75}K^{0.25}\,dK = \frac{5^{1.75}}{17.5}\left[\frac{K^{1.25}}{1.25}\right]\Big|_0^2 = \frac{5^{.75}}{17.5}\left[\frac{2^{1.25}}{1.25}\right]$$

$$= \frac{5^{7/4}2^{5/4}}{\frac{175}{8}} = \frac{8}{175}\sqrt[4]{5^7}\ \sqrt[4]{2^5} = \frac{8}{175}\left(5\sqrt[4]{5^3}\right)\left(2\sqrt[4]{2}\right)$$

$$= \frac{80}{175}\left(\sqrt[4]{250}\right) = \frac{16}{35}\sqrt[4]{250} \approx 1.81776166$$

8.6 Correlation and Least Squares Applications

1. yes **3.** no **5.** yes **7.** no **9.** no **11.** no

13. $n = 4,\ \Sigma x = 10,\ \Sigma y = 27,\ \Sigma^2 = 30,\ \Sigma y^2 = 259,\ \Sigma xy = 87$

$$r = \frac{4(87) - 10(27)}{\sqrt{4(30) - 10^2}\sqrt{4(259) - 27^2}} \approx .995$$

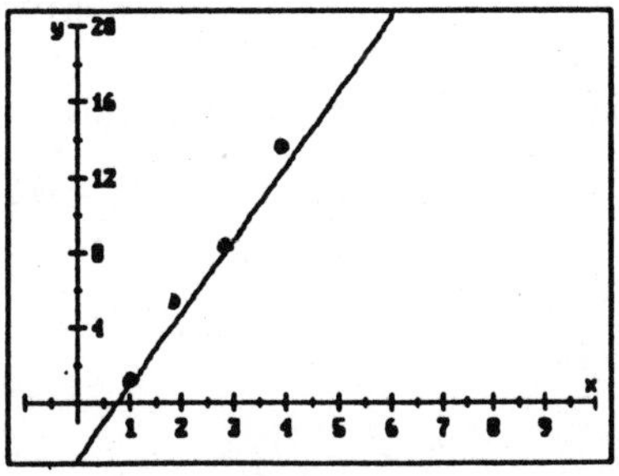

There is a linear correlation at 5%.

15. $n = 5$, $\Sigma x = 10$, $\Sigma y = 82$, $\Sigma x^2 = 30$, $\Sigma y^2 = 1486$, $\Sigma xy = 127$

$$r = \frac{5(127) - 10(82)}{\sqrt{5(30) - (10)^2}\sqrt{5(1486) - (82)^2}} \approx -.9847$$

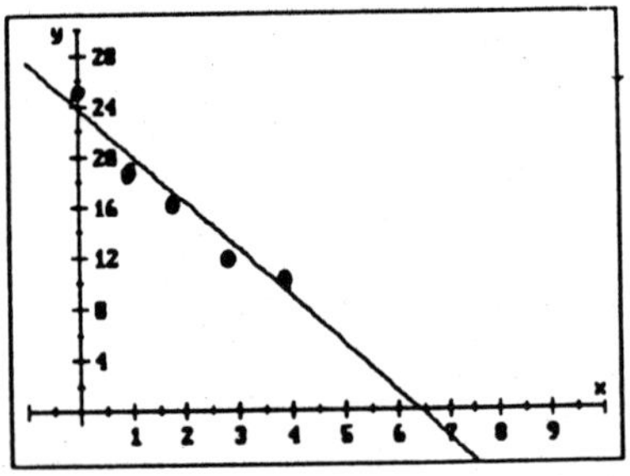

There is a linear correlation at 1%.

17. $n = 6$, $\Sigma x = 200$, $\Sigma y = 361$, $\Sigma x^2 = 8400$, $\Sigma y^2 = 22{,}293$, $\Sigma xy = 13{,}000$

$$r = \frac{6(13{,}000) - 200(361)}{\sqrt{6(8400) - 200^2}\sqrt{6(22{,}293) - 361^2}} \approx .970$$

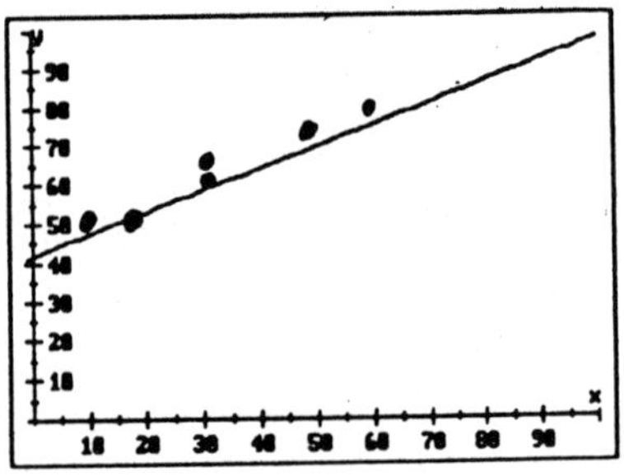

There is a linear correlation at 1%.

19. See Problem 13. $m = \dfrac{4(87) - 10(27)}{4(30) - 10^2} = 3.9$, $b = \dfrac{27 - 3.9(10)}{4} = -3$,

$y = 3.9x - 3$

21. See Problem 17. $m = \dfrac{6(13{,}000) - 200(361)}{6(8400) - 200^2} \approx .5577$, $b = \dfrac{361 - .5577(200)}{6}$

≈ 41.58 Rounding to two decimals, $y = .56x + 41.58$

23. $n = 5$, $\Sigma x = 20$, $\Sigma y = 96$, $\Sigma x^2 = 108$, $\Sigma y^2 = 2070$, $\Sigma xy = 308$;

$$m = \frac{5(308) - 20(96)}{5(108) - 20^2} \approx -2.71, \; b = \frac{96 - (-2.17)(20)}{5} \approx 30.06,$$

$y = -2.71x + 30.06$

25. $n = 8$, $\Sigma x = 27$, $\Sigma y = 115$, $\Sigma x^2 = 109$, $\Sigma y^2 = 1805$, $\Sigma xy = 432$;

$$r = \frac{8(432) - 27(115)}{\sqrt{8(109) - 27^2}\sqrt{8(1805) - 115^2}} \approx .842$$

This is significant at 1%.

27. See Problem 25. $m = \dfrac{8(432) - 27(115)}{8(109) - 27^2} \approx 2.4545,$

$b = \dfrac{115 - 2.4545(27)}{8} \approx 6.0909; \;\; y = 2.45x + 6.09$

29. $r = .358$, no significant correlation

31. $r = -.936$, significant at 1%

33. $y = .468x + 19.235$

35. $y = -6.186x + 87.179$

8.7 Review

Objective Questions

1. $f(5, -12) = 3(5)^2 - 2(5)(-12) + (-12)^2 = 339$

3. $K(5, -12) = (5)(-12) = -60$

5.

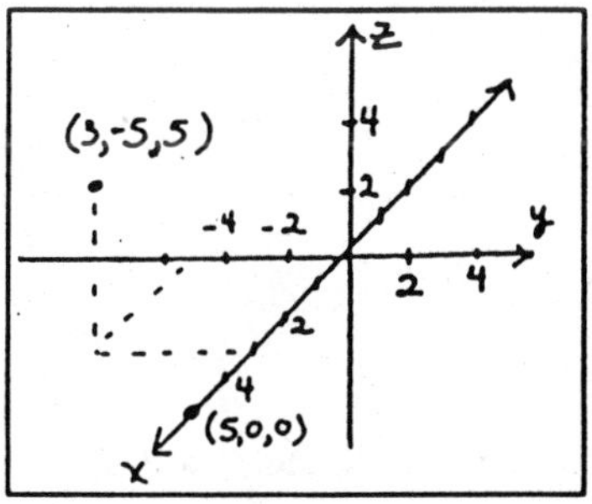

7.

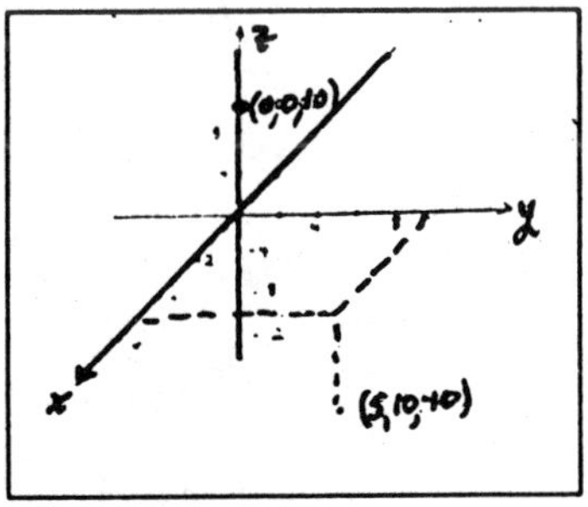

9.

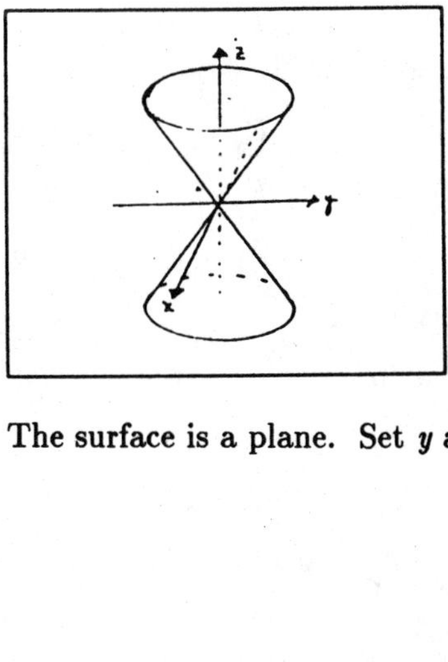

11. The surface is a plane. Set y and z equal to zero to get $x = 10$, the x-intercept.

Set x and y equal to zero to get $z = 5$, the z-intercept. Set x and z equal to zero to get $y = 10$, the y-intercept.

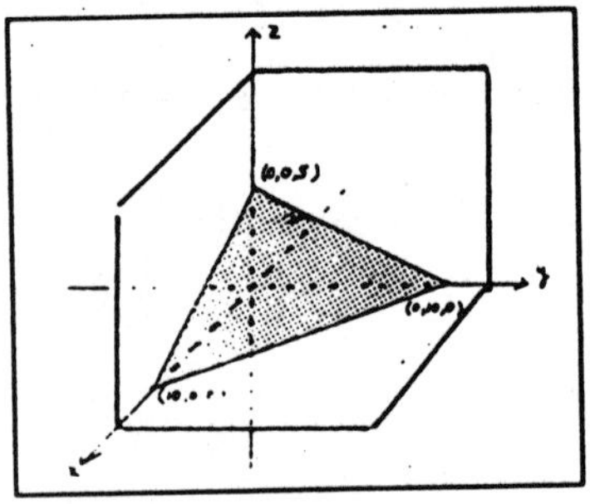

13. $f_x(x, y) = 12(2x - 5)^{11}(2) = 24(2x - 5y)^{11}$

15. Treat x as a constant: $\dfrac{\partial z}{\partial y} = 1000[-\frac{1}{2}(x^2 + y^2)^{-3/2} \cdot \dfrac{\partial}{\partial y}(x^2 + y^2)] =$

$-500(x^2 + y^2)^{-3/2}(2y) = -1{,}000y(x^2 + y^2)^{-3/2}$

17. $f_x(x, y) = 15x^4 - 20x^3y^3 + 4x$

$f_{xx}(x, y) = 60x^3 - 60x^2y^3 + 4$

19. $\dfrac{\partial z}{\partial y} = -15x^4y^2;\ \dfrac{\partial^2 z}{\partial y \partial x} = \dfrac{\partial^2 z}{\partial x \partial y} = \dfrac{\partial^2 z}{\partial x}(-15x^4y^2) = -60x^3y^2$

21. No relative maximums or minimums.

23. $f_x = 4x - 3y,\ f_y = -3x + 2y$. Solve $\begin{cases} 4x - 3y = 0 \\ -3x + 2y = 0 \end{cases}$

The only solution is $x = 0$ and $y = 0$.

$f_{xx} = 4,\ f_{xy} = -3,\ f_{yy} = 2;\ \ D = (-3)^2 - (4)(2) = 1 > 0$; Since $D > 0$, there is a saddle point at (0, 0, 0).

25. Maximize $f(x, y) = 12 - x^2 - y^2$ subject to $x + y = 10$

$F(x, y, \lambda) = 12 - x^2 - y^2 + \lambda(x + y - 10)$. Solve $\begin{cases} F_x = -2x + \lambda = 0 \\ F_y = -2y + \lambda = 0 \\ F_\lambda = x + y - 10 = 0 \end{cases}$

The solution is $x = 5,\ y = 5\ (\lambda = 10)$. Maximum is -38.

27. Letting x and y be the numbers, we want to maximize

$f(x, y) = xy$ subject to $x + y = 250$

$F(x, y, \lambda) = xy + \lambda(x + y - 250)$. Solve $\begin{cases} F_x = y + \lambda = 0 \\ F_y = x + \lambda = 0 \\ F_\lambda = x + y - 250 = 0 \end{cases}$

The solution is $x = 125$, $y = 125$ $(\lambda = -125)$. Maximum is 15,625.

29. $\displaystyle\int_1^3 x^2y^3\,dx = \frac{x^3y^3}{3}\Big|_1^3 = \frac{26y^3}{3}$

31. $\displaystyle y^2\int_0^1 x^2\sqrt{x^3+8}\,dx = y^2\int_0^1 (x^3+8)^{1/2}x^2\,dx$. Let $u = x^3 + 8$, then $du = 3x^2\,dx$:

$$y^2\int_{x=0}^{x=1} u^{1/2}\frac{du}{3} = \frac{y^2}{3}\left[\tfrac{2}{3}u^{3/2}\right]\Big|_{x=0}^{x=1}$$

$$= \frac{2y^2}{9}\left[(x^3+8)^{3/2}\right]\Big|_0^1$$

$$= \frac{2y^2}{9}(27 - 8^{3/2})$$

$$= \left(6 - \tfrac{32}{9}\sqrt{2}\right)y^2$$

33. $\displaystyle\int_0^9\int_{\sqrt{x}}^{\sqrt{x}+2} y\,dy\,dx = \int_0^9 \frac{y^2}{2}\Big|_{\sqrt{x}}^{\sqrt{x}+2} dx$

$$= \tfrac{1}{2}\int_0^9 (x + 4x + 4 - x)\,dx$$

$$= \tfrac{1}{2}\left(\tfrac{2}{3}\right)x^{3/2} + 2\text{x}\Big|_0^9$$

$$= 54$$

35. $\displaystyle\int_1^2\int_0^{1/y} y^3\,dxdy = \int_1^2 y^3x\,\Big|_{x=0}^{x=\frac{1}{y}} dy$

$$= \int_1^2 y^2\,dy$$

$$= \frac{y^3}{3}\Big|_1^2 = \frac{7}{3}$$

37. $\displaystyle\int_1^8\int_{2/y}^{\sqrt{2y}} dx\,dy = \int_1^8 x|_{2/y}^{\sqrt{2y}}\,dy$

$$= \int_1^8\left(\sqrt{2y} - \tfrac{2}{y}\right)dy$$

$$= \left[\sqrt{2}\left(\frac{2}{3}\right)y^{3/2} \; - \; 2 \ln y\right]\Bigg|_1^8$$

$$= \frac{64 \; - \; 2\sqrt{2}}{3} \; - \; 6 \ln 2 \; \approx \; 16.2$$

39. The region of integration is bounded below by the parabola $y = x^2$, above the line $y = 4$ and by the y-axis. To reverse the order, x must go from $x = 0$ to the parabola $x = \sqrt{y}$:

$$\int_0^2 \int_{x^2}^4 dx\; dy = \int_0^4 \int_0^{\sqrt{y}} dx\; dy = \frac{16}{3}$$

41. $$\int_0^5 \int_{x^2}^{5x} (x + 2y)\, dy\; dx = \int_0^5 \left[xy + y^2\right]_{x^2}^{5x} dx$$

$$= \int_0^5 (5x^2 + 25x^2 \; - \; x^3 \; - \; x^4)\, dx$$

$$= (10x^3 \; - \; \frac{x^4}{4} \; - \; \frac{x^5}{5} \,\Bigg|_0^5$$

$$= \frac{1{,}875}{4}$$

$$\approx \; 468.75$$

43. $$\int_0^1 \int_0^2 x^2 y^3\; dx\; dy = \int_0^1 \frac{x^3}{3} y^3 \,\Bigg|_0^2\; dy = \int_0^1 \frac{8}{3} y^3\; dy = \frac{2y^4}{3}\Bigg|_0^1 = \frac{2}{3}$$

45.

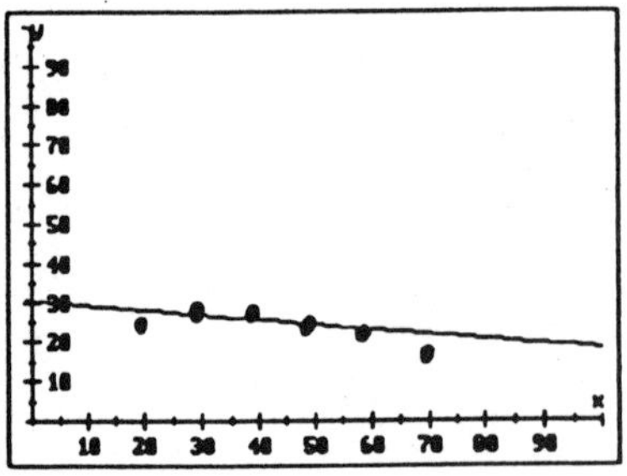

47.

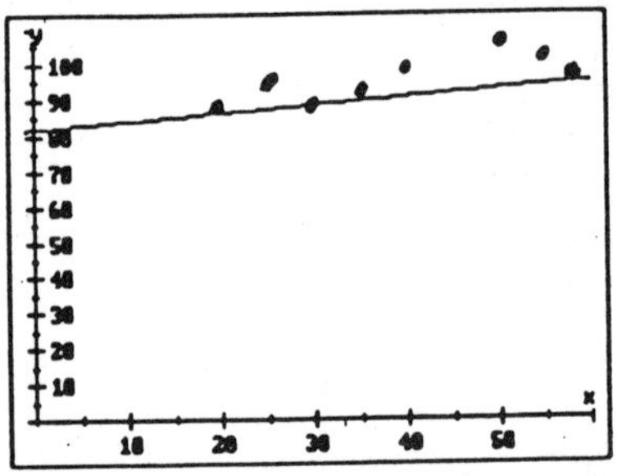

49. $r = \; -.914$; significant at both 1% and 5%

51. $y = \; -.12x + 30.4$ **53.** $y = .218x + 81.821$

55. $$\frac{\partial P}{\partial x}\Bigg|_{(2,4)} = 4y \; - \; 6x\Big|_{(2,4)} = 16 \; - \; 12 = 4$$

When the labor force is 2 (in hundreds) and the capital investment is 4 (in millions of dollars), then the rate of change of productivity with respect to labor is 4 per employee per day.

57. The maximum productivity is .35111 when $x = \frac{16}{27}$ [labor force is $\frac{16}{27}(100) \approx 59$ workers] with a capital investment of $y = \frac{8}{9}$ [in millions $\approx$ \$888,889].

Test Questions

1. $P(10, 50) = 10^{.4}50^{.2} \approx 5.49$

3. Since all variables are of first degree, the surface is a plane. The intercepts are:

$x = 0,\ y = 0 \Rightarrow z = 8$

$x = 0,\ z = 0 \Rightarrow y = 8$

$y = 0,\ z = 0 \Rightarrow x = 4$

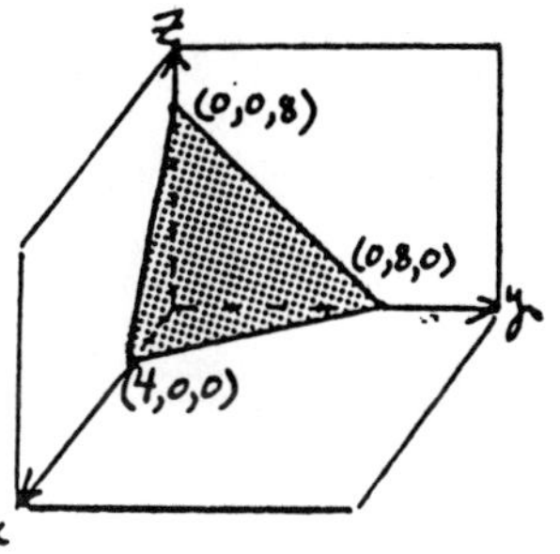

5. $f_x = 10(x - 3y)^9 \cdot \frac{\partial}{\partial x}(x - 3y) = 10(x - 3y)^9$

7. $f_x = 16x^3 - 9x^2y^2 + 4x;$

$f_{xx} = 48x^2 - 18xy^2 + 4$

9. $\int_0^9 \int_{\sqrt{x}}^{\sqrt{x}+2} dy\, dx = \int_0^9 y \Big|_{\sqrt{x}}^{\sqrt{x}+2} dx = \int_0^9 [\sqrt{x} + 2 - \sqrt{x}]\, dx = \int_0^9 (x + 2\ dx)dx$

$= 2x \Big|_0^9 = 18$

11. $f_x = 4x - 3y,\ f_y = -3x + 2y$ To find critical points, solve

$$\begin{cases} 4x - 3y = 0 \\ -3x + 2y = 0 \end{cases}$$

The solution is (0, 0). Using the second derivative test,

$A = f_{xx} = 4$: $B = f_{xy} = -3$; $C = f_{yy} = 2$, $D = B^2 - AC = 1$

Since $D > 0$, (0, 0, 0) is a saddle point.

13. $\int_0^2 \int_1^{e^x} dy\, dx = e^2 - 3 \approx 4.39$

15. Revenue $= R(x, y) = (50 - .05x + .001y)x + (130 + .01x - .04y)y$

$$= 50x - .05x^2 + .011xy + 130y - .04y^2$$

$$R(5, 8) = 50(5) - .05(5^2) + .011(5)(8) + 130(8) - .04(8^2) = 1{,}286.63$$

$$C(5, 8) = 90 + 20(5) + 90(8) = 910$$

CHAPTER 9

DIFFERENTIAL EQUATIONS

9.1 First Order Differential Equations

1. $dy = x^2\,dx$

$$\int dy = \int x^2\,dx$$

$$y = \frac{x^3}{3} + C$$

3. $dy = (8x - 10)\,dx$

$$\int dy = \int (8x - 10)\,dx$$

$$y = 4x^2 - 10x + C$$

5. $dy = (4x^3 - 3x^2 - 5)\,dx$

$$\int dy = \int (4x^3 - 3x^2 - 5)\,dx$$

$$y = x^4 - x^3 - 5x + C$$

7. $dy = \frac{1}{5}e^x\,dx$

$$\int dy = \int \tfrac{1}{5}e^x\,dx$$

$$y = \tfrac{1}{5}e^x + C$$

9. $dy = \frac{1}{12}(5x + 1)^{1/2}\,dx$

$$\int dy = \tfrac{1}{12}\int (5x + 1)^{1/2}\,dx$$

$$y = \tfrac{1}{12}\left[\tfrac{2}{15}(5x + 1)^{3/2}\right] + C$$

$$y = \tfrac{1}{90}(5x + 1)^{3/2} + C$$

11. $ydy = xdx$

$$\int ydy = \int xdx$$

$$\frac{y^2}{2} = \frac{x^2}{2} + C_1$$

$$y^2 = x^2 + C \quad (C = 2C_1)$$

13. $y^2 dy = (x^3 - 3)dx$

$$\int y^2 dy = \int (x^3 - 3)dx$$

$$\frac{y^3}{3} = \frac{x^4}{4} - 3x + C_1$$

$$y^3 = \frac{3x^4}{4} - 9x + C \quad (C = 3C_1)$$

15. $\frac{dy}{y} = 2xdx$

$$\int \frac{dy}{y} = \int 2xdx$$

$$\ln|y| = x^2 + C_1$$

We could isolate y:

$$|y| = e^{x^2 + C_1} = e^{x^2}e^{C_1}$$

$$|y| = C_2 e^{x^2} \quad (C_2 = e^{C_1})$$

$$\pm C_2 e^{x^2}$$

$$y = Ce^{x^2} \quad (C = \pm C_2)$$

17. Using Theorem 3, where $K = .02$, $P = Me^{.02t}$. Without Theorem 3:

$$\frac{dP}{P} = .02dt$$

$$\ln|P| = .02t + C_1$$

$$50 \ln|P| = t + C \quad (C = 50C_1).$$

If we want to solve for P, we would get $P = Me^{.02t}$.

19. Using Theorem 3, where $K = .001$, $N = Me^{.001t}$. Without Theorem 3:

$$\frac{dN}{N} = .001dt$$

$$\ln|N| = .001t + C_1$$

$1000 \ln|N| = t + C \quad (C = 1000C_1)$. This can also be written $N = Me^{.001t}$.

21. $y^2\,dy = x^2\,dx$

$$\int y^2\,dy = \int x^2\,dx$$

$$\frac{y^3}{3} = \frac{x^3}{3} + C_1$$

$$y^3 = x^3 + C \quad (C = 3C_1).$$

Substitute $y = 5$, $x = 0$ and solve for C:

$$5^3 = 0^3 + C$$

$$C = 125$$

The particular solution is $y^3 = x^3 + 125$

23. See Problem 17. $P = Me^{.02t}$. Substitute $P = e^3$, $t = 0$ and solve for the constant

M: $e^3 = Me^0$

$$M = e^3.$$

The particular solution is $P = e^3 e^{.02t}$. An equivalent form is $P = e^{3\,+\,.02t}$ or $\ln P = 3 + .02t$.

25. Multiply both sides by $\frac{dx}{xy}$: $\frac{dy}{y} - \frac{dx}{\sqrt{x}} = 0$

$$\int \frac{dy}{y} - \int x^{-1/2}\,dx = C_1$$

$$\ln|y| - 2x^{1/2} = C$$

$$\ln|y| = 2\sqrt{x} + C.$$

Substitute $y = 1$, $x = 1$ and solve for C:

$$\ln 1 = 2\sqrt{1} + C$$

$$0 = 2 + C$$

$$C = -2.$$

The particular solution is $\ln|y| = 2\sqrt{x} - 2$.

27. $\frac{dy}{y} = \frac{x\,dx}{1 + x^2}$

$$\ln|y| = \tfrac{1}{2}\ln\left|1 + x^2\right| + C$$

Since $1 + x^2$ is always positive, $\ln|y| = \frac{1}{2}\ln(1 + x^2) + C$. Substitute $x = -1$, $y = 2$ and solve for C:

$$\ln 2 = \tfrac{1}{2}\ln 2 + C$$

$\frac{1}{2}\ln 2 = C$

The particular solution is $\ln|y| = \frac{1}{2}\ln(1 + x^2) + \frac{1}{2}\ln 2$.

29. If N is the number of crimes and t represents the number of years after 1986, $\frac{dN}{dt} = .03t$. By Theorem 3, $N = Me^{.03t}$. The number of crimes was 3,500 when the year was 1986 $\Rightarrow N = 3{,}500$ when $t = 0$. Substitute this into $N = Me^{.03t}$ and solve for the constant M: $3{,}500 = Me^0 \quad M = 3{,}500$. The particular solution is $N = 3{,}500e^{.03t}$. The number of crimes, N, 1996 ($N = 10$) will be $N = 3{,}500e^{.03(10)} = 3{,}500e^{.3} \approx 4{,}725$.

31. If N is the number of bacteria and t is the number of hours, $\frac{dN}{dt} = .10t$. By Theorem 3, $N = Me^{.10t}$. Initially there are 12 (million), so substitute $N = 12$ when $t = 0$ and solve for constant M: $12 = Me^0 = M$. So the particular solution is $N = 12e^{.10t}$. Next, let $N = 1{,}000$ and solve for t: $1{,}000 = 12e^{.10t}$

$$\frac{1{,}000}{12} = e^{.10t}$$

$$.10t = \ln\left(\frac{1{,}000}{12}\right)$$

$$t = 10\ln\left(\frac{1{,}000}{12}\right)$$

$$\approx 44.2 \text{ hours.}$$

9.2 Applications - Growth Models

1. This uninhibited growth with a growth rate of 12%, $P_0 = 25{,}000$ and $t = 5$ (see Example 1). We have $P = 25{,}000e^{.12(5)} \approx 45{,}553$.

3. See Example 2. Since 28% is present: $-.0001205479t = \ln\left(\frac{P}{P^0}\right) = \ln .28$

$t \approx \frac{\ln .28}{-.0001205479} \approx 10{,}560$ years.

5. Using $P'(x) = 6x^2 - 10$

$$P(x) = 2x^3 - 10x + C.$$

Since $P(3) = 1{,}000$:

$$1{,}000 = P(3) = 2(3)^3 - 10(3) + C$$

$$C = 976.$$

Thus, $P(x) = 2x^3 - 10x + 976$. Now $P(5) = 2(5)^3 - 10(5) + 976 = 1{,}176$.

The increase is $P(5) - P(3) = 1{,}176 - 1{,}000 = 176$ items per month.

7. Using the uninhibited growth model, $P = P_0 e^{.0715t}$ where t is the number of years after 1990 and $P_0 = 5{,}200.8$ (billion dollars). In 2000, $t = 11$:
$P = 5{,}200.8^{.0715t(11)} \approx \$11{,}419.4$ billion.

9. Use $P = P_0 e^{rt}$ where t is the number of years after 1980 and $P_0 = 226{,}545{,}805$
$P = 226{,}545{,}805 e^{rt}$. Since $P = 249{,}632{,}692$ when $t = 10$: $\dfrac{249{,}632{,}692}{226{,}545{,}805} = e^{10r}$ or

$$r = \frac{\ln \dfrac{249{,}632{,}692}{226{,}545{,}805}}{10} \approx .0097043451.$$ So in 2000, $t = 20$:

$$P = 226{,}545{,}805 e^{.0097043451(20)} \approx 275{,}072{,}323$$

11. Use $P = P_0 e^{rt}$ where t is the number of years after 1980 and $P_0 = 14{,}609{,}000$:

$P = 14{,}609{,}000 e^{rt}$. Since $P = 15{,}575{,}000$ when $t = 4$: $\dfrac{15{,}575{,}000}{14{,}609{,}000} = e^{r(4)}$ or

$$r = \frac{\ln \dfrac{15{,}575{,}000}{14{,}609{,}000}}{4} \approx .0160073219.$$ So in 1990, $t = 10$:

$$P = 14{,}609{,}000 e^{.0160073219(10)} \approx 17{,}145{,}076$$

13. Use $P = P_0 e^{rt}$ where t is the number of years after 1990 and r is .97%:

$$P = 2{,}423{,}000 e^{.0097t}$$

$$P(10) = 2{,}423{,}000 e^{.0097t}$$

$$\approx 2{,}669{,}808$$

15.

$$\frac{dy}{dt} = .7 - .001y$$

$$\frac{dy}{.7 - .001y} = dt$$

$$\int (.7 - .001y)^{-1} dy = \int dt$$

$$-1{,}000 \ln |.7 - .001y| = t + C$$

If $t = 0$ then $y = 200$ so

$$-1{,}000 \ln |.7 - .001(200)| = C$$

$$693.147 \approx C$$

Thus, $-1{,}000 \ln |.7 - .001y| = t + 693.147$

When $t = 168$ we have

$$-1{,}000 \ln |.7 - .001y| = 168 + 693.147$$
$$\ln |.7 - .001y| = -.861147$$
$$.7 - .001y = e^{.861147}$$
$$.2773230063 = .001y$$
$$277 \approx y$$

17. Using the limited growth model, $P(t) = .5(1 - e^{-rt})$.

$$.2 = P(10) = .5[1 - e^{-r(10)}]$$
$$r = \frac{\ln\left(\frac{.2}{.5}\right)}{-10} \approx .05108256$$

We have $P(t) = .5(1 - e^{.05108256t})$. We want to find t such that $P = 30\%$:

$$.30 = .5(1 - e^{.05108256t}).$$
$$\frac{.30}{.5} = 1 - e^{.05108256t}$$
$$e^{.05108256t} = .4$$
$$t = \frac{\ln(.4)}{-.05108256} \approx 18 \text{ times per week.}$$

19. $$\frac{dP}{L - P} = Kdt$$

$$\int \frac{dP}{L - P} = \int Kdt$$
$$-\ln |L - P| = Kt + C$$

Since $L = 1$ and $L - P \geq 0$, we have $-\ln(1 - P) = Kt + C$. When $t = 0$, $P = 0$:

$$-\ln(1 - 0) = K(0) + C$$
$$C = 0$$

This gives us $-\ln(1 - P) = Kt$

$$\ln(1 - P) = -Kt$$
$$1 - P = e^{-Kt}$$
$$P = 1 - e^{-Kt}.$$

9.3 Special Second Order Differential Equations

1. $\frac{dy}{dx} = \int -24x\,dx = -12x^2 + C_1$

$$y = \int (-12x^2 + C_1)dx = -4x^3 + C_1x + C_2$$

Using $y = 5$ when $x = 0$: $5 = -4(0)^3 + C_1(0) + C_2 \Rightarrow C_2 = 5$

We have $y = -4x^3 = C_1x + 5$. Using $y = 29$ when $x = 2$:

$$29 = -4(2)^3 + C_1(2) + 5$$

$$C_1 = 28$$

The particular solution is $y = -4x^3 + 28x + 5$.

3. $$\frac{dy}{dx} = \int (10 - 6x)dx = 10x - 3x^2 + C_1$$
$$y = \int (10x - 3x^2 + C_1)dx = 5x^2 - x^3 + C_1x + C_2$$

Using $y = 8$ when $x = 0$:

$$8 = 5(0)^2 - 0^3 + C_1(0) + C_2$$

$$C_2 = 8$$

We have $y = 5x^2 - x^3 + C_1x + 8$. Using $y = 52$ when $x = 4$:

$$52 = 5(4)^2 - 4^3 + C_1(4) + 8$$

$$C_1 = 7$$

The particular solution is $y = 5x^2 - x^3 + 7x + 8$

5. $$\frac{dy}{dx} = \int (2x^2 - 3x + 5)dx = \frac{2x^3}{3} - \frac{3x^2}{2} + 5x + C_1$$
$$y = \int (\frac{2x^3}{3} - \frac{3x^2}{2} + 5x + C_1)dx = \frac{x^4}{6} - \frac{x^3}{2} + \frac{5x^2}{2} + C_1x + C_2$$

Using $y = 3$ when $x = 0$:

$$3 = \frac{0^4}{6} - \frac{0^3}{2} + \frac{5(0)^2}{2} + C_1(0) + C_2$$

We have $y = \frac{x^4}{6} - \frac{x^3}{2} + \frac{5x^2}{2} + C_1x + 3$. Using $y = 12$ when $x = 2$:

$$12 = \frac{2^4}{6} - \frac{2^3}{2} + \frac{5(2)^2}{2} + C_1(2) + 3$$

$$C_1 = \frac{1}{6}$$

The particular solution is $y = \frac{x^4}{6} - \frac{x^3}{2} + \frac{5x^2}{2} + \frac{x}{6} + 3$

7. $y'' = 4 - 3x$

$$y' = \int (4 - 3x)\,dx = 4x - \frac{3x^2}{2} + C_1$$

Using $y = 4{,}500$ when $x = 0$:

$$4{,}500 = 2(0)^2 - \frac{0^3}{2} + C_1(0) + C_2$$

We have $y = 2x^2 - \frac{x^3}{2} + C_1x + 4{,}500$. Using $y = 50$ when $x = 10$:

$$50 = 20(10)^2 - \frac{10^3}{2} + C_1(10) + 4{,}500$$

$$C_1 = -415$$

The particular solution is $y = 2x^2 - \frac{x^3}{2} - 415x + 4{,}500$.

9. See Example 2. Let $y' = p$ so that $y'' = \frac{dp}{dx}$:

$$xy'' = 2y'$$

$$x\frac{dp}{dx} = 2p$$

Separate the variables:

$$\frac{dp}{p} = 2\frac{dx}{x}$$

$$\ln|p| = 2\ln|x| + k$$

Let $k = \ln C_1$:

$$\ln|p| = 2\ln|x| + \ln C_1 = \ln x^2 + \ln C_1$$

$$\ln|p| = \ln C_1x^2$$

$$p = \pm C_1x^2$$

Substituting $p = \frac{dy}{dx}$: $\frac{dy}{dx} = \pm C_1x^2$

$$y = \int \pm C_1x^2\,dx = \pm\frac{C_1x^3}{3} + C_2$$

Using $y = -4$ when $x = 0$: $-4 = \pm\frac{C_1(0)^3}{3} + C_2$.

We have $y = \frac{\pm C_1x^3}{3} - 4$. Using $y = 8$ when $x = 3$:

$$8 = \pm\frac{C_1(3)^3}{3} - 4$$

$$C_1 = \pm\frac{4}{3}$$

The particular solution is $y = \pm\frac{4}{9}x^3 - 4$.

11. Subtract y from both sides to get the form $ay'' + by' + cy = 0$:

$$2y'' - y' - y = 0$$

Using Theorem 4, the auxiliary equation is:

$$2m^2 - m - 1 = 0$$
$$(2m + 1)(m - 1) = 0$$
$$m = -\frac{1}{2} \text{ or } 1$$

The general solution is $y = Ae^{-1/2x} + Be^x$.

Using $y = 0$ when $x = 0$: $0 = Ae^0 + Be^0 = A + B$.

Using $y = 1$ when $x = 4$: $1 = Ae^{-2} + Be^4$. We need to solve the system

$$\begin{cases} A + B = 0 \\ e^{-2}A + e^4B = 1 \end{cases}$$ The solution is $A = \dfrac{e^2}{1 - e^6}$ and $B = \dfrac{e^2}{e^6 - 1}$

The particular solution is $y = \left(\dfrac{e^2}{1 - e^6}\right) e^{-x/2} + \left(\dfrac{e^2}{e^6 - 1}\right) e^x$.

13. $\dfrac{dy}{dx} = \displaystyle\int e^x dx = e^x + A$, where A is a constant.

$y = \displaystyle\int (e^x + A)dx = e^x + Ax + B$, where B is a constant.

15. $\dfrac{dy}{dx} = \displaystyle\int (1 - e^{-x})dx = x + e^{-x} = A$, where A is a constant.

$y = \displaystyle\int (x + e^{-x} + A)dx = \frac{x^2}{2} - e^x + Ax + B$, where B is a constant.

17. $y'' = 5x^2 - 2$

$y' = (5x^2 - 2)dx = \dfrac{5x^3}{3} - 2x + A$

$y = \displaystyle\int\left(\frac{5x^3}{3} - 2x + A\right)dx = \frac{5x^4}{12} - x^2 + Ax + B$, where A and B are constants.

19. $y'' = 2 - x$

$y' = \displaystyle\int (2 - x)dx = 2x - \frac{x^2}{2} + A$

$y = \displaystyle\int\left(2x - \frac{x^2}{2} + A\right)dx = x^2 - \frac{x^3}{6} + Ax + B$, where A and B are constants.

21. Using Theorem 4, the auxiliary equation is $6m^2 + 23m - 4m = 0$. The solutions are $m_1 = \frac{1}{6}$ and $m^2 = -4$. The general solution is $y = Ae^{m_1x} + Be^{m_2x} = Ae^{3x} + Be^{-3x}$.

23. Using Theorem 4, the auxiliary equation is $m^2 - 9 = 0$. The solutions are $m_1 = 3$ and $m^2 = -3$.

The general solution is $y = Ae^{m_1x} + Be^{m_2x} = Ae^{3x} + Be^{-3x}$.

25. Since the variable y is missing, let $p = \frac{dy}{dx}$ so that $\frac{dp}{dx} = \frac{d^2y}{dx^2}$.

$$xy'' + 2y' = 0$$

$$x\frac{dp}{dx} + 2p = 0$$

Separating variables, $\frac{dp}{p} = -\frac{2dx}{x}$

$$\ln|p| = -2\ln|x| + k$$

Let $k = \ln C_1$ and use properties of logarithms:

$$\ln|p| = \ln x^{-2} + \ln C_1 = \ln C_1x^{-2}$$

$$p = \pm C_1x^{-2}$$

$$\frac{dy}{dx} = \pm C_1x^{-2}$$

$$y = \int \pm C_1x^{-2}dx = \pm C_1x^{-1} + B$$

Letting $A = \pm C_1$ we have $y = Ax^{-1} + B$.

27. Let $p = \frac{dy}{dx}$ so the differential equation becomes

$$2\frac{dp}{dx} + (x^3)\frac{dp}{dx} - 3x^2p = 0$$

$$(2 + x^3)\frac{dp}{dx} = 3x^2p$$

$$\frac{dp}{p} = \frac{3x^2}{2 + x^3}dx.$$

Integrating both sides we get $\ln|p| = \ln|2 + x^3| + C_1$.

Letting $C_1 = \ln C_2$,

$$\ln|p| = \ln|2 + x^3| + \ln C_2 = \ln C_2|2 + x^3|.$$

Therefore, $p = \pm C_2(2 + x^3)$. Substituting $\frac{dy}{dx}$ for p,

$$\frac{dy}{dx} = \pm C_2(2 + x^3)$$

$$y = \int \pm C_2(2 + x^3)dx = \pm C_2(2x + \tfrac{x^4}{4}) + C_3 = \pm\tfrac{1}{4}C_2(8x + x^4) + C_3.$$

Letting $A = \pm\frac{1}{4}C_2$ and $B = C_3$, we have $y = A(8x + x^4) + B$.

29. Using Theorem 4, write the equation as $D'' - MD = 0$. The auxiliary equation is $m^2 - M = 0$. The solutions are $m_1 = \sqrt{M}$ and $m_2 = -\sqrt{M}$. The general solution is: $D = Ae^{m_1 t} + Be^{m_2 t} = Ae^{\sqrt{M}t} + Be^{-\sqrt{M}t}$.

9.4 Review

Objective Questions

1. $\frac{dy}{dx} = 8x^3 - 2x^2 + 1$

$$\int dy = \int (8x^3 - 2x^2 + 1)dx$$

$$y = 2x^4 - \tfrac{2}{3}x^3 + x + C$$

3. $\int 3dy = \int (2x - 5)dx$

$$3y = x^2 - 5x + C_1$$

$$y = \tfrac{1}{3}x^2 - \tfrac{5}{3}x + C$$

5. $\int 4y\, dy = \int (x + 5)dx$

$$2y^2 = \tfrac{1}{2}x^2 + 5x + C$$

$$y^2 = \tfrac{1}{4}x^2 + \tfrac{5}{2}x + C$$

7. $\int y^{-1}dy = \int 5x\, dx$

$$\ln|y| = \tfrac{5}{2}x^2 + C_1$$

$$y = Me^{5x^2/2}$$

9. $\int y^2\, dy = \int x^3\, dx$

$$\frac{y^3}{3} = \frac{x^4}{4} + C_1$$

For $(0, -5)$ $y^3 = \tfrac{3}{4}x^4 + C$

$$-125 = C$$

$$y^3 = \tfrac{3}{4}x^4 - 125$$

11. $2xy = y'$

$2x\,dx = y^{-1}\,dy$

$\int 2x\,dx = \int y^{-1}\,dy$

$x^2 + C = \ln|y|$

For $(1, e^2)$ $y = Me^{x^2}$

$e^2 = Me$

$e = M$

$y = ee^{x^2} = e^{x^2+1}$

13. $y' = \int (15 - 3x^2)dx$

$= 15x - x^3 + C$

$y = \int (15x - x^3 + C)dx$

$y = \frac{15x^2}{2} - \frac{x^4}{4} + C_x + D$

If $x = 0$ then $y = 3$

$3 = 0 - 0 + 0 + D$

If $x = 2$ then $y = 9$

$9 = 15 \cdot 2 - 4 + 2C + 3$

$-10 = C$

$y = -\frac{1}{4}x^4 + \frac{15}{2}x^2 - 10x + 3$

15. $y'' = 2x^3 - 4$

$y' = \int (2x^3 - 4)dx$

$= \frac{1}{2}x^4 - 4x + A$

$y = \int \left(\frac{1}{2}x^4 - 4x + A\right)dx$

$= \frac{1}{10}x^5 - 2x^2 + Ax + B$

17. From Theorem 4, $y = Ae^{4x} + Be^{-4x}$

19. $y = \frac{-2e^2}{(1 - e^2)^2}e^x + \frac{2e^2}{(1 - e^2)^2}e^{-x}$

21. $y = Me^{.05t}$ from Theorem 3

If $t = 0$ then $y = 1{,}250$ so

$1{,}250 = M$

In 3 years

$$y = 1{,}250e^{.05(3)}$$

$$\approx 1452.29$$

The solution is 1,452 scooters.

23. $$\frac{dV}{dt} = k(100 - V)$$

$$\int (100 - V) = \int k\, dt$$

$$-\ln|100 - V| = kt + C$$

If $t = 0$, $V = 10$ so

$$-\ln|100 - 10| = C$$

$$-4.49980967 = C$$

If $t = 18$, $V = 40$

$$-\ln|100 - 40| = 18k - 4.49980967$$

$$k \approx .022528393$$

Test Questions

1. $$dy = (12x^3 - 6x^2 + 5)dx$$

$$\int dy = \int (12x^3 - 6x^2 + 5)dx$$

$$y = 3x^4 - 2x^3 + 5x + C$$

3. $$9\frac{dy}{dx} = 14x - 18$$

$$dy = \left(\frac{14}{9}x - 2\right)dx$$

$$y = \frac{7}{9}x^2 - 2x + C$$

5. $$8y\,dy = (x + 3)\ dx$$

$$\int 8y\,dy = \int (x + 3)dx$$

$$4y^2 = \frac{x^2}{2} + 3x + C_1$$

$$8y^2 = x^2 + 6x + C_2$$ This can also be written as $y^2 = \frac{1}{8}(x + 3)^2 + C$.

7. $\frac{dy}{y} = 10x\,dx$

$$\int \frac{dy}{y} = \int 10x\,dx$$

$$\ln|y| = 5x^2 + C_1$$

$$|y| = e^{5x^2 + C_1} = e^{C_1}e^{5x^2} = C_2e^{5x^2} \quad (e^{C_1} = C_2)$$

$$|y| = C_2e^{5x^2}$$

$$y = Me^{5x^2} \qquad (M = \pm C_2)$$

9. $y^2\,dy = x^4\,dx$

$$\int y^2\,dy = \int x^4\,dx$$

$\frac{y^3}{3} = \frac{x^5}{5} + C.$ Substituting $y = -1$ and $x = 0$,

$-\frac{1}{3} = 0 + C.$ So

$$\frac{y^3}{3} = \frac{x^5}{5} - \frac{1}{3}$$

$$5y^3 = 3x^5 - 5$$

11. $\frac{(4 - y^2)dy}{y} = x\,dx$

$$\int \left(\frac{4}{y} - y\right)dy = \int x\,dx$$

$4\ln|y| - \frac{y^2}{2} = \frac{x^2}{2} + C.$ Substituting $y = 1$ and $x = 1$,

$$4\ln 1 - \frac{1}{2} = \frac{1}{2} + C$$

$$0 - \frac{1}{2} = \frac{1}{2} + C$$

$$-1 = C$$

Thus,

$$4\ln|y| - \frac{y^2}{2} = \frac{x^2}{2} - 1$$

$$\ln|y| = \frac{x^2}{8} + \frac{y^2}{8} - \frac{1}{4}$$

$$y^8 = e^{x^2+y^2-2}$$

13. $y'' = x^3 - 8$

$$y' = \frac{x^4}{4} - 8x + C_1$$

$$y = \frac{x^5}{20} - 4x^2 + C_1x + C^2$$

$$8 \ln|y| = x^2 + y^2 - 2$$

$$\ln y^8 = x^2 + y^2 - 2$$

$$y^8 = e^{x^2 + y^2 - 2}$$

15. $8y'' - 17y' + 2y = 0$

Auxiliary equation is $8m^2 - 17m + 2 = 0$

$$(m - 2)(8m - 1) = 0$$

$$m = \tfrac{1}{8} \text{ or } 2$$

The solution to the differential equation is $y = Ae^{x/8} + Be^{2x}$.

17. The auxiliary equation is $m^2 - m - 6 = 0$

$$(m - 3)(m + 2) = 0$$

$$m = 3 \text{ or } -2$$

The solution to the differential equation is $y = Ae^{3x} + Be^{-2x}$.

First substitute $x = 0$ and $y = 0$, then $x = 1$ and $y = 4$: $\begin{cases} 0 = A + B \\ 4 = Ae^3 + Be^{-2} \end{cases}$

Solving this system for A and B gives $A = \dfrac{4e^2}{e^5 - 1}$, $B = \dfrac{-4e^2}{e^5 - 1}$. The solution of the differential equation is $y = \dfrac{4e^2}{e^5 - 1}e^{x/8} - \dfrac{4e^2}{e^5 - 1}e^{2x}$.

19. $\dfrac{dP}{1 - P} = kdt$

$$\int \frac{dP}{1 - P} = \int kdt$$

$$-\ln|1 - P| = kt + C_1$$

$$\ln|1 - P| = -kt - C_1$$

$$|1 - P| = e^{-kt - C_1} - C_2e^{-kt} \qquad (C_2 = e^{-C_1})$$

$$1 - P = C_3e^{-kt} \qquad (C_3 = \pm C_2)$$

$$P = 1 - C_3e^{-kt}$$

a. Initially 0% are aware of the new proof, which means $P = 0$ when $t = 0$:

$$0 = 1 - C_3$$

$$C_3 = 1 \text{ So } P = 1 - e^{-Kt}$$

We are given $P = 25\%$ when $t = 6$:

$$.25 = 1 - e^{-6K}$$
$$.75 = e^{-6K}$$
$$K = \frac{\ln .75}{-6}$$
$$\approx .047947$$
$$P = 1 - e^{-.047947t}$$

b. If $P = 75\%$, $.75 = 1 - e^{-.047947t}$

$$.25 = e^{-.047947t}$$
$$\frac{\ln .25}{-.047947} = t$$
$$t \approx 29 \text{ (months)}$$

CUMULATIVE REVIEW

1. Answers vary. The limit of a function $f(x)$, as x approaches a, is the number L if the value of $f(x)$ approaches L as x gets closer to a. The values of x get closer and closer (but never reach) a, the values of $f(x)$ get closer and closer (and may or may not) reach L.

3. Answers vary. Integration is the inverse process of differentiation; it anti-differentiation.

$\int f(x)\,dx = F(x)$ means that $F'(x) = f(x)$. That is when looking for the integral of $f(x)$ we are seeking a function, $F(x)$, whose derivative is $f(x)$. The definite integral is used to find areas, as well as to sum or total functional values. Finding areas under a curve can be used in calculating consumer and producer surplus, as well as certain probabilities if we are given a probability density function.

5.
$$y = 5x^{-1} + 3\sqrt{x}$$
$$\frac{dy}{dx} = -5x^{-2} + \tfrac{3}{2}x^{-1/2}$$

7.
$$y = \ln|1 - 11x|$$
$$\frac{dy}{dx} = \frac{1}{1 - 11x}(-11) = \frac{-11}{1 - 11x}$$

9. $$\int_0^2 4x\,dx = \left.\frac{4x^2}{2}\right|_0^2 = 8$$

11. $\displaystyle\int_0^1\int_{e^x}^{e}\frac{1}{\ln y}\,dydx = \int_1^e\int_0^{\ln y}\frac{1}{\ln y}\,dxdy$

$$= \int_1^e \frac{x}{\ln y}\Big|_0^{\ln y} dy$$

$$= \int_1^e \frac{\ln y}{\ln y}\,dy$$

$$= \int_1^e dy$$

$$= y\Big|_0^e$$

$$= e - 1$$

$$\approx 1.718281828$$

13. **a.** Break even occurs when

Revenue = Cost.

$130x = 5x + 8{,}000$

$130x = 5x + 8{,}000$

$x = 64$ units.

b. $R(64) = 130(64) = \$8{,}320$

15. $P'(x) = 3x^2 - 26x + 40$ $P'(x) = 0 \Rightarrow x = 2\frac{20}{3}$, $P''(x) = 6x - 26 - 20$

The maximum $P''(2) < 0$ max and $P''\left(\frac{20}{3}\right) > 0$ min.

17. Profit is $P(2) = 36$ (thousand)

The area is $\displaystyle\int_{-2}^{4} \left(\tfrac{1}{2}y + 2 - \tfrac{1}{4}y^2\right)dy = \frac{y^2}{4} + 2y - \frac{1}{12}y^3\Big|_{-2}^{4}$

$$= \left(4 + 8 - \frac{16}{3}\right) - \left(1 - 4 + \frac{2}{3}\right)$$

$$= 9$$

19. $f'(x) = x^3 - 3x$ Setting this equal to zero and solving gives

$x(x^2 - 3) = 0$

$x = 0$ or $x = \pm\sqrt{3}$;

$f''(x) = 3x^2 - 3$. Using the Second-Derivate Test, $f''(0) = -3$, $f''(\sqrt{3}) = 6$, and $f''(-\sqrt{3}) = 6$, so the graph has local minimums at $x = \pm\sqrt{3}$ and a local maximum at $x = 0$.

Set $f''(x) = 3x^2 - 3 = 0 \Rightarrow x = \pm 1$. For $x < -1$ or $x > 1$, $f''(x) > 0$ so the graph is concave up for $x < -1$ or $x > 1$. Similarly, the graph is concave down for $-1 < x < 1$ since $f''(x) < 0$ in that interval. Thus, $x = \pm 1$ are where points of inflection occur. Some key values are:

$x = \pm\sqrt{3}$, $f(\pm\sqrt{3}) = -\frac{9}{4}$

$x = \pm 1$, $f(\pm 1) = -\frac{5}{4}$

$x = 0$, $f(0) = 0$

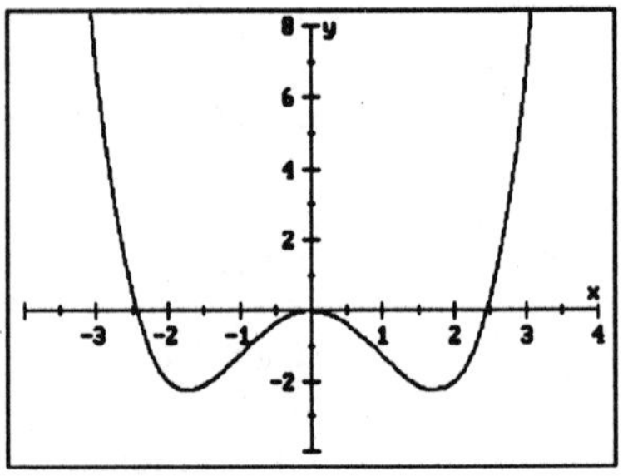

APPENDIX

A.1 Algebra Pretests

PART I

1. E; Even though the 5 is also a factor of $5x^2y$, the best answer is the numerical coefficient since 5 is not a *literal* factor.

3. A; $(-2)(-2)(-2) = -8$

5. C; $a^2 - 2ab + b^2 = (-3)^2 - 2(-3)(5) + 5^2 = 9 + 30 + 25 = 64$

7. D; $3x - 5x = -2x$

9. D; $(-3y^2)^3 = (-3)^3(y^2)^3 = -27y^6$

11. C;

$$\begin{aligned} 3x + 12 &= 6 & \\ 3x &= -6 & \textit{Subtract 12 from both sides} \\ x &= -2 & \textit{Divide both sides by 3} \end{aligned}$$

13. B;

$$\begin{aligned} 2x - 16 &= 3x - 9 & \\ -16 &= x - 9 & \textit{Subtract 2x from both sides} \\ -7 &= x & \end{aligned}$$

15. D; $3(x + 2) - (x - 4y) + (x + y) = 3x + 6 - x + 4y + x + y = 3x + 5y + 6$

17. F; The left side of the equation is $2xy$; the right side of the equation is $4xy$.

19. F; This is "almost" true. Since $0^2 = 0$, x^2 could be 0. A *true* statement would be, "x^2 is non-negative."

21. F; Do not "cancel" terms. Had the "+" been "$\cdot$" a true statement would result, $\frac{A \cdot C}{B \cdot C} = \frac{A}{B}$.

23. F; $\frac{0}{5}$ is defined and is equal to 0, but $\frac{5}{0}$ is not defined.

25. T; A third expression equal to both sides could be $\frac{A + C}{B}$

27. T; One way to see this is simply choose values for x and y and observe that the square of a number is the same as the square of the number's opposite. Another way to see the validity is to expand each side:

$(x - y)^2 = x^2 - 2xy + y^2 = y^2 - 2xy + x^2 = (y - x)^2$

29. T; $\frac{1}{x}\,y\,\frac{1}{z} = \frac{1}{x} \cdot \frac{y}{1} \cdot \frac{1}{z} = \frac{1 \cdot y \cdot 1}{x \cdot 1 \cdot z} = \frac{y}{xz}$

PART II

1. B; Because $2\pi - 7$ is negative $|2\pi - 7| = 7 - 2\pi$

3. C; The distance between any two points x and y is $|x - y|$, which must be non-negative. The distance between 5 and $\sqrt{10}$ is *exactly* $|5 - \sqrt{10}|$. An approximation is listed for B, but the *exact* answer is C.

5. C; $(x + y)^2 = (x + y)(x + y) = x^2 + 2xy + y^2$

7. A; Factor each; $10x^3 = 2 \cdot 5 \cdot x^3$, $5x^2 = 5 \cdot x^2$, $25x = 5^2 \cdot x$. All three have $5x$ in common and no more.

9. B; $15x^3 - 15xy^2 = 15x(x^2 - y^2) = 15x(x + y)(x - y)$.

11. D; Remember to reverse the direction of an inequality when multiplying or dividing both sides by a negative number:

$$-2x \leq 8$$

$$-\frac{2x}{-2} \geq \frac{8}{-2} \qquad \textit{Divide both sides by } -2$$

$$x \geq -4$$

This solution includes all *real* numbers greater than or equal to -4, not just integers as listed in B.

13. C;
$$\begin{aligned} x^2 - 7x + 12 &= 0 \\ (x - 3)(x - 4) &= 0 \\ x &= 3,\ 4 \end{aligned}$$

15. A; Critical values are $x = -1$ and $x = 2$; Use a number line to find that the solution is all numbers to the left of -1 or to the right of 2, namely $x < -1$ or $x > 2$; this is also written as $(-\infty, -1) \cup (2, \infty)$

17. D;
$$\begin{aligned} \frac{x^2 - 5x + 4}{x + 3} \cdot \frac{x^2 + 2x - 3}{x - 4} &= \frac{(x - 1)(x - 4)}{(x + 3)} \cdot \frac{(x + 3)(x - 1)}{x - 4} \\ &= \frac{(x - 1)(x - 4)(x + 3)(x - 1)}{(x + 3)(x - 4)} \\ &= (x - 1)(x - 1) \\ &= (x - 1)^2 \end{aligned}$$

A.2 Real Numbers

1. a. -9 is not a natural number or a whole number, but is an integer. Answer is *I*, *Q*, *R*. **b.** $\sqrt{30}$ is not a rational number. Answer is Q', *R*.

3. **a.** 5 is a natural number. Answer is N, W, I, Q, R. **b.** $\sqrt{4} = 2$ which is a natural number. Answer is N, W, I, Q, R.

5. **a.** Since all repeating decimals are rational numbers, $.\overline{4}$ is a rational number. Answer is Q, R. **b.** Interpreting $.5252\cdots$ as a repeating decimal, $.\overline{52}$, it is a rational number. Answer is Q, R.

7. **a.** Since $.381 = \frac{381}{1000}$, it is a rational number. Answer is Q, R. **b.** Since π is irrational, $\frac{\pi}{6}$ is also irrational. Answer is Q', R.

9. **a.** $\frac{16}{0}$ is undefined and does not belong to any set of numbers **b.** $\frac{0}{16} = 0$ which is not a natural number, but is a whole number. Answer is W, I, Q, R.

11. Since all prime numbers are natural numbers P is shown inside N.

13. All even numbers are whole numbers, but all are not natural numbers; (0 is even, but not natural) so E is shown inside W but not entirely inside N.

15. False **17.** False **19.** True **21.** True; $|5| = 5$ **23.** True

25. False; counterexample is $9 + 8 = 17$ **27.** True **29.** True **31.** True

33. False; $-.\overline{3}$, for example **35.** False; counterexample is 9, which is an odd integer but not prime. **37.** False **39.** True **41.** 9 **43.** 19 **45.** $-\pi$

47. Since $2 - \pi$ is negative, $|2 - \pi| = \pi - 2$. **49.** Since $\pi - 10$ is negative, $|\pi - 10| = 10 - \pi$. **51.** Since $\sqrt{20} - 4$ is positive, $|\sqrt{20} - 4| = \sqrt{20} - 4$

53. Since $\sqrt{50} - 8$ is negative, $|\sqrt{50} - 8| = 8 - \sqrt{50}$.

55. Since $2\pi - 7$ is negative, $|2\pi - 7| = 7 - 2\pi$.

57. If $x \leq -3$ then $x + 3$ is non-positive. Therefore, $|x + 3| = -(x + 3)$.

59. If $y \geq 5$ then $y - 5$ is non-negative. Therefore, $|y - 5| = y - 5$.

61. If $t < -10$ then $4 + 3t$ is negative. Therefore, $|4 + 3t| = -(4 + 3t)$.

63. $|(-15) - (-5)| = 10$ **65.** $|(9) - (-2)| = 11$

67. $|2 - \pi| = \pi - 2$ **69.** $|2 - \sqrt{3}| = 2 - \sqrt{3}$

A.3 Algebraic Expressions

1. $3 + 2 \cdot 5 = 3 + 10 = 13$

3. $5 + 3 \cdot 2 = 5 + 6 = 11$

5. $(-6)^2 = (-6)(-6) = 36$

7. $-5^2 = -5 \cdot 5 = -25$; do not confuse this with $(-5)^2 = 25$

9. $-7^2 = -7 \cdot 7 = -49$; do not confuse this with $(-7)^2 = 49$

11. $8 - 5(2 - 7) = 8 - 5(-5) = 8 - (-25) = 33$

13. $\frac{(-2)6 + (-4)}{-4} + \frac{(-3)(-4)}{6} = \frac{-12 + (-4)}{-4} + \frac{12}{6} = \frac{-16}{-4} + \frac{12}{6} = 4 + 2 = 6$

15. $x^2 = (-3)^2 = 9$ 17. $y^2 = (2)^2 = 4$ 19. $z^2 = (-1)^2 = 1$

21. $x + y - z = -3 + 2 - (-1) = 0$

23. $$\begin{aligned}(xy)^2 + xy^2 + x^2y &= [(-3)(2)]^2 + (-3)(2)^2 + (-3)^2(2)\\ &= (-6)^2 + (-3)(4) + (9)(2)\\ &= 36 + (-12) + 18\\ &= 42\end{aligned}$$

25. $$\begin{aligned}x^2 - w^2(y^2 + z^2) &= (-3)^2 - (-4)^2[(2)^2 + (-1)^2]\\ &= 9 - 16(4 + 1)\\ &= 9 - 16(5)\\ &= 9 - 80\\ &= -71\end{aligned}$$

27. $\frac{x - z}{y} = \frac{(-3) - (-1)}{2} = \frac{-3 + 1}{2} = \frac{-2}{2} = -1$

Note: For Problems 29 - 41 the middle step is shown for you. According to the directions, you should do this step *mentally.*

29. a. $(x - 2)(x + 6) = x^2 + 6x - 2x - 12 = x^2 + 4x - 12$

b. $(x + 5)(x - 4) = x^2 - 4x + 5x - 20 = x^2 + x - 20$

31. a. $(x - 5)(x - 3) = x^2 - 3x - 5x + 15 = x^2 - 8x + 15$

b. $(x + 3)(x - 4) = x^2 - 4x + 3x - 12 = x^2 - x - 12$

33. a. $(2y + 1)(y - 1) = 2y^2 - 2y + y - 1 = 2y^2 - y - 1$

b. $(2y - 3)(y - 1) = 2y^2 - 2y - 3y + 3 = 2y^2 - 5y + 3$

35. a. $(2y + 3)(3y - 2) = 6y^2 - 4y + 9y - 6 = 6y^2 + 5y - 6$

b. $(2y + 3)(3y + 2) = 6y^2 + 4y + 9y + y = 6y^2 + 13y + 6$

37. a. $(x + y)(x - y) = x^2 - xy + xy - y^2 = x^2 - y^2$

b. $(a + b)(a - b) = a^2 - ab + ab - b^2 = a^2 - b^2$

39. a. $(x + 2)^2 = (x + 2)(x + 2) = x^2 + 2x + 2x + 4 = x^2 + 4x + 4$

b. $(x - 2)^2 = (x - 2)(x - 2) = x^2 - 2x - 2x + 4 = x^2 - 4x + 4$

41. a. $(a+b)^2 = (a+b)(a+b) = a^2 + ab + ab + b^2 = a^2 + 2ab + b^2$

b. $(a-b)^2 = (a-b)(a-b) = a^2 - ab - ab + b^2 = a^2 - 2ab + b^2$

43. $(x-y-z) + (2x+y-3z) = x + 2x - y + y - z - 3z = 3x - 4z$

45. $$(x+3y-2z)+(3x-5y+3z) = x+3x+3y-5y-2z+3z$$
$$= 4x - 2y + z$$

47. $(2x-y)-(2x+y) = 2x - y - 2x - y = -2y$

49. $(6x-4y)-(4x-6y) = 6x - 4y - 4x + 6y = 2x + 2y$

51. $$(x+y-5)-(2x-3y+4) = x+y-5-2x+3y-4$$
$$= x - 2x + y + 3y - 5 - 4$$
$$= -x + 4y - 9$$

53. $$(6x^2-3x+2)-(2x^2+5x+3) = 6x^2-3x+2-2x^2-5x-3$$
$$= 6x^2 - 2x^2 - 3x - 5x + 2 - 3$$
$$= 4x^2 - 8x - 1$$

55. $$(2x^2-3x-5)-(5x^2-6x+4) = 2x^2-3x-5-5x^2+6x-4$$
$$= 2x^2 - 5x^2 - 3x + 6x - 5 - 4$$
$$= -3x^2 + 3x - 9$$

57. $$(x^2-x)+(x-3)-(x-x^2) = x^2 - x + x - 3 - x + x^2$$
$$= x^2 + x^2 - x + x - x - 3$$
$$= 2x^2 - x - 3$$

59. $$(x^2-7)-(3x+4)-(x-2x^2) = x^2 - 7 - 3x - 4 - x + 2x^2$$
$$= x^2 + 2x^2 - 3x - x - 7 - 4$$
$$= 3x^2 - 4x - 11$$

61. $$(x+1)^3 = (x+1)(x+1)(x+1)$$
$$= (x+1)(x^2+2x+1)$$
$$= x^3 + 2x^2 + x + x^2 + 2x + 1$$
$$= x^3 + 3x^2 + 3x + 1$$

63. $$(3x^2-5x+2)+(x^3-4x^2+x-4) = x^3 + 3x^2 - 4x^2 - 5x + x + 2 - 4$$
$$= x^3 - x^2 - 4x - 2$$

65. $(3x^2-5x+2)-(5x+1) = 3x^2 - 5x + 2 - 5x - 1 = 3x^2 - 10x + 1$

67. $$(5x+1)(3x^2-5x+2) = 15x^3 - 25x^2 + 10x + 3x^2 - 5x + 2$$
$$= 15x^3 - 22x^2 + 5x + 2$$

A.4 Factoring

1. $20xy - 12x = 4x(5y - 3)$

3. $6x - 2 = 2(3x - 1)$

5. $xy + xz^2 + 3x = x(y + z^2 + 3)$

7. $a^2 + b^2$ does not factor

9. $m^2 - 2mn + n^2 = (m - n)(m - n) = (m - n)^2$

11. Common factor of x^{-3}; $x^{-3}(1 + x^2 + x^5)$; remember to add exponents.

13. Both terms have the quantity $(4x - 1)$; factor this common factor to obtain $(4x - 1)(x + 3)$

15. Try pairs of numbers whose product is -15; $2x^2 + 7x - 15 = (2x - 3)(x + 5)$

17. $3x^2 - 5x - 2 = (3x + 1)(x - 2)$

19. Common factor first; $2x^{-1} - 10x^{-2} - 48x^{-3} = 2x^{-3}(x^2 - 5x - 24)$
$$= 2x^{-3}(x - 8)(x + 3)$$

21. $2 - 5x^{-1} + 3x^{-2} = x^{-2}(2x^2 - 5x + 3) = x^{-2}(2x - 3)(x - 1)$

23. $x^{1/3} + x^{4/3} + x^{7/3} = x^{1/3}(1 + x + x^2) = x^{1/3}(x^2 + x + 1)$

25. $(x - y)^2 - 1 = [(x - y) - 1][(x - y) + 1] = (x - y - 1)(x - y + 1)$

27. $x^{3/2} + 6x^{1/2} + 9x^{-1/2} = x^{-1/2}(x^2 + 6x + 9) = x^{-1/2}(x + 3)^2$

29. This is a difference of squares;
$$(a + b)^2 - (x + y)^2 = [(a + b) - (x + y)][(a + b) + (x + y)]$$
$$= (a + b - x - y)(a + b + x + y)$$

31. $2x^2 + x - 6 = (2x - 3)(x + 2)$

33. $6x^2 + 47x - 8 = (6x - 1)(x + 8)$

35. $6x^2 + 49x + 8 = (6x + 1)(x + 8)$

37. $4x^2 + 13x - 12 = (4x - 3)(x + 4)$

39. $9x^2 - 56x + 12 = (9x - 2)(x - 6)$

41. $10x^2 - 9 - x^4 = -(x^4 - 10x^2 + 9)$
$$= -(x^2 - 1)(x^2 - 9)$$
$$= -(x - 1)(x + 1)(x - 3)(x + 3)$$

43. $(x - \frac{1}{4})(x^2 - \frac{1}{9}) = \frac{1}{4}(4x^2 - 1)\frac{1}{9}(9x^2 - 1)$
$$= \frac{1}{36}(2x - 1)(2x + 1)(3x - 1)(3x + 1)$$

45. $(x^2 - 3x - 6)^2 - 2^2 = [(x^2 - 3x - 6) - 2][(x^2 - 3x - 6) + 2]$

$= (x^2 - 3x - 8)(x^2 - 3x - 4)$ *Difference of squares*

$= (x^2 - 3x - 8)(x - 4)(x + 1)$

47. Let $s = (x + y)$ and $t = (a + b)$; then we need to factor $2s^2 - 5st - 3t^2 = (2s + t)(s - 3t)$ Therefore,

$$2(x + y)^2 - 5(x + y)(a + b) - 3(a + b)^2 = [2(x + y) + (a + b)][(x + y - 3(a + b)]$$
$$= (2x + 2y + a + b)(x + y - 3a - 3b)$$

49. Common factor is $18x^2(x - 5)^2$:

$$18x^2(x - 5)^2[(x - 5) + x] = 18x^2(x - 5)^2(2x - 5)$$

51. Common factor is $5x(3x + 1)^2$:

$$5x(3x + 1)^2[2(3x + 1) + 9x] = 5x(3x + 1)^2(15x + 2)$$

53. Common factor is $2x(4x + 3)^2$:

$$2x(4x + 3)^2[(4x + 3) + 3x(2)] = 2x(4x + 3)^2(4x + 3 + 6x) = 2x(4x + 3)^2(10x + 3)$$

55. Common factor is $(x + 1)^2(x - 2)^3$:

$$(x + 1)^2(x - 2)^3[3(x - 2) + 4(x + 1)] = (x + 1)^2(x - 2)^3(3x - 6 + 4x + 4)$$
$$= (x + 1)^2(x - 2)^3(7x - 2)$$

57. Common factor is $(3)(2)(2x - 1)^2(3x + 2)$:

$$6(2x - 1)^2(3x + 2)[(3x + 2) + (2x - 1)] = 6(2x - 1)^2(3x + 2)(5x + 1)$$

59. Common factor is $2(x + 5)^3(x^2 - 2)^2$:

$$2(x + 5)^3(x^2 - 2)^2[2(x^2 - 2) + 3x(x + 5)]$$
$$= 2(x + 5)^3(x^2 - 2)^2(2x^2 - 4 + 3x^2 + 15x)$$
$$= 2(x + 5)^2(x^2 - 2)^2(5x^2 + 15x - 4)$$

A.5 Linear Equations and Inequalities

1. $3x = 5x - 4$ *Subtract 5x from both sides*

$-2x = -4$ *Divide both sides by -2*

$x = 2$

3. $7x + 10 = 5x$ *Subtract 7x from both sides*

$10 = -2x$ *Divide both sides by -2*

$-5 = x$

5. $3x + 22 = 1 - 4x$ *Add 4x to both sides*

$7x + 22 = 1$ *Subtract 22 from both sides*

$7x = -21$ *Divide both sides by 7*

$x = -3$

7. $2x - 13 = 7x + 2$ *Subtract 2x from both sides*

$-13 = 5x + 2$ *Subtract 2 from both sides*

$-15 = 5x$ *Divide both sides by 5*

$-3 = x$

9. $7x + 18 = -2x$ *Subtract 7x from both sides*

$18 = -9x$ *Divide both sides by − 9*

$-2 = x$

11. $8x - 3 = 15 - x$ *Add x to both sides*

$9x - 3 = 15$ *Add 3 to both sides*

$9x = 18$ *Divide both sides by 9*

$x = 2$

13. $2(x - 1) = 1 + 3(x - 2)$ *Remove parentheses and simplify*

$2x - 2 = 3x - 5$ *Subtract 2x from both sides*

$-2 = x - 5$ *Add 5 to both sides*

$3 = x$

15. $5 - 2x = 1 + 3(x - 2)$ *Remove parentheses and simplify*

$5 - 2x = 3x - 5$ *Add 2x to both sides*

$5 = 5x - 5$ *Add 5 to both sides*

$10 = 5x$ *Divide both sides by 5*

$2 = x$

17. $3x - 2 \leq 7$ *Add 2 to both sides*

$3x \leq 9$ *Divide both sides by 3*

$x \leq 3$ or $(-\infty, 3]$

19. $4 - 5x > 29$ *Subtract 4 from both sides*

$-5x > 25$ *Divide both sides by − 5*

$x < -5$ or $(-\infty, -5)$ *(reverse inequality)*

21. $9x + 7 \geq 5x - 9$ — *Subtract 5x from both sides*

$4x + 7 \geq -9$ — *Subtract 7 from both sides*

$4x \geq -16$ — *Divides both sides by 4*

$x \geq -4$ or $[-4, \infty)$

23. $2(3 - 4x) < 30$ — *Eliminate parentheses*

$6 - 8x < 30$ — *Subtract 6 from both sides*

$-8x < 24$ — *Divide both sides by − 8 (reverse inequality)*

$x > -3$ or $(-3, \infty)$

25. $7 < x + 2 < 11$ — *Subtract 2 from all three sides*

$5 < x < 9$ or $(5, 9)$

27. $-2 < x - 1 < 3$ — *Add 1 to all three sides*

$-1 < x < 4$ or $(-1, 4)$

29. $9 < 1 - 2x < 15$ — *Subtract 1 from both sides*

$8 < -2x < 14$ — *Divide both sides by − 2 (reverse inequality)*

$-4 > x > -7$ — *This should be rewritten as:*

$-7 < x < -4$ or $(-7, -4)$

31. $5x^4 - 5y^4y' = 0$

$$5x^4 = 5y^4y'$$

$$y' = \frac{x^4}{y^4}$$

33. $4x^3y^2 + 2x^4yy' = 0$

$$y' = \frac{-4x^3y^2}{2x^4y}$$

$$= \frac{-2y}{x}$$

35. $2x - (xy' + y) + 2yy' = 0$

$$2x - y = xy' - 2yy'$$

$$2x - y = (x - 2y)y'$$

$$y' = \frac{2x - y}{x - 2y}$$

37. $3x^2 + 4x + (xy' + y) = 0$

$$xy' = -3x^2 - 4x - y$$

$$y' = \frac{-3x^2 - 4x - y}{x}$$

39. $\frac{1}{25}[2(x + 1)] + \frac{1}{4}[2(y - 1)y'] = 0$

$$4(x + 1) + 25(y - 1)y' = 0 \qquad \textit{Multiply both sides by 50}$$

$$25(y - 1)y' = -4(x + 1)$$

$$y' = \frac{-4(x + 1)}{25(y - 1)}$$

41. $3[x^2(3y^2y') + 2xy^3] - 3[x(2yy') + y^2] + 5(xy' + y) = 0$

$$9x^2y^2y' + 6xy^3 - 6xyy' - 3y^2 + 5xy' + 5y = 0$$

$$9x^2y^2y' - 6xyy' + 5xy' = 3y^2 - 6xy^3 - 5y$$

$$(9x^2y^2 - 6xy + 5x)\, y' = 3y^2 - 6xy^3 - 5y$$

$$y' = \frac{3y^2 - 6xy^3 - 5y}{9x^2y^2 - 6xy + 5x}$$

43.

$$\begin{aligned} 2(x - 3) - 5x &= 3(1 - 2x) \\ 2x - 6 - 5x &= 3 - 6x \\ -3x - 6 &= 3 - 6x \\ 3x - 6 &= 3 \\ 3x &= 9 \\ x &= 3 \end{aligned}$$

45.

$$\begin{aligned} 5(x - 1) + 3(2 - 4x) &> 8 \\ 5x - 5 + 6 - 12x &> 8 \\ -7x + 1 &> 8 \\ -7x &> 7 \\ x &< -1 \text{ or } (-\infty, -1) \end{aligned}$$

47.

$$\begin{aligned} 4(x - 2) + 1 &\le 3(x + 1) \\ 4x - 8 + 1 &\le 3x + 3 \\ 4x - 7 &\le 3x + 3 \\ x - 7 &\le 3 \\ x &\le 10 \text{ or } (-\infty, 10] \end{aligned}$$

49.

$$\begin{aligned} 2(4 - 3x) &> 4(3 - x) \\ 8 - 6x &> 12 - 4x \\ 8 &> 12 + 2x \\ -4 &> 2x \\ -2 &> x \text{ or } (-\infty, -2) \end{aligned}$$

51.
$$\begin{aligned} 2(3-7x) &\geq -4-(5-x) \\ 6-14x &\geq -4-5+x \\ 6-14x &\geq -9+x \\ 6 &\geq -9+15x \\ 15 &\geq 15x \\ 1 &\geq x \text{ or } (-\infty, 1] \end{aligned}$$

53.
$$\begin{aligned} 3(1-x)-5(x-2) &= 5 \\ 3-3x-5x+10 &= 5 \\ 13-8x &= 5 \\ -8x &= -8 \\ x &= 1 \end{aligned}$$

55.
$$\begin{aligned} 6(2x+5) &= 4(3x+1) \\ 12x+30 &= 12x+4 \\ 30 &= 4 \end{aligned}$$
No solution

57.
$$\begin{aligned} 3(2x-5) &= 5(4x-3) \\ 6x-15 &= 20x-15 \\ -15 &= 14x-15 \\ 0 &= 14x \\ 0 &= x \end{aligned}$$

59.
$$\begin{aligned} 5(1-2x) &= 3(x-4)-(13x-17) \\ 5-10x &= 3x-12-13x+17 \\ 5-10x &= 10x+5 \end{aligned}$$
Any real number is a solution.

A.6 Quadratic Equations and Inequalities

1.
$$\begin{aligned} x^2+2x-15 &= 0 \\ (x+5)(x-3) &= 0 \\ x+5=0 \text{ or } x-3 &= 0 \\ x=-5 \text{ or } x &= 3 \end{aligned}$$

3.
$$\begin{aligned} x^2+7x-18 &= 0 \\ (x+9)(x-2) &= 0 \\ x+9=0 \text{ or } x-2 &= 0 \\ x=-9 \text{ or } x &= 2 \end{aligned}$$

5.
$$\begin{aligned} 10x^2-3x-4 &= 0 \\ (2x+1)(5x-4) &= 0 \\ 2x+1=0 \text{ or } 5x-4 &= 0 \\ x=-\tfrac{1}{2} \text{ or } x &= \tfrac{4}{5} \end{aligned}$$

7.
$$x^2 + 5x - 6 = 0$$
$$(x - 1)(x + 6) = 0$$
$$x - 1 = 0 \text{ or } x + 6 = 0$$
$$x = 1 \text{ or } x = -6$$

9.
$$x^2 - 10x + 25 = 0$$
$$(x - 5)(x - 5) = 0$$
$$x - 5 = 0$$
$$x = 5$$

11.
$$12x^2 + 5x - 2 = 0$$
$$(3x + 2)(4x - 1) = 0$$
$$3x + 2 = 0 \text{ or } 4x - 1 = 0$$
$$x = -\tfrac{2}{3} \text{ or } x = \tfrac{1}{4}$$

13.
$$5x^2 - 4x + 1 = 0$$

$5x^2 - 4x + 1$ *does not factor.*

$$x = \frac{-b \pm \sqrt{b^2 - 4ac}}{2a}$$
$$= \frac{4 \pm \sqrt{16 - 20}}{10}$$
$$= \frac{4 \pm \sqrt{-4}}{10}$$

No real solutions since $\sqrt{-4}$ is not real.

15.
$$4x^2 - 5 = 0$$
$$x^2 = \tfrac{5}{4}$$
$$x = \pm \sqrt{\tfrac{5}{4}}$$
$$= \pm \tfrac{\sqrt{5}}{2}$$

17.
$$3x^2 = 7x$$
$$3x^2 - 7x = 0$$
$$x(3x - 7) = 0$$
$$x = 0 \text{ or } x = \tfrac{7}{3}$$

19.
$$3x^2 = 5x + 2$$
$$3x^2 - 5x - 2 = 0$$
$$(3x + 1)(x - 2) = 0$$
$$3x + 1 = 0 \text{ or } x - 2 = 0$$
$$x = -\tfrac{1}{3} \text{ or } x = 2$$

21. Critical points are 6 and 2.

$x - 6$:	− − −		− − −		+ + +
$x - 2$:	− − −		− − −		+ + +
	pos	2	neg	6	pos
			↑		

desired value where the product is less than or equal to zero (neg)

The answer is [2, 6].

23. Critical points are 0 and -3.

x:	− − −		− − −		+ + +
$x + 3$:	− − −		+ + +		+ + +
	pos	-3	neg	0	pos
			↑		

desired value where the product is less than zero (neg)

The answer is $(-3, 0)$.

25. Critical points are -2 and 8.

$x + 2$:	− − −		+ + +		+ + +
$8 - x$:	+ + +		+ + +		− − −
	neg	-2	pos	8	neg
	↑				↑

desired value where the product is less than or equal to zero (neg)

The answer is $x \leq -2$ or $x \geq 8$; that is $(-\infty, -2] \cup [8, \infty)$.

27. Critical points are $\frac{1}{3}$ and 4.

$1 - 3x$:	+ + +		− − −		− − −
$x - 4$:	− − −		− − −		+ + +
	neg	$\frac{1}{3}$	pos	4	neg
	↑				↑

desired value where the product is less than zero (neg)

The answer is $(-\infty, \frac{1}{3}) \cup (4, \infty)$.

29. $x^2 - 9 \geq 0$

$(x + 3)(x - 3) \geq 0$. Critical points are $x = -3$ and $x = 3$.

$x + 3$:	− − −		+ + +		+ + +
$x - 3$:	− − −		− − −		+ + +
	pos	-3	neg	3	pos
	↑				↑

desired value where the product is greater than or equal to zero (pos)

The answer is $[-\infty, -3] \cup [3, \infty)$.

31. Since there are no solutions to $x^2 + 9 = 0$, there are no critical points.
Pick any number, say $x = 1$: $(1)^2 + 9 \leq 0$ false. There is no real solution.

33. $x^2 - x - 6 > 0$
$(x - 3)(x + 2) > 0$. Critical points are $x = 3$ and $x = -2$.

$x - 3$:	$- - -$		$- - -$		$+ + +$
$x + 2$:	$- - -$		$+ + +$		$+ + +$
	pos	-2	neg	3	pos
	↑				↑

desired value where the product is greater than to zero (pos)
The answer is $(-\infty, -2) \cup (3, \infty)$.

35. $5x - 6 \geq x^2$
$0 \geq x^2 - 5x + 6$
$0 \geq (x - 2)(x - 3)$. Critical points are $x = 2$ and $x = 3$.
You are less likely to make a mistake if you rewrite this as
$(x - 2)(x - 3) \leq 0$

$x - 2$:	$- - -$		$+ + +$		$+ + +$
$x - 3$:	$- - -$		$- - -$		$+ + +$
	pos	2	neg	3	pos
			↑		

desired value where the product is less than or equal to zero (neg)
The answer is $[2, 3]$.

37. $5 - 4x \geq x^2$
$0 \geq x^2 + 4x - 5$
$0 \geq (x + 5)(x - 1)$. Critical points are $x = -5$ and $x = 1$.
You are less likely to make a mistake if you rewrite this as
$(x + 5)(x - 1) \leq 0$

$x + 5$:	$- - -$		$+ + +$		$+ + +$
$x - 1$:	$- - -$		$- - -$		$+ + +$
	pos	-5	neg	1	pos
			↑		

desired value where the product is less than or equal to zero (neg)
The answer is $[-5, 1]$.

39. $6x^2 - 10 > 59x$
$6x^2 - 59x - 10 > 0$
$(6x + 1)(x - 10) > 0$

$6x + 1$:	$- - -$		$+ + +$		$+ + +$
$x - 10$:	$- - -$		$- - -$		$+ + +$
	pos	$-\frac{1}{6}$	neg	10	pos
	↑				↑

desired value where the product is greater than zero (pos)
The answer is $(-\infty, -\frac{1}{6}) \cup (10, \infty)$.

A.7 Rational Expressions

1. $\dfrac{2}{x+y}+3=\dfrac{2}{x+y}+\dfrac{3(x+y)}{x+y}=\dfrac{2+3x+3y}{x+y}$

3. $\dfrac{x^2-y^2}{2x+2y}=\dfrac{(x+y)(x-y)}{2(x+y)}=\dfrac{x-y}{2}$

5. $\dfrac{3x^2-4x-4}{x^2-4}=\dfrac{(3x+2)(x-2)}{(x+2)(x-2)}=\dfrac{3x+2}{x+2}$

7. $\dfrac{3}{x+y}+\dfrac{5}{2x+2y}=\dfrac{3}{x+y}\cdot\dfrac{2}{2}+\dfrac{5}{2(x+y)}=\dfrac{6}{2(x+y)}+\dfrac{5}{2(x+y)}=\dfrac{11}{2(x+y)}$

9. Any non-zero number raised to the power of 0 is 1.

11. $(x^2-36)\left(\dfrac{3x+1}{x+6}\right)=(x+6)(x-6)\left(\dfrac{3x+1}{x+6}\right)=(x-6)(3x+1)$

$=3x^2-17x-6$

13. $\dfrac{x+3}{x}+\dfrac{3-x}{x^2}=\dfrac{x(x+3)}{x^2}+\dfrac{3-x}{x^2}=\dfrac{x^2+3x+3-x}{x^2}=\dfrac{x^2+2x+3}{x^2}$

15. $\dfrac{x}{x-1}+\dfrac{x-3}{1-x}=\dfrac{x}{x-1}+\dfrac{3-x}{x-1}=\dfrac{3}{x-1}$

17. $\dfrac{2x+3}{x^2}+\dfrac{3-x}{x}=\dfrac{2x+3}{x^2}+\dfrac{x(3-x)}{x^2}=\dfrac{2x+3}{x^2}+\dfrac{3x-x^2}{x^2}=\dfrac{3+5x-x^2}{x^2}$

19. $\frac{x}{y}+2+\frac{y}{x}=\frac{x^2}{xy}+\frac{2xy}{xy}+\frac{y^2}{xy}=\dfrac{x^2+2xy+y^2}{xy}=\dfrac{(x+y)^2}{xy}$

21. $\frac{x}{y}-2+\frac{y}{x}=\frac{x^2}{xy}-\frac{2xy}{xy}+\frac{y^2}{xy}=\dfrac{x^2-2xy+y^2}{xy}=\dfrac{(x-y)^2}{xy}$

23. $\dfrac{1}{3xy^2}+xy+\dfrac{1}{x^3y}=\dfrac{x^2}{3x^3y^2}+\dfrac{3x^4y^3}{3x^3y^2}+\dfrac{3y}{3x^3y^2}=\dfrac{x^2+3x^4y^3+3y}{3x^3y^2}$

25. $\dfrac{4x-12}{x^2-49}\div\dfrac{18-2x^2}{x^2-4x-21}=\dfrac{4x-12}{x^2-49}\cdot\dfrac{x^2-4x-21}{18-2x^2}$

$$=\frac{4(x-3)}{(x+7)(x-7)}\cdot\frac{(x-7)(x+3)}{-2(x+3)(x-3)}$$

$$=\frac{-2}{x+7}$$

27. $\dfrac{1}{x^2+1}-\dfrac{x^2}{x^2+1}=\dfrac{1-x^2}{x^2+1}=\dfrac{(1-x)(1+x)}{x^2+1}$

29. $\dfrac{4x^2}{x^4-2x^3}+\dfrac{8}{4x-x^3}-\dfrac{-4}{x+2}=\dfrac{4x^2}{x^3(x-2)}+\dfrac{8}{-x(x+2)(x-2)}+\dfrac{4}{x+2}$

$$= \frac{4}{x(x-2)} - \frac{8}{x(x+2)(x-2)} + \frac{4}{x+2}$$
$$= \frac{4(x+2)}{x(x+2)(x-2)} - \frac{8}{x(x+2)(x-2)} + \frac{4x(x-2)}{x(x+2)(x-2)}$$
$$= \frac{4x+8-8+4x^2-8x}{x(x+2)(x-2)}$$
$$= \frac{4x^2-4x}{x(x+2)(x-2)}$$
$$= \frac{4x(x-1)}{x(x+2)(x-2)}$$
$$= \frac{4(x-1)}{(x+2)(x-2)}$$

A.8 Chapter Review

Objective Questions

1.

Number	*Natural*	*Whole*	*Integer*	*Set* *Rational*	*Irrational*	*Real*
-8	No	No	Yes	Yes	No	Yes
$\frac{5}{6}$	No	No	No	Yes	No	Yes
$.2\overline{3}$	No	No	No	Yes	No	Yes
$\sqrt{169}$	Yes	Yes	Yes	Yes	Yes	Yes
$\frac{3}{0}$	No	No	No	No	No	No

3. $-|-5| = -5$

5. If $x < -2$ then $x - 4$ is negative so $(x-4) = -|x-4| = 4 - x$

7. $|-4-2\pi| = -(-4-2\pi) = 2\pi + 4$

9. **a.** -64 **b.** 64 **c.** -64 **d.** -1

11. $\dfrac{6+4(-3)}{-2} = \dfrac{6-12}{-2} = 3$

13. $-1 - [2(-3) - 2(-3)] = -1 - 0 = -1$

15. **a.** $-(-1)^2 = -1$ **b.** $[-(-1)]^2 = 1^2 = 1$

17. **a.** $x^2 + 5x - 14$ **b.** $x^2 - 9$

19. **a.** $x^2 + xy - 2y^2$ **b.** $x^2 - 9y^2$

21. $(x - y - z) + (2x + y - 3z) = 3x - 4z$

23. $2(3 - x^2) - 3(5 - x) - (3x - 2) = 6 - 2x^2 - 15 + 3x - 3x + 2$
$= -2x^2 - 7$

25. $x^2 - 5x + 6 = (x - 2)(x - 3)$

27. $25 - x^2 = (5 - x)(5 + x)$

29. $9x^4 - 40x^2 + 16 = (9x^2 - 4)(x^2 - 4) = (3x + 2)(3x - 2)(x + 2)(x - 2)$

31.
$$90 = 5x - 10$$
$$100 = 5x$$
$$20 = x$$

33.
$$3 - 4x = 3(5 - 2x)$$
$$3 - 4x = 15 - 6x$$
$$2x = 12$$
$$x = 6$$

35.
$$3x - 5 > 16$$
$$3x > 21$$
$$x > 7$$
$$(7, \infty)$$

37.
$$x < 3(2 + x)$$
$$x < 6 + 3x$$
$$-2x < 6$$
$$x > -3$$
$$(-3, \infty)$$

39.
$$-8 \leq x + 5 \leq -3$$
$$-13 \leq x \leq -8$$
$$[-13, -8]$$

41.
$$6 < -x < 10$$
$$-6 > x > -10$$
$$-10 < x < -6$$
$$(-10, -6)$$

43. $(3x - 1)(5x + 2) = 0$
$$x = \tfrac{1}{3} \text{ or } x = -\tfrac{2}{5}$$

45.
$$6x^2 + 19x = -15$$
$$6x^2 + 19x + 15 = 0$$
$$(3x + 5)(2x + 3) = 0$$
$$x = -\tfrac{5}{3}, -\tfrac{3}{2}$$

47.
$$x^2 - 6x + 7 = 0$$
$$x = \frac{6 \pm \sqrt{36 - 4(7)}}{2}$$
$$= \frac{6 \pm 2\sqrt{2}}{2}$$
$$= 3 \pm \sqrt{2}$$

49. $(x - 5)(x - 7) < 0$

$x - 5$:	$- - -$	$+ + +$	$+ + +$
$x - 7$:	$- - -$	$- - -$	$+ + +$
	pos	5 neg 7	pos
	↑		↑

desired value where the product is negative

The answer is (5, 7).

51. $6x^2 > 11x + 10$

$6x^2 - 11x - 10 > 0$

$(3x + 2)(2x - 5) > 0$

$3x + 2$:	$- - -$		$+ + +$		$+ + +$
$2x - 5$:	$- - -$		$- - -$		$+ + +$
	pos	$-\frac{2}{3}$	neg	$\frac{5}{2}$	pos
	↑				↑

desired value where the product is greater than zero (pos)

The answer is $(-\infty, -\frac{2}{3}) \cup (\frac{5}{2}, \infty)$.

53. $\frac{2x}{15} - \frac{y}{12} = \frac{2x}{15}(\frac{4}{4}) - \frac{y}{12}(\frac{5}{5}) = \frac{8x - 5y}{60}$

55.
$$\frac{3x + 2}{9x^2 - 6x + 1} + \frac{1}{3x - 1} = \frac{3x + 2}{(3x - 1)(3x - 1)} + \frac{1}{3x - 1}$$
$$= \frac{3x + 2 + (3x - 1)}{(3x - 1)(3x - 1)}$$
$$= \frac{6x + 1}{(3x - 1)^2}$$

Sample Test

1.

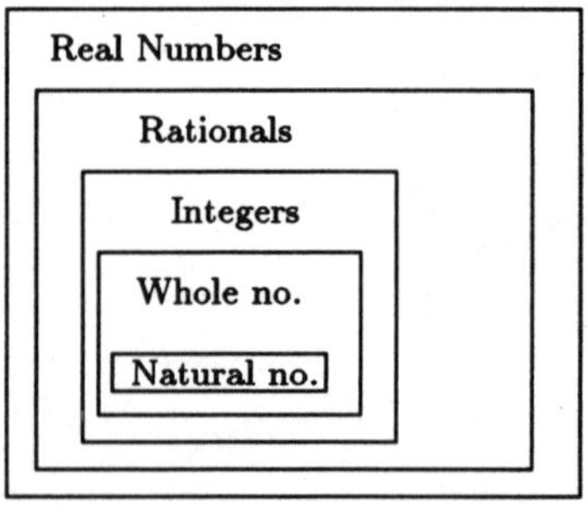

3. $|5 - \sqrt{10}| = 5 - \sqrt{10}$

5. $-(-6)^2 + 5(-6)[3 - 2(-3)] = -36 + 5(-6)(9) = -306$

7. $(x + y)(x + y) = x^2 + 2xy + y^2$

9. $6x^2 - 29x - 5 = (6x + 1)(x - 5)$

11.

$4 - 3x = 2(5 - 2x)$	*Remove parentheses*
$4 - 3x = 10 - 4x$	*Add 4x to both sides*
$4 + x = 10$	*Subtract 4*
$x = 6$	

13.

$2(x + 1) + 5 < 3(1 - x)$	*Remove parentheses*
$2x + 2 + 5 < 3 - 3x$	*Add 3x and simplify*
$5x + 7 < 3$	*Subtract 7 and divide by 5*
$x < -\frac{4}{5}$ or $(-\infty, -\frac{4}{5})$	

15.

$$x = 5x^2$$
$$x - 5x^2 = 0$$
$$x(1 - 5x) = 0$$
$$x = 0 \text{ or } 1 - 5x = 0$$
$$x = 0 \text{ or } x = \frac{1}{5}$$

17. The critical values are $x = 1$ and $x = \frac{2}{3}$. Use a number line to find the solution $(\frac{2}{3}, 1)$.

19. $\frac{7x}{30} - \frac{y}{12} = \frac{14x}{60} - \frac{5y}{60} = \frac{14x - 5y}{60}$